Prostaglandin and Lipid Metabolism in Radiation Injury

Prostaglandin and Lipid Metabolism in Radiation Injury

Edited by
Thomas L. Walden, Jr.
and
Haywood N. Hughes

Armed Forces Radiobiology Research Institute
Bethesda, Maryland

Plenum Press • New York and London

Library of Congress Cataloging in Publication Data

International Conference on Prostaglandins and Lipid Metabolism in Radiation Injury
(1986: Rockville, Md.)
    Prostaglandin and lipid metabolism in radiation injury / edited by Thomas L.
Walden, Jr. and Haywood N. Hughes.
        p.   cm.
    "Proceedings of the International Conference on Prostaglandins and Lipid
Metabolism in Radiation Injury, held October 2–3, 1986, in Rockville, Maryland"—
T.p. verso.
    Includes bibliographies and index.
    ISBN 0-306-42793-1
    1. Radiation—Physiological effect—Congresses. 2. Radiation injuries—Congresses.
3. Prostaglandins—Effect of radiation on—Congresses. 4. Lipids—Effect of radiation
on—Congresses. 5. Radiation-protective agents—Congresses. I. Walden, Thomas L.
II. Hughes, Haywood N. III. Title. [DNLM: 1. Lipids—metabolism—congresses. 2.
Prostaglandins—metabolism—congresses. 3. Radiation Injuries—congresses. WN
610 I63p 1986]
QP82.2.R3I57   1986   87-32663
616.98'9707—dc19   CIP

Proceedings of the International Conference on Prostaglandins and Lipid
Metabolism in Radiation Injury, held October 2–3, 1986, in Rockville, Maryland

© 1987 Plenum Press, New York
A Division of Plenum Publishing Corporation
233 Spring Street, New York, N.Y. 10013

Printed in the United States of America

PREFACE

This volume contains the proceedings of the International Conference on Prostaglandins and Lipid Metabolism in Radiation Injury held in Rockville, Maryland, on October 2-3, 1986. Over 200 persons from eight countries attended the program, which consisted of 24 oral presentations and 38 poster presentations. Forty-two of those presentations have been included in this volume. The conference was sponsored by the Armed Forces Radiobiology Research Institute, located in Bethesda, Maryland.

The effects of radiation on lipid synthesis and membrane damage are aptly summarized in the first five chapters. These chapters describe the effects of radiation on lipid peroxidation of model membranes, and the role of lipid composition in mammalian cell death and in bacterial radio- and thermosensitivity. In bacteria, lipid peroxidation is not essential for radiation-induced cell death.

One of the key points of the conference was the paradoxical nature of the radiobiology of eicosanoids. On the one hand, eicosanoids are mediators of damage; on the other, they are radioprotective agents. It is clear from the literature and from the data presented at the conference that both of these properties may also be observed as a consequence of radiotherapy. Some studies indicate that nonsteroidal anti-inflammatory drugs may minimize or prevent certain radiation-induced damage, but other studies show no positive effect. It was pointed out at the conference that care should be exercised in manipulating the eicosanoid response of tumors to favor radiotherapy because in blocking either the cyclooxygenase pathway or the lipoxygenase pathway, arachidonic acid may be shunted through the other pathway and elicit similar radiation-induced damage or tumor protection. Two additional chapters are devoted to prostaglandins and cancer, showing the potential uses of nonsteroidal anti-inflammatory drugs in therapy and also demonstrating that prostaglandins are radioprotective to tumor cells. Both the prostaglandins and the leukotrienes have been shown to exhibit radioprotective qualities. In addition, the latest information has been included on the alteration of prostaglandin levels in humans following total-body irradiation for bone marrow transplants. The increase in urinary 6-keto $PGF_{1\alpha}$ has now been observed in humans and in laboratory animals postirradiation.

The conference highlighted a number of methods for the detection and quantitation of lipid peroxidation and eicosanoids. The proceedings is organized to reflect the methods of detecting eicosanoids and lipid peroxidation, and summarizes the state of the art in those fields. The definitive detection is represented by the extreme sensitivity and characteristic spectral analysis of the GC-mass spectrometer, with which femtogram quantities can be detected. The rapidly developing field of enzyme-labeled immunoassay (ELISA) is summarized, including the detection of eicosanoids using acetylcholine esterase. This technique permits the detection of eicosanoids down to one picogram per well and does not require the use of radiolabeled compounds. A relatively new technique for the introduction of nonvolatile samples for mass spectroscopic analysis is the thermospray interface. As new techniques and lower limits of sensitivity have been acquired, the list of eicosanoid metabolites and their biological properties that have been identified will most certainly grow. The picture becomes more complicated in accounting for the metabolites, the distributions, and their contributions to either radiation injury or radioprotection.

The effects of ionizing and nonionizing radiations on eicosanoid and lipid metabolism are presented on experimental systems ranging from liposomes and bacteria to man. Very little work has been done with either the platelet-activating factor or leukotrienes regarding their roles in radiation injury. Reports in this volume demonstrate that the production of both compounds is increased in response to radiation exposure. The conference has summarized much of the complexity within the field and pointed out the need for more research and, perhaps, the need for a future conference.

The editors gratefully acknowledge the support and assistance of Junith A. Van Deusen, Janet B. Gillette, Beth A. Roberts, Modeste E. Greenville, and Nancy F. Brown in producing this volume. We also thank Beverly S. Geisbert, Sonia Jones, Carolyn B. Wooden, and Sandra L. Work for preparing the manuscripts.

Thomas L. Walden, Jr.
Haywood N. Hughes
Editors

The articles in this volume have been accepted by a peer review process. The editors express their sincere appreciation to the following members of the Editorial Board:

CONTENTS

RADIATION EFFECTS ON LIPID SYNTHESIS AND MEMBRANE DAMAGE

RADIATION-INDUCED CHANGES
IN EICOSANOID SYNTHESIS

PROSTAGLANDINS IN RADIATION PROTECTION AND RECOVERY

CLINICAL IMPLICATION

DETECTION AND QUANTITATION OF EICOSANOIDS AND LIPID PEROXIDES

RADIATION EFFECTS ON LIPID SYNTHESIS AND MEMBRANE DAMAGE

RADIATION PEROXIDATION IN MODEL MEMBRANES

J. A. Raleigh

Department of Radiobiology
Cross Cancer Institute
Edmonton, Alberta, Canada T6G 1Z2

ABSTRACT

The potential for chain reaction decomposition puts unsaturated lipids among the most radiosensitive of biological molecules. The possibility that radiation peroxidation of unsaturated lipids in biological membranes might underlie some of the biological effects of ionizing radiation has prompted a number of studies of radiation-induced lipid peroxidation. In model membranes the hydroxyl radical (OH) from water radiolysis initiates peroxidation which proceeds without a lag phase. The process is dose-rate dependent, with peroxidation increasing with decreasing dose rate. In spite of the fact that OH initiates peroxidation, not all OH scavengers inhibit peroxidation. Secondary peroxy radicals from the OH scavengers might in some cases (e.g., dimethylsulfoxide, DMSO) initiate peroxidation as efficiently as OH radicals. An additional important feature of radiation-induced lipid peroxidation is that approximately 60% of the peroxides formed in linoleic acid (18:2) are diene-conjugated hydroperoxides, whereas in linolenic acid (18:3) these are minor products even though overall oxygen consumptions are similar for the two unsaturated fatty acids. The difference in product distribution can be accounted for, in part, by the formation of malonaldehyde-type products in linolenic acid. The formation of these products is also subject to the paradoxical effect of OH scavengers. Antioxidants such as vitamin E remove the paradoxical OH scavenger effect. Studies like these can be instructive in unravelling the effects of ionizing radiation on biological membranes. For example, we have concluded from these studies that the paradoxical effect of the OH scavenger dimethylsulfoxide on erythrocyte hemolysis arises in the lipid component of the plasma membrane and is due to secondary peroxy radicals formed from dimethylsulfoxide. The effect appears to occur in the initial stages of radiation damage and precedes the extensive autoxidative degradation of membrane lipids that parallels hemolysis during the postirradiation incubation period.

INTRODUCTION

The potential for chain reaction peroxidation puts unsaturated lipids among the most radiosensitive of biological molecules. This has led to speculation that lipid peroxidation may initiate some of the biological effects of ionizing radiation, which in turn has stimulated investigations of the basic process of peroxidation in biological and chemical systems. Among the objectives of these investigations has been a characterization of the responses of lipid peroxidation to physical and chemical modifiers. These responses might be used as indirect probes of lipid peroxidation in biological membranes where direct measurement of low levels of peroxidation is often not feasible. Indeed, a recurring observation in irradiated biological membranes at less than massive doses (i.e., less than 200 Gy) is that the amount of lipid peroxide immediately after irradiation is not measurably different from that in unirradiated controls (1-5). Postirradiation incubation does lead to easily measurable levels of peroxidation in membranes, indicating that radiation-induced changes have occurred that make the membrane susceptible to autoxidative change after the radiation has ceased. The general assumption appears to be that lipid peroxidation in the postirradiation period is the cause of membrane disintegration. However, evidence for lipid peroxidation as an initiating event at low doses is lacking.

Wills (3) has proposed three possible mechanisms for the radiation-induced promotion of postirradiation lipid peroxidation in biological membranes. (a) The lipids of the membranes may be altered, possibly by a conformation change so that they are more sensitive to peroxidation; (b) protective antioxidants may be destroyed by irradiation; or (c) some component of the membrane structure may be altered so that it becomes a more effective catalyst for postirradiation peroxidation. The activation of oxygenases by low levels of radiation-induced hydroperoxides leading to unscheduled, naturally occurring peroxidative processes involved in prostaglandin synthesis (6,7) could conceivably be a special case of the last mechanism. If lipid peroxidation can be identified as the initiating event in membrane damage, this mechanism could provide a link between radiation damage and prostaglandin biochemistry.

However, the challenge is to establish whether lipid peroxidation as such can initiate membrane damage. Although the approach of using the response of radiation-induced peroxidation to chemical and physical modifiers as a probe is indirect, it has shown some promise (8). Basic studies of peroxidation in model lipid membranes form the basis of this approach.

Micelles of polyunsaturated fatty acids have proven to be a simple yet useful system in which to characterize the peroxidation process. The micelles provide an association of lipid molecules in contact with an aqueous phase, which can be readily manipulated with respect to both the analysis of damage and the study of the effects of physical and chemical modifiers on radiation peroxidation. In addition, unsaturated fatty acids can be obtained in highly purified form, which can be expected to be the condition of lipids in biological membranes. The high purity minimizes a concern that radiation effects in the micelles might be due to the interaction of water radiolysis products (reactions 1 and 2) with oxidative impurities in the lipid, such as epoxides, carbonyl compounds, and peroxides (9). Phospholipid liposomes are useful models for the lipid bilayer structure in biological membranes and for studies of radiation effects on membrane permeability (e.g., 10-12). Both models have been useful in establishing the basic radiation response of peroxidation in lipid membranes, and the results from both systems are presented in the following.

INITIATION OF RADIATION PEROXIDATION

In dilute micelle solutions of unsaturated fatty acids, most of the energy of the ionizing radiation is absorbed by the water, leading to the formation of hydrated electrons (e^-_{aq}), hydroxyl radicals (OH), molecular hydrogen, and hydrogen peroxide (reaction 1). In the presence of molecular oxygen, e^-_{aq} is rapidly converted to superoxide radical anion ($k[e^-_{aq} + O_2] = 1.88$ x 10^{10} $dm^3 mol^{-1} s^{-1}$) (13), which is in equilibrium with its conjugate acid, the hydroperoxyl radical ($pK_a = 4.88$; reaction 2) (14).

Radical scavenging experiments have established that the hydroxyl radical but neither e^-_{aq} nor O_2^- can initiate lipid hydroperoxide formation in pure lipids (15,16). Subsequent studies (17) have shown that under appropriate pH conditions, HO_2^- can also initiate lipid peroxidation (reaction 3 where LH is an unsaturated lipid molecule). In the case of HO_2^-, hydrogen abstraction at the doubly allylic methylene position in polyunsaturated lipids is estimated to be exothermic by approximately 10 kcal/mole with reaction rate constants of 1.18 to 3.05 x 10^3 $dm^3 mol^{-1} s^{-1}$ reported for a series of polyunsaturated fatty acids (18).

$$H_2O \xrightarrow{\quad\sim\sim\sim\quad} e^-_{aq}, HO^{\cdot}, H_2, H_2O_2 \tag{1}$$

$$e^-_{aq} + O_2 \rightarrow O_2^- \underset{\rightleftarrows}{\overset{H^+}{}} HO_2^{\cdot} \tag{2}$$

Initiation

$$LH + HO^{\cdot} \text{ (or } HO_2^{\cdot}) \rightarrow L^{\cdot} + H_2O \text{ (or } H_2O_2) \tag{3}$$

Propagation

$$L^{\cdot} + O_2 \rightarrow LOO^{\cdot} \tag{4}$$

$$LOO^{\cdot} + LH \rightarrow L^{\cdot} + LOOH \tag{5}$$

Termination

$$\left. \begin{array}{l} LOO^{\cdot} + LOO^{\cdot} \\ \\ L^{\cdot} + LOO^{\cdot} \\ \\ L^{\cdot} + L^{\cdot} \end{array} \right\} \longrightarrow \quad \text{non-radical products} \tag{6}$$

There are reports that superoxide radical anion can initiate lipid peroxidation. Although the reaction rate of O_2^- with pure, unsaturated lipids is vanishingly small ($k = 0.01$ to 0.1 $dm^3mol^{-1}s^{-1}$) (18), there are two possibilities for the involvement of O_2^- in the initiation of peroxidation. Superoxide radical anions might react with lipid peroxides present in the lipid before irradiation, which is often the case in lipids isolated from natural sources. The products of this reaction are lipoxyl radicals (LO'), which can initiate peroxidation. The direct reaction between O_2^- and LOOH (reaction 7) is now considered unlikely (19), but one mediated by trace metal catalysis is possible (reactions 8 and 9). This topic has been reviewed recently (20).

$$O_2^- + LOOH \rightarrow LO^{\cdot} + OH^- + O_2 \tag{7}$$

$$O_2^- + Fe^{+3} \rightarrow Fe^{+2} + O_2 \tag{8}$$

$$LOOH + Fe^{+2} \rightarrow LO^{\cdot} + OH^- + Fe^{+3} \tag{9}$$

It has been shown that alkoxy radicals, of which the lipoxyl radical is an example, have a marked specificity for attacking the allylic or doubly allylic site in polyunsaturated lipids (21). Electron spin resonance studies, in which the pentadienyl radical has been identified, indicate that alkoxy radicals attack the doubly allylic site at least two to three times faster than other sites in polyunsaturated fatty acids (22).

A second possibility arises from the effect of membrane surface potential on pH. A drop of several pH units at the membrane surface would lead to a higher $HO_2^{\cdot}$ concentration due to protonation of the O_2^- diffusing or being drawn into the membrane (10,18,23). A similar proposal has been made in connection with $HO_2^{\cdot}$ formed from α-hydroxyalkylperoxy radicals (24).

In some elegant pulse radiolysis studies, Patterson and his co-workers (25) have established that OH radicals can attack randomly along the fatty acid chain by hydrogen atom abstraction at each of the methylene (CH_2) positions. More recent studies by Heijman et al. (26,27) have established that OH can also attack the unsaturated fatty acids at the double bonds by addition. It has been estimated that hydrogen abstraction predominates over addition to the extent of approximately 10 to 1 (28). The overall second-order reaction rate constant for OH attack on micellar linoleate molecules is reported to be 1.3×10^9 $dm^3mol^{-1}s^{-1}$ (15). This reaction rate constant is an order of magnitude lower than that for monomeric linoleate, which indicates that unsaturated fatty acids are less susceptible to attack by aqueous oxidizing species when located in the close packing of micelles (15).

In the absence of oxygen, extensive secondary hydrogen atom abstraction by the initial lipid radicals occurs at the doubly allylic position to produce the penta-dienyl radical ($-CH_2-CH=CH-CH=CH-\overset{\cdot}{C}H-$). In the close packing of lipids in micelles and liposomes, the secondary abstraction is most likely to occur intermolecularly with a reaction rate constant on the order of 10^3 $dm^3mol^{-1}s^{-1}$ (26,28). Although the dienyl radical can be produced in the absence of oxygen, either it is formed in low yield or it collapses in a way that does not produce the conjugated diene structure formed in the presence of oxygen since few or no diene products are observed in the absence of oxygen (16). Dimerization reactions of the sort observed by Howton (29) in irradiated oleate micelles may be involved.

In the presence of oxygen, carbon radicals in the lipid (L·) react rapidly to produce LOO· (reaction 4; $k_4 \sim 3 \times 10^8$ $dm^3 mol^{-1} s^{-1}$) (30). The hydroperoxy radicals are approximately 10^7 times less reactive than OH and 10^5 times less reactive than LO· so that LOO· will react almost exclusively at the doubly allylic positions (C-H bond energy ~ 76 kcal/mole) (31) in the fatty acid. The overall effect is that, following the initial OH attack in the presence or absence of oxygen, the damage is quickly transferred to the labile allylic and doubly allylic positions in the unsaturated fatty acids, from which propagation of the initial damage by chain reaction peroxidation can proceed.

It is assumed that the propagation and termination steps in radiation-induced lipid peroxidation are generally similar to those in autoxidation (reactions 4, 5, and 6). The length of the chain reaction is determined by the probability that two lipid radicals will react in a termination reaction to produce non-radical products (reaction 6). In radiation studies, this has major ramifications because the concentration of lipid radicals such as LOO· is dependent on the rate at which the radiation is delivered to the system. As this can be controlled for both ionizing radiation and ultraviolet light (32), the dose rate becomes a useful variable in the study of radiation peroxidation, as discussed later.

FATTY ACID ASSOCIATION AND RADIATION PEROXIDATION

The importance of lipid association on radiation-induced peroxidation was first clearly demonstrated by Gebicki and Allen (33). The efficiency of peroxidation increases dramatically as the concentration of sodium linoleate increases above its critical micelle concentration (shown in Figure 1). Gebicki and Allen also identified an ionic strength effect. At high salt concentrations, the critical micelle concentration of fatty acids shifts downward so that the sharp increase in lipid peroxide formation occurs at a lower concentration of unsaturated fatty acid (Figure 2). Although, in general, increasing ionic strength also leads to a decrease in the maximum G value for radiation peroxidation, interesting exceptions were discovered with rubidium and cesium salts (34). The ionic strength effect was observed with these salts, but it was found that they exhibited a specific ion effect in that they promoted the formation of lipid hydroperoxides in linoleate and linolenate micelles in a way that sodium or potassium salts did not (Figure 2). Analysis of the various salts showed that the effect was not due to contamination by transition metal salts. The effect was largely independent of the anion, although this is not true of anions such as thiocynate, bromide, and formate, which react with OH at significant rates under the conditions of the experiment (Figure 3). In fact, bromide (Br⁻) and thiocyanate ions (SCN⁻) fall into the class of OH scavengers, which do not inhibit lipid peroxidation in polyunsaturated lipids (24).

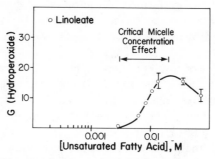

Fig. 1. Effect of linoleate association on yield of lipid hydroperoxide, the critical micelle effect. 10 Gy at 4 Gy/min.

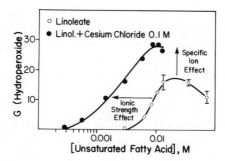

Fig. 2. Yield of linoleate hydroperoxide as a function of fatty acid concentration in presence (●) and absence (○) of 0.1 M CsCl. Ionic strength effect decreases critical micelle concentration. Specific ion effect increases maximum yield of LOOH. 10 Gy at 4 Gy/min.

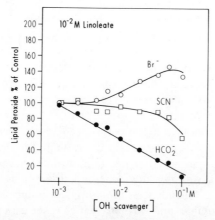

Fig. 3. Effect of bromide (○), thiocyanate (□), and formate (●) on diene conjugated lipid hydroperoxide formation in linoleate micelles. Formate protects as expected but neither bromide nor thiocyanate protects.

The ionic strength and specific ion effects do not appear to be related to increases in bulk viscosity of the micellar solutions (Figure 4). The drop in lipid hydroperoxide yield follows the decreased critical micelle concentration for linoleate solutions, but these changes precede major bulk viscosity changes. Viscosity changes occurring at high salt concentration are not accompanied by additional changes in the G value for hydroperoxide (Figure 4).

However, the specific ion effect does seem to be related to fluidity in the micelles. The nuclear magnetic resonance spectra of linolenate micelles in solutions containing high concentrations of sodium chloride showed that the absorption band due to the non-allylic methylene protons on carbon atoms 3 through 7 was greatly broadened with an attendant drop in peak height. This was not the case for micelles in the presence of high cesium chloride concentrations (Figure 5). The increased broadening of the methylene protons, which is dependent on the sodium chloride concentration (Figure 6), is associated with a loss of freedom of motion of the linolenate monomers within the micelle, presumably caused by a tighter packing of the carboxylate head groups in the presence of increasing concentrations of sodium ions. This loss of fluidity could be the basis of the ionic strength effect of lowering the maximum G value for hydroperoxide formation (33). The NMR signal for the terminal methyl group is not visibly affected by the presence of either sodium or cesium salt (Figure 5). It appears that the restricted motion near the head of the fatty acid molecule seen in the presence of high concentrations of sodium chloride is not transmitted along the whole length of the molecule. ^{13}C NMR indicates that water molecules penetrate the micelles at least as far as the first seven carbon atoms (35), and it is conceivable that the salts are affecting water/fatty acid association in this region. A similar broadening of CH_2 NMR resonances in cetyltrimethylammonium micelles containing benzene molecules was interpreted in this way (36).

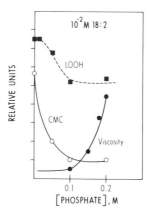

Fig. 4. Response of lipid hydroperoxide formation (HPLC) in linoleate micelles to increasing concentrations of sodium dihydrogen phosphate. Hydroperoxide yield drops in response to ionic strength effect on CMC but does not change with overall increase in solution viscosity.

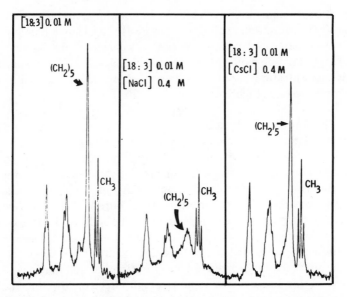

Fig. 5. 'H NMR spectra of linolenate micelles in absence of added salt (left panel), in presence of 0.4 M NaCl (middle panel), and in presence of 0.4 M CsCl (left panel). Note broadening in presence of NaCl of signal for methylene protons near carboxylate end of molecule.

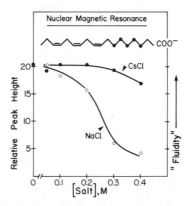

Fig. 6. Effect of increasing salt concentration on broadening of 'H NMR signal for methylene protons near carboxylate group in linolenate micelles. Broadening is expressed on left ordinate as peak height and on right as fluidity in arbitrary units.

Maintenance of fluidity within the micelle in the presence of cesium salts appears to be compatible with a sustained yield of hydroperoxide formation. Unfortunately, the NMR experiment as performed does not reveal whether cesium ions actually increase the fluidity beyond that in fatty acid micelles in the absence of salts. Therefore, the basis of the specific ion effect in raising the maximum G value of lipid hydroperoxides is not resolved by these studies. Attempts to investigate the role of cesium ions on micellar size by freeze-etch electron microscopy were not successful due to the extensive crystallization of the salts in the freeze etch preparation (Raleigh and Copps, unpublished).

A final observation on the specific ion effect concerns a comparison of linoleate and linolenate micelles. The yield of fatty acid hydroperoxides possessing a conjugated diene chromophore is much lower in irradiated linolenate (18:3) micelles. On the other hand, the specific ion effect of cesium ions is very much greater in linolenate micelles in that the yield of linolenate hydroperoxide formation is increased much more than in linoleate micelles (34). Oxygen consumption studies (see below) indicate that linoleate and linolenate are essentially equally sensitive to overall radiation peroxidation but the distribution of peroxidation products is very different in the two fatty acids. The specific ion effect on linolenate and possibly on linoleate as well appears to be a shift in product distribution toward the formation of lipid hydroperoxides, as distinct from other types of peroxidation products.

DOSE RATE EFFECT AND RADIATION PEROXIDATION

Mead (37) appears to be the first to report that radiation peroxidation in unsaturated lipids, notably linoleate micelles, is dose-rate dependent. Petkau reinvestigated this phenomenon in phospholipid black lipid membranes (38) and in phospholipid liposomes (10). A number of other authors have also investigated this phenomenon in various model systems (15,16,32,39,40). The basic observation is that the yield of lipid peroxide at a fixed dose of radiation increases in the membranes as the dose rate decreases (Figure 7). As mentioned above, this is due to the fact that the yield of peroxidation is dependent on a termination step involving two lipid radicals whose instantaneous concentration is dependent on the dose rate of irradiation. The expected relationship is that lipid peroxide is proportional to (dose rate)$^{-0.5}$ (41). This relationship holds very well for the data for pure linoleate and linolenate micelles (34), and for the egg lecithin liposomes shown in Figure 7. Incidentally, the specific ion effect does not alter this relationship (34). When plotted in the manner of Figure 7, the data suggest a pseudo "threshold" dose rate below which lipid peroxidation rapidly increases. For linoleate micelles, this threshold is around 1 Gy/min (100 rad/min), while for the phospholipid liposomes it is around 0.15 Gy/min (15 rad/min). The scales on the abscissa in Figure 7 differ by a factor of ten for the two model membrane systems. One interpretation of these very different threshold dose rates is that in linoleate micelles where LOO˙ radicals have many sensitive sites to attack, termination reactions are relatively less frequent than would be the case in the lecithin liposomes, so that the drop in instantaneous LOO˙ concentration represented by a dose rate of around 1 Gy/min for linoleate micelles is effective in tripping the balance increasingly in favor of propagation (reaction 5) over termination (reaction 6). The implication is that all other things being equal, a lower degree of lipid unsaturation should lead to a lower dose-rate threshold for radiation peroxidation.

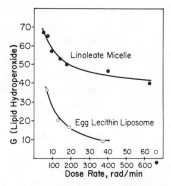

Fig. 7. Dose-rate effect for lipid peroxidation in linoleate micelles (0.01 M fatty acid + 0.5 M CsCl) (●) and in sonicated egg lecithin liposomes (o). 10 Gy at various dose rates. Products measured by high performance liquid chromatography (HPLC).

HYDROXYL RADICAL SCAVENGERS AND RADIATION PEROXIDATION

Patterson and Redpath (15) and Raleigh et al. (16) investigated in some detail the effect of hydroxyl radical scavengers on radiation-induced peroxidation in unsaturated fatty acid micelles and reached similar conclusions. Sodium formate was found to be an effective inhibitor of radiation peroxidation (Figure 3). From these data, a second-order reaction rate constant for OH reaction with micellar linoleate molecules of 3.3×10^9 $dm^3mol^{-1}s^{-1}$ can be calculated in reasonable agreement with a value of 2×10^9 $dm^3mol^{-1}s^{-1}$ calculated on the basis of the benzoate result in Figure 3. Patterson and Redpath (15) calculated a lower value of 1.3×10^9 $dm^3mol^{-1}s^{-1}$, which might be due in part to a higher ionic strength in their micellar solutions, which contained 20 mM phosphate buffer. Although formate and sodium benzoate effectively inhibit peroxidation, both groups of investigators reported that some OH scavengers did not inhibit peroxidation to the extent expected. Perhaps the most dramatic effect (16,42) was seen with dimethylsulfoxide [DMSO; $k_{OH} = 5.8 \times 10^9$ $dm^3mol^{-1}s^{-1}$ (43)], which showed very little if any inhibition of peroxidation. In addition, at concentrations above that necessary to guarantee that essentially all OH radicals were scavenged by DMSO, peroxidation was promoted (Figure 8). The unusual effect of high concentrations of DMSO in promoting lipid hydroperoxide formation in linoleate and linolenate micelles was reminiscent of the specific ion effect of cesium and rubidium ions (34), and suggested that DMSO was acting to change micellar structure. Freeze etch electron microscopy of 10^{-2} M solutions of sodium linoleate revealed that the micelles were highly aggregated in rod-like structures extending up to and, in some cases, beyond 1000 nm in length. This is the picture expected on the basis of known micellar behavior (44). In the same solutions containing 1.0 M DMSO, the free etching was of poorer quality, but it seemed clear that the micelles were smaller (200-500 A diameter) and more uniformly spherical in shape (Raleigh and Copps, unpublished). It may well be that the promotional effect of high concentrations of DMSO arises in part from a more efficient lipid hydroperoxide formation in the smaller micelles for which the water lipid interphase is greatly increased in a situation where OH scavenging by DMSO does not inhibit lipid hydroperoxide formation for reasons discussed below. A similar promotional effect on peroxidation of increasing lipid/water interphase in phospholipid liposomes by sonication (45) has been reported (10,39,46). However, the disproportionate increase in linolenate hydroperoxide formation cannot be accounted for by this mechanism.

Although DMSO is somewhat unique in promoting lipid hydroperoxide formation at high concentration, a number of other OH scavengers were also decidedly less effective than expected at inhibiting lipid hydroperoxide formation in fatty acid micelles. Mannitol, for example (Figure 9), with a reaction rate constant with the hydroxyl radical ($k_{OH} = 1.8 \pm 0.4 \times 10^9$ $dm^3mcl^{-1}s^{-1}$) (47) not much different from formate ($k_{OH} = 2.8 \times 10^9$ $dm^3mol^{-1}s^{-1}$) (48), is very much less effective as an inhibitor of radiation peroxidation. A series of OH scavengers has been classified (24,49) with respect to the ability to inhibit radiation peroxidation in linoleate micelles.

Although a number of possible explanations for the diminished effectiveness of many of the OH scavengers were considered (15,16), the most likely explanation is that secondary radicals formed in the OH scavengers, although less reactive than OH, are capable of initiating peroxidation in the lipids. For example, in the case of ethanol, OH attack produces two types of radical (reaction 10).

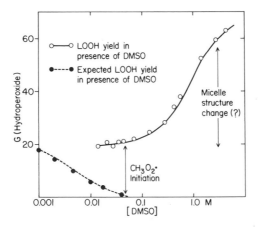

Fig. 8. Effect of increasing DMSO concentration on yield of linoleate hydroperoxide (o) compared to expected response (●). 20 Gy at 4 Gy/min.

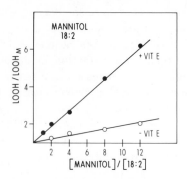

Fig. 9. Effect of a supplement (10^{-5} M) of vitamin E on efficiency of mannitol, a hydroxyl radical scavenger, in inhibiting lipid hydroperoxide formation in linoleate micelles (0.012 M). Slope of curve in presence of vitamin E is identical with theory. 20 Gy at 2 Gy/ min.

$$
\begin{array}{c}
\overset{\displaystyle OO^{\cdot}}{\underset{\displaystyle}{|}} \\
\longrightarrow CH_3CHOH
\end{array}
$$

$$
CH_3CH_2OH \xrightarrow{\quad OH \quad}
\begin{cases}
\xrightarrow{85\%} CH_3\overset{\cdot}{C}HOH \quad \alpha \\
\\
\xrightarrow{15\%} {}^{\cdot}CH_2CH_2OH \xrightarrow{\quad O_2 \quad} {}^{\cdot}OOCH_2CH_2OH \quad \beta
\end{cases}
\tag{10}
$$

In the presence of molecular oxygen, both intermediate α and β radicals will react rapidly ($k[R^{\cdot} + O_2] = 10^8$ - 10^9 dm^{3}mol^{-1}s^{-1}) (50) to produce the corresponding alkyl peroxy radicals. The peroxy radical derived from the α radical can undergo a further transformation to form acetaldehyde and a hydroperoxy radical (reaction 11) (51).

$$
\begin{array}{c}
\overset{\displaystyle OO^{\cdot}}{\underset{\displaystyle}{|}} \\
CH_3CHOH \longrightarrow CH_3CHO + HO_2^{\cdot}
\end{array}
\tag{11}
$$

Both $HO_2^{\cdot}$ and ${}^{\cdot}OOCH_2CH_2OH$ would be capable of initiating peroxidation in the lipids, although a pK_a of 4.88 for $HO_2^{\cdot}$ ensures that its conjugate base O_2^{-} would be present in the pH 10.5 aqueous phase of the micellar solution. As discussed, O_2^{-} is not capable of initiating peroxidation in pure lipids. Reactions analogous to 11 can occur for most of the alcohols in Table 1. The estimated distributions of α-hydroxyalkylperoxy [RCH(OO$^{\cdot}$)OH] and alkylperoxy radicals [RCH(OO$^{\cdot}$)R$'$] for the straight chain alcohols have been calculated from known partial rate factors for OH reactions with organic compounds (52) and listed in Table 1. While reaction 11 will be common to each of the alcohols, it is clear that there is a direct relationship

Table 1. Calculated Peroxy Radical Distribution in Alcohols Used as OH Scavengers

| | | Peroxy Radical Distribution (Percent) | | |
| | | $\overset{\displaystyle OO^{\cdot}}{\underset{\displaystyle RCHOH}{\mid}}$ | $\overset{\displaystyle OO^{\cdot}}{\underset{\displaystyle RCHR'}{\mid}}$ | Deviation[1] |
Alcohol	k_{OH} ($10^9 M^{-1}s^{-1}$)	RCHOH	RCHR'	Factor
CH_3CH_2OH	1.9	85	15	4
$CH_3(CH_2)_2CH_2OH$	4.0	43	57	11
$CH_3(CH_2)_3CH_2OH$	4.9	34	66	16
$CH_3(CH_2)_4CH_2OH$	5.9	29	71	21

[1]Factor by which alcohol is less effective than expected on basis of k_{OH}; determined by slopes in competition plots such as Figure 9.

between the percent of secondary alkyl peroxy radicals and the decreasing effectiveness of the alcohols as inhibitors of lipid hydroperoxide formation. Evidence that the secondary peroxy radicals and not some other factor is involved is provided in the next section.

ANTIOXIDANTS AND RADIATION PEROXIDATION

Polister and Mead in their early studies (53) showed that α-tocopherol (vitamin E) inhibits radiation peroxidation in polyunsaturated fatty acids, and many investigators have confirmed this result. It is clear that very low concentrations of α-tocopherol are capable of inhibiting lipid peroxidation (15). However, radiation destroys the antioxidant, and at the stage where, on average, two hydroxyl radicals have been produced for each α-tocopherol molecule present, the molecule is changed from an anti- to a pro-oxidant.

The mechanism of action of α-tocopherol is that of a chain-breaking antioxidant. Chain-carrying LOO$^{\cdot}$ radicals react with α-tocopherol at a rate ($k_{LOO^{\cdot} + vit E} = 8 \times 10^4$ $dm^3mol^{-1}s^{-1}$) (30) that exceeds that of the rate-determining propagation step in radiation peroxidation of linoleate [reaction 5, $k_5 \sim 60$ (54)] by many orders of magnitude, which accounts for the high efficiency of the process. Rather larger reaction rate constants have been reported for the reaction of cyclohexylperoxy and linolenate peroxy radicals with α-tocopherol in nonaqueous systems [k = 7.9 $\times 10^6$ $dm^3mol^{-1}s^{-1}$ (55) and 5.1×10^5 $dm^3mol^{-1}s^{-1}$ (56), respectively]. This difference between micellar and nonaqueous systems may reflect the sensitivity of the reaction to stereoelectronic factors (56).

The lipid-soluble α-tocopherol is not the only antioxidant capable of inhibiting radiation peroxidation. For example, water-soluble NADH, catechol, and cysteamine are all effective inhibitors of radiation peroxidation in micellar fatty acids and in the same concentration range as vitamin E (49). However, at high concentration, cysteamine converts to a pro-oxidant (16). NADH and vitamin E are additive in their effects (Raleigh and Shum, unpublished), with no indication of the synergistic effect proposed for other antioxidant couples such as vitamin E and glutathione (8). The fact that an antioxidant need not be lipid soluble to be effective combined with the observation that water soluble OH scavengers can inhibit lipid peroxidation suggests very strongly that in micelles of fatty acids, at least, much of the radical chemistry is occurring in an environment easily accessible to the aqueous phase. Similar studies with liposomes show that α-tocopherol is the most efficient in inhibiting radiation peroxidation and in sparing the most highly unsaturated fatty acids from degradation. However, water-soluble antioxidants such as cysteamine and glutathione are also effective (39).

The ability of antioxidants to quench chain-carrying LOO˙ radicals in competition with the propagation step in lipid peroxidation suggested an experiment to test the idea that secondary peroxy radicals from OH scavengers were accounting for their diminished efficiency. A supplement of α-tocopherol that in itself did not completely inhibit peroxidation should scavenge secondary peroxy radicals from the OH scavengers and turn them into perfectly good inhibitors of radiation peroxidation. This was found to be the case (24,49), as demonstrated for mannitol in Figure 9. In the absence of vitamin E, the slope of the inhibition curve for mannitol is much too shallow, which indicates an apparently low reaction rate constant for the reaction of mannitol with OH. However, in the presence of a supplement of vitamin E (which itself gives less than 50 percent protection), the slope of the inhibition curve for mannitol is restored to the theoretical value. This effect of supplemental vitamin E was observed with all the less-than-fully effective OH scavengers. (Br_2^- and SCN_2^- were not included in the study.) This provides strong evidence that secondary peroxy radicals are the primary source of the deviations observed.

DISTRIBUTION OF PRODUCTS IN RADIATION PEROXIDATION

A number of investigators have been struck by the observation that in the radiation peroxidation of polyunsaturated fatty acid micelles, the more unsaturated linolenate molecule, although more susceptible to autoxidation than linoleate (Table 2), forms very much less lipid hydroperoxide possessing a conjugated diene chromophore. This is true whether the analysis is done spectrophotometrically (15,16,58) or chromatographically (16). This apparent anomaly was resolved when oxygen consumption in the irradiated micelles was used as a measure of radiation peroxidation (Table 2) (59) or when overall destruction of the fatty acid was measured (12). A G value for O_2 consumption of ~30 is essentially the same for linoleate, linolenate, and 1:1 mixed micelles of these two fatty acids. However, it is clear that the distribution of products is very different in these micelles. For linoleate, approximately 70 percent of the oxygen consumption can be accounted for in terms of diene conjugated lipid hydroperoxide, which was identified as a mixture of 9- and 13-hydroperoxides by HPLC (16). On the other hand, only 13 percent

Table 2. Comparison of Autoxidation and Radiation Peroxidation in Unsaturated Fatty Acids

| Fatty Acid | Autoxidation Rel. Rate[2] | Radiation Peroxidation[1] | | |
		G(LOOH)	G(TBA)	G(O$_2$ Consumption)
Oleic (18:1)	1.0	-	-	-
Linoleic (18:2)	40	24	0	33
Linolenic (18:3)	98	4	3	30
18:2/18:3 = 1/1	-	4.5	1.5	36

[1]0.012 M fatty acid: 20 Gy at 4 Gy/min
[2]See ref. 57.

of the oxygen consumption shows up as diene lipid hydroperoxide in linolenate micelles. Peroxidation processes in linolenate are known to be complex (Figure 10) and, in particular, malondialdehyde type products seemed a likely alternative to diene hydroperoxides. Using standard procedures and authentic malondialdehyde as a standard, the yield of the barbituric acid-positive materials in irradiated linolenate micelles was measured. This type of product accounted for only an additional 10 percent of the yield of oxygen consumption, leaving 77 percent of the products of O$_2$ consumption unaccounted for. Even allowing for the possibility of the consumption of two molecules of O$_2$ for every lipid radical formed according to Figure 10 (9), approximately 50 percent of the O$_2$ consumption would still be unaccounted for in terms of specific products. The specific ion effect and the effect of high concentrations of DMSO on the product distribution in linolenate micelles is perhaps interesting in this regard. In both cases, there is a proportionately much larger increase in yield of lipid hydroperoxides containing the diene chromophore than is seen in linoleate micelles. It seems possible at this stage that the fluidity or some other parameter of micelle structure can have a profound effect on the distribution of peroxidation products, particularly in linolenate micelles. In a study spanning a narrower concentration range of DMSO than earlier studies, it was found that the yield of malonaldehyde type products also increased at DMSO concentrations greater than 0.1 M in 0.012 M micellar linolenate solutions (Figure 11). Whether these products follow the pattern of the lipid hydroperoxides at DMSO concentrations up to 3.0 M (42) is not known.

MIXED MICELLES AND RADIATION PEROXIDATION

The specific ion effect and the effect of high DMSO concentrations in increasing diene lipid hydroperoxide disproportionately in linolenate micelles are in contrast to the effect of forming mixed micelles of linolenate (18:3) and linoleate (18:2) with oleate (18:1) molecules (15). The effect of oleate is to disproportionately decrease the yield of diene conjugation in linolenate compared to linoleate molecules. It

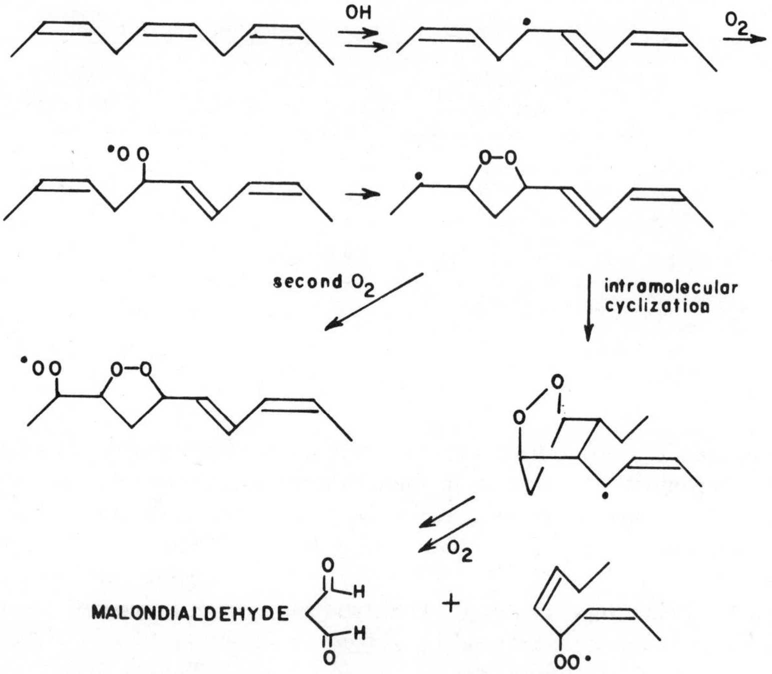

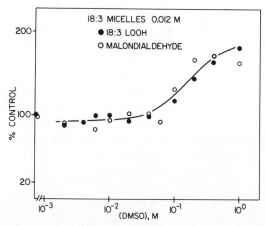

Fig. 10. Possible mechanism for extra oxygen consumption in irradiated linolenate leading to formation of malondialdehyde but not diene hydroperoxide (9)

Fig. 11. Effect of increasing DMSO concentration on formation of lipid hydroperoxides (o) and malondialdehyde-type products (●) in irradiated linolenate micelles. Yields have been normalized for comparison purposes. 30 Gy at 2.9 Gy/min.

was postulated that the effect might be related to a negative influence of oleate on linolenate packing with respect to chain propagation in linolenate molecules. However, a possible alternative is that the overall degree of radiation peroxidation is unchanged in linolenate molecules but that a shift in product distribution away from diene lipid hydroperoxide has occurred in the linolenate/oleate mixed micelles, which would be the opposite of the effect of cesium ions and of high concentrations of DMSO.

The yield of overall O_2 consumption does not appear to be affected in mixed micelles, which are 1:1 in linoleate and linolenate compared to micelles containing single components (Table 2). The yield of linoleate diene hydroperoxide decreases slightly more than expected on the basis of relative concentrations and reactivities of linolenate (four doubly allylic CH bonds) and linoleate (two doubly allylic CH bonds) in mixed micelles (Figure 12). In the case of the products of linolenate peroxidation (that is, conjugated diene hydroperoxide and malondialdehyde), there is indication of a small excess yield of malonaldehyde but an expected yield of the hydroperoxide in mixed micelles containing low concentrations of linolenate (Figure 12). Therefore, even though all the products of linolenate peroxidation have not been identified, there is little evidence for a major distribution of products among 18:2 and 18:3 fatty acids in these mixed micelles. This is consistent with the results of a study of linoleate and linolenate mixed micelles in which the overall yield of fatty degradation was measured (12). At a total fatty acid concentration (10^{-2} M) the same as that for the mixed micelles (from which the data in Figures 12 and 13 were obtained), a similar small effect on linolenate destruction, albeit

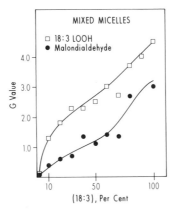

Fig. 12. Lipid hydroperoxide (HPLC) and malonaldehyde (colorimetric) yields from linolenate molecules in mixed micelles with linoleate. Note relatively low G values for these products compared to an overall G value for O_2 consumption of approximately 30.

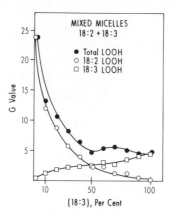

Fig. 13. Linoleate and linolenate diene conjugated hydroperoxides in mixed micelles. At high linolenate concentrations, lipid hydroperoxide is a relatively minor peroxidation product. 20 Gy at 4 Gy/min.

in the opposite direction, was observed (12). However, when a competition kinetic analysis of the sort used by Patterson and Redpath (15) in their study of mixed micelles containing oleate is applied to the 10^{-2} M data of Mooibroek et al., a trend emerges. The least unsaturated fatty acid, linoleate, inhibits decomposition of the more unsaturated fatty acids linolenate (18:3), arachidonate (20:4), and clupanadonate (22:6). In terms of overall decomposition, linoleate protects 18:3, 20:4, and 22:6 to approximately the same extent. However, in terms of the expected relative sensitivity of these fatty acids on the basis of their number of doubly allylic bonds, linoleate is increasingly effective as a protector as the degree of unsaturation increases 22:6 > 20:4 > 18:3. This is accompanied by a larger-than-expected decomposition of 18:2 in these mixed micelles. A similar result has been obtained in studies (60) of autoxidizing mixtures of unsaturated fatty esters, where again a less unsaturated oleate molecule is oxidized more efficiently than expected in the presence of an excess of linoleate molecules.

Patterson and Redpath (15) proposed that the effect of oleate in decreasing conjugated diene formation in the more unsaturated linolenate molecules in mixed micelles is due to a steric effect of the oleate on the rate-determining propagation step (reactions 4 and 5) but not the initiation step of peroxidation. Oleate was seen to inhibit close packing of linolenate molecules with a concomitant decrease in the efficiency of propagation step which, on general principles, was assumed to be enhanced by close packing. However, diene conjugation in linoleate was affected to a lesser extent than that in linolenate, which is reminiscent of the specific ion effect of cesium ion (Figure 2). If the physical basis of the effect in both cases is similar, then the NMR results (Figures 5 and 6) indicate that close packing of linolenate molecules accompanied by a decrease in freedom of motion of the micellar fatty acids would not lead to an increase in the yield of diene conjugation as proposed (15). The effect of oleate with one cis double bond might, instead, be to force linolenate with three cis double bonds to self-associate to an unusual extent in the mixed micelle, with a resultant decrease in freedom of motion and in the efficiency of conjugated diene hydroperoxide formation. A similar effect of linoleate on linolenate, arachidonate, and clupanadonate might account for the results reported for Mooibroek et al. (12). Increased packing with loss of motional freedom might also underlie the decrease in maximum G value for hydroperoxide formation in homogeneous micelles in the presence of high salt concentration (Figure 4).

RADIOBIOLOGY AND RADIATION PEROXIDATION

The heuristic value of physical and chemical modifications of radiation response is well established in radiation biology. The phenomenon of the oxygen effect, for example, has been a guiding light to the establishment of a broadly acceptable picture of the likely chemical events occurring in an irradiated cell (61,62). It has also led to investigations of the role of membrane damage in radiobiology following Alper's N (DNA) and O (membrane) classification of radiation damage (63).

Dose-rate studies in repair-competent, proliferating cells have led to (a) the observation that cell survival at a fixed dose of radiation generally increases at low dose rate and (b) the conclusion that the rate of repair outstrips the rate of damage formation at a dose rate below 1 Gy/min (64). It is not surprising that the inverse to this normal dose-rate effect, which has a "threshold" of less than 1 Gy/min in model membranes (Figure 7), would be masked in repair-proficient cells. However, under appropriate circumstances, the inverse dose-rate effect might be a useful probe of chain reaction degradation of membranes. Konings has tested this concept in biological membrane damage and has shown increased dose-rate responses in erythrocyte permeability (65) and lymphocyte interphase death (8). While these results are consistent with lipid peroxidation's being the initiating event in permeability changes, there is much evidence to suggest that membrane thiols are important targets in radiation-induced changes in cation permeability (review, 66). Thiols themselves can be subject to radiation-induced chain reaction decomposition (67), so that if appropriately arranged in the protein lipid mosaic of biological membranes, they could participate in an inverse dose-rate response. This might be particularly true if close spatial and functional relationships of thiols and unsaturated lipids in biological membranes (68) were to permit a cooxidation of thiols and unsaturated lipids. At this stage the observed inverse dose rate suggests the involvement of a chain reaction and is an important clue to the initiating event. Whether this is confined to the lipid component of the membrane is an open question.

In the absence of chain reactions, thiol destruction could still lead to lipid peroxidation in the postirradiation period since these compounds are known to be effective *in vitro* and *in vivo* antioxidants against lipid peroxidation, and their destruction could represent a specific example of Will's second mechanism proposed to account for the stimulation of postirradiation peroxidation.

The observation that secondary peroxy radicals from OH scavengers, including carbohydrate molecules like glucose and mannitol, can initiate peroxidation in micellar fatty acids may have relevance to the situation in biological membranes that have glycoproteins attached to their surface. Scavenging of OH radicals by these proteins may in fact not protect membrane lipids. The ultimate sensitivity of biological membranes may be determined by their complement of chain-terminating antioxidants such as thiols, reduced pyridine coenzymes, or α-tocopherol. For example, in a study of erythrocyte hemolysis, it was found that human erythrocytes (which appeared to lack a full complement of α-tocopherol) were not protected by DMSO whereas bovine erythrocytes (for which additional α-tocopherol added no further radioprotection) were fully protected by DMSO (69). This DMSO effect might be developed into a probe of the antioxidant status of irradiated membranes. For example, it should be possible to establish if other human blood cells also lack a full complement of antioxidant as it appears that erythrocytes do.

The mixed micelle results combined with the ionic strength effect, the specific ion effect, and the DMSO effect all point to the importance of steric factors in peroxidation of membrane lipids. There is the definite indication in these studies that the most unsaturated lipid need not be the most radiosensitive, even though there is little doubt that the most unsaturated lipids will be subject to the greatest degree of peroxidative destruction in the postirradiation incubation period (39). Wills and Wilkinson (2) had reported many years ago that the amount of lipid but not its degree of unsaturation determined the radiosensitivity of subcellular organelles with respect to postirradiation peroxidation. Recent attempts to implicate membrane lipid damage in radiobiological end points by lipid replacement studies (review, 66) are consistent with this finding. Generally, it is not the sensitivity of the membrane fatty acids to autoxidation that is important but their effect on membrane fluidity or packing. Permeability changes and cell surface effects, although transient, are among the most radiosensitive. These permeability changes, associated with apparently normal physiological responses to stress, may very well predispose membrane lipids to autoxidation by changing the conformation and packing of unsaturated lipids or by increasing the aqueous/lipid interface. Hence the lipids are exposed to metal ion catalyzed autoxidation in a manner analogous to the effect of chaotropic ions in stimulating autoxidation in the membranes of subcellular organelles (70). A third possibility involves the interesting observation that very low concentrations of lipid hydroperoxides (10^{-8} M) are capable of activating the normal physiological, but peroxidative, processes involved with prostaglandin synthesis (6,7). This relationship may ultimately provide a link between radiation damage and prostaglandin biochemistry and form the basis of a rationale for developing radioprotectors that inhibit radiation-induced, unscheduled peroxidation.

ACKNOWLEDGMENTS

The author thanks W. Kremers and F. Y. Shum for excellent technical assistance. Dr. C. Koch measured O_2 consumption in the irradiated micelles and has been a source of stimulating discussions over a number of years. Ms. G. Kennedy and the members of the Audiovisual Department at the Cross Cancer Institute headed by K. Liesner have been of great assistance in the preparation of this manuscript. This work has been supported by the Atomic Energy of Canada Limited, the Alberta Cancer Board, and the Alberta Heritage Savings Trust Fund.

REFERENCES

1. Myers, D. K., and Bide, R. W. Biochemical effects of X-irradiation on erythrocytes. Radiat. Res. 27: 250-263, 1966.
2. Wills, E. D., and Wilkinson, A. E. The effect of irradiation on lipid peroxide formation in subcellular fractions. Radiat. Res. 31: 732-747, 1967.
3. Wills, E. D. Effects of irradiation on subcellular components. I. Lipid peroxide formation in the endoplasmic reticulum. Int. J. Radiat. Biol. 17: 217-228, 1970.
4. Konings, A. W. T., and Oosterloo, S. K. Radiation effects on membranes. II. A comparison of the effects of X-irradiation and ozone exposure with respect to the relation of antioxidant concentration and the capacity for lipid peroxidation. Radiat. Res. 81: 200-207, 1980.
5. Fonck, K., Scherphof, G. L., and Konings, A. W. T. The effect of X-irradiation on membrane lipids of lymphosarcoma cells *in vivo* and *in vitro*. J. Radiat. Res. 23: 371-384, 1982.
6. Lands, W. E. M. Interactions of lipid hydroperoxides with eicosanoid biosynthesis. J. Free Rad. Biol. Med. 1: 97-101, 1985.
7. Marshall, P. J., Kulmacz, R. J., and Lands, W. E. M. Hydroperoxides, free radicals, and prostaglandin synthesis. In: "Oxygen Radicals in Chemistry and Biology." W. Bors, M. Saran, and D. Tait, eds. de Gruyter, New York, 1984, pp. 299-307.
8. Konings, A. W. T. The involvement of polyunsaturated fatty acyl chains of membrane phospholipids in radiation induced cell death of mammalian cells. In: "Oxygen Radicals in Chemistry and Biology." W. Bors, M. Saran, and D. Tait, eds. de Gruyter, New York, 1984, pp. 593-602.
9. Frankel, E. N. Chemistry of free radical and singlet oxidation of lipids. Prog. Lipid Res. 23: 197-221, 1985.
10. Petkau, A., and Chelak, W. S. Radioprotective effect of superoxide dismutase on model phospholipid membranes. Biochim. Biophys. Acta 433: 445-456, 1976.
11. Nakazawa, T., and Nagatsuka, S. Radiation-induced lipid peroxidation and membrane permeability in liposomes. Int. J. Radiat. Biol. 38: 537-544, 1980.

12. Mooibroek, J., Trieling, W. B., and Konings, A. W. T. Comparison of the radiosensitivity of unsaturated fatty acids, structured as micelles or liposomes, under different experimental conditions. Int. J. Radiat. Biol. 42: 601-609, 1982.

13. Anbar, M., and Neta, P. A compilation of specific bimolecular rate constants for the reactions of hydrated electrons, hydrogen atoms and hydroxyl radicals with inorganic and organic compounds in aqueous solutions. Int. J. Appl. Rad. Isotop. 18: 493-523, 1967.

14. Behar, D., Czapski, G., Robani, J., Dorfman, L. M., and Schwarz, H. A. The acid dissociation constant and decay kinetics of the perhydroxyl radical. J. Phys. Chem. 74: 3209-3213, 1970.

15. Patterson, L. K., and Redpath, J. L. Radiation-induced peroxidation in fatty acid soap micelles. In: "Micellization, Solubilization and Microemulsions," Vol. 2. K. L. Mittal, ed. Plenum Press, New York, 1977, pp. 589-601.

16. Raleigh, J. A., Kremers, W., and Gaboury, B. Dose-rate and oxygen effects in models of lipid membranes: Linoleic acid. Int. J. Radiat. Biol. 31: 203-213, 1977.

17. Gebicki, J. M., and Bielski, B. H. J. Comparison of the capacities of the perhydroxyl and the superoxide radicals to initiate chain oxidation of linoleic acid. J. Amer. Chem. Soc. 103: 7020-7023, 1981.

18. Bielski, B. H., Arudi, R. L., and Sutherland, M. W. A study of the reactivity of HO_2/O_2^- with unsaturated fatty acids. J. Biol. Chem. 258: 4759-4761, 1983.

19. Thomas, M. J., Sutherland, M. W., Arudi, R. L., and Bielski, B. H. J. Studies of the reactivity of HO_2/O_2^- with unsaturated hydroperoxides in ethanolic solutions. Arch. Biochem. Biophys. 233: 772-775, 1984.

20. Girotti, A. W. Mechanisms of lipid peroxidation. J. Free Rad. Biol. Med. 1: 87-95, 1985.

21. Small, R. D., Jr., Scaiano, J. C., and Patterson, L. K. Radical processes in lipids. A laser photolysis study of t-butoxy radical reactivity toward fatty acids. Photochem. Photobiol. 29: 49-51, 1979.

22. Bascetta, E., Gunstone, F. D., Scrimgeour, C. M., and Walton, J. C. E.S.R. observation of pentadienyl and allyl radicals on hydrogen abstraction from unsaturated lipids. J. Chem. Soc., Chem. Commun. 110-112, 1982.

23. Fukuzawa, K., and Gebicki, J. M. Oxidation of α-tocopherol in micelles and liposomes by the hydroxyl, perhydroxyl, and superoxide free radicals. Arch. Biochem. Biophys. 226: 242-251, 1983.

24. Raleigh, J. A., and Shum, F. Y. Radioprotection in model lipid membranes by hydroxyl radical scavengers: Supplementary role for α-tocopherol in scavenging secondary peroxy radicals. In: "Radioprotectors and Anticarcinogens." O. F. Nygaard and M. G. Simic, eds. Academic Press, New York, 1983, pp. 87-102.

25. Patterson, L. K., and Hasegawa, K. Pulse radiolysis studies in model lipid systems. The influence of aggregation on kinetic behavior of OH induced radicals in aqueous sodium linoleate. Ber. Bunsenges. Phys. Chem. 82: 951-956, 1978.

26. Heijman, M. G. J., Nauta, H., and Levine, Y. K. A pulse radiolysis study of the dienyl radical in oxygen-free linoleate solutions: Time and linoleate concentration dependence. Radiat. Phys. Chem. 26: 73-82, 1985.

27. Heijman, M. G. J., Heitzman, A. J. P., Nauta, H., and Levine, Y. K. A pulse radiolysis study of the reactions of OH/O⁻ with linoleic acid in oxygen-free aqueous solution. Radiat. Phys. Chem. 26: 83-88, 1985.

28. Patterson, L. K. Studies of radiation induced peroxidation in fatty acid micelles. In: "Oxygen and Oxy-Radicals in Chemistry and Biology." M. A. J. Rodgers and E. L. Powers, eds. Academic Press, New York, 1981, pp. 89-95.

29. Howton, D. R. Nature of the products formed by gamma irradiation of deaerated aqueous potassium oleate. Radiat. Res. 20: 161-186, 1963.

30. Patterson, L. K. Investigation of micellar behavior by pulse radiolysis. In: "Solution Behavior of Surfactants," Vol. 1. K. L. Mittal and E. J. Fendler, eds. Plenum Press, New York, 1982, pp. 285-297.

31. Korcek, S., Chenier, J. H. B., Howard, J. A., and Ingold, K. U. Absolute rate constants for hydrocarbon autoxidation. XXI. Activation energies for propagation and the correlation of propagation rate constants with carbon-hydrogen bond strengths. Can. J. Chem. 50: 2285-2297, 1972.

32. Mandal, T. K., and Chatterjee, S. N. Ultraviolet and sunlight-induced lipid peroxidation in liposomal membrane. Radiat. Res. 83: 290-302, 1980.

33. Gebicki, J. M., and Allen, A. O. Relationship between critical micelle concentration and rate of radiolysis of aqueous sodium linoleate. J. Phys. Chem. 73: 2443-2445, 1969.

34. Raleigh, J. A., and Kremers, W. Promotion of radiation peroxidation in models of lipid membrane by cesium and rubidium counter-ions: Micellar linoleic and linolenic acids. Int. J. Radiat. Biol. 34: 439-447, 1978.

35. Menger, F. M. On the structure of micelles. Accts. Chem. Res. 12: 111-117, 1979.

36. Eriksson, J. C., and Gillberg, G. NMR-studies of the solubilization of aromatic compounds in cetyltrimethylammonium bromide solution. Surface Chemistry 148-156, 1965.

37. Mead, J. F. The irradiation-induced autoxidation of linoleic acid. Science 115: 470-472, 1952.

38. Petkau, A. Effect of ²²Na⁺ on a phospholipid membrane. Health Physics 22: 239-244, 1972.

39. Konings, A. W. T., Damen, J., and Trieling, W. B. Protection of liposomal lipids against radiation induced oxidative damage. Int. J. Radiat. Biol. 35: 343-350, 1979.

40. Chatterjee, S. N., and Agarwal, S. Lipid peroxidation by ultraviolet light and high energy α particles from a cyclotron. Radiat. Environ. Biophys. 21: 275-280, 1983.

41. Spinks, J. W. T., and Woods, R. J. In: "An Introduction to Radiation Chemistry," second edition. Wiley-Interscience, New York, 1976, pp. 187-191.

42. Raleigh, J. A., and Kremers, W. DMSO does not protect against hydroxyl radical induced peroxidation in model membranes. Int. J. Radiat. Biol. 39: 441-444, 1981.

43. Reuvers, A. P., Greenstock, C. L., Borsa, J., and Chapman, J. D. Studies on the mechanism of chemical radioprotection by dimethylsulfoxide. Int. J. Radiat. Biol. 24: 533-536, 1973.

44. Mittal, K. L., and Mukerjee, P. The wide world of micelles. In: "Micellization, Solubilization, and Microemulsions," Vol. 1. K. L. Mittal, ed. Plenum Press, New York, 1977, pp. 1-21.

45. Chapman, D., Fluck, D. J., Penkett, S. A., and Shipley, G. G. Physical studies of phospholipids X. The effect of sonication of aqueous dispersions of egg yolk lecithin. Biochim. Biophys. Acta 163: 255-261, 1968.

46. Edwards, J. C., and Quinn, P. J. The structure of unsaturated lipid dispersions in aqueous systems influences susceptibility to oxidation. Biochim. Biophys. Acta 710: 502-505, 1982.

47. Goldstein, S., and Czapski, G. Mannitol as an OH scavenger in aqueous solution and in biological systems. Int. J. Radiat. Biol. 46: 725-729, 1984.

48. Willson, R. L., Greenstock, C. L., Adams, G. E., Wageman, R., and Dorfman, L. M. The standardization of hydroxyl radical rate data from radiation chemistry. Int. J. Radiat. Phys. Chem. 3: 211-220, 1971.

49. Raleigh, J. A., and Shum, F. Y. Radiation peroxidation in micellar fatty acids. In: "Oxygen Radicals in Chemistry and Biology." W. Bors, M. Saran, and D. Tait, eds. de Gruyter, New York, 1984, pp. 581-591.

50. Ross, A. B., and Neta, P. Rate constants for radiation of aliphatic carbon-centered radicals in aqueous solution. U.S. National Bureau of Standards (NSRDS-NBS 70), Gaithersburg, MD, 1982, pp. 25-27.

51. Bothe, E., Schuchman, M. N., Schulte-Frohlinde, D., and von Sonntag, C. HO_2 elimination from α-hydroxyalkyl-peroxyl radicals in aqueous solutions. Photochem. Photobiol. 28: 639-644, 1978.

52. Anbar, M., Meyerstein, D., and Neta, P. Reactivity of aliphatic compounds towards hydroxyl radicals. J. Chem. Soc. B: 742-747, 1966.

53. Polister, B. H., and Mead, J. F. Effect of certain vitamins and antioxidants on irradiation-induced autoxidation of methyl linoleate. Agric. Food Chem. 2: 199-202, 1954.

54. Howard, J. A., and Ingold, K. U. Absolute rate constants for hydrocarbon autoxidation. VI. Alkyl, aromatic, and olefinic hydrocarbons. Can. J. Chem. 45: 793-802, 1967.

55. Hunter, E. P. L., and Simic, M. G. Kinetics of peroxy radical reactions with antioxidants. In: "Oxy-Radicals and Their Scavenging Systems," Vol. 1. G. Cohen and R. A. Greenwald, eds. Elsevier, New York, 1983, pp. 32-37.

56. Burton, G. W., Doba, T., Gobe, E. J., Hughes, L., Lee, F. L., Prasad, L., and Ingold, K. U. Autoxidation of biological molecules. 4. Maximizing the antioxidant activity of phenols. J. Amer. Chem. Soc. 107: 7053-7065, 1985.

57. Gebicki, J. M. Linoleate micelles as models for radiobiological effects. In: "Biophysical Aspects of Radiation Quality." International Atomic Energy Agency, Vienna, 1971, pp. 229-238.

58. Raleigh, J. A., Shum, F. Y., and Koch, C. J. Radiation chemistry of membrane damage. Distribution of oxygenated products in linoleate and linolenate micelles irradiated at low doses. Proc. 7th I.C.R.R., A4-31. Amsterdam, 1983.

59. Holman, R. T., and Elmer, D. C. The rates of oxidation of unsaturated fatty acids and esters. J. Amer. Oil Chem. Soc. 24: 127-129, 1947.

60. Wong, W-S.D., and Hammond, E. G. Analysis of oleate and linoleate hydroperoxides in oxidized ester mixtures. Lipids 12: 475-479, 1977.

61. Howard-Flanders, P., and Alper, T. The sensitivity of microorganisms to irradiation under controlled gas conditions. Radiat. Res. 7: 518-540, 1957.
62. Koch, C. J. Competition between radiation protectors and radiation sensitizers in mammalian cells. In: "Radioprotectors and Anticarcinogens." O. F. Nygaard and M. G. Simic, eds. Academic Press, New York, 1983, pp. 275-295.
63. Alper, T. "Cellular Radiobiology." Cambridge University Press, New York, 1979.
64. Hall, E. J. "Radiobiology for the Radiologist," second edition. Harper and Row, New York, 1978, pp. 148-150.
65. Konings, A. W. T. Radiation-induced efflux of potassium ions and haemoglobin in bovine erythrocytes at low doses and low dose rates. Int. J. Radiat. Biol. 40: 441-444, 1981.
66. Edwards, J. C., Chapman, D., Cramp, W. A., and Yatvin, M. B. The effects of ionizing radiation on biomembrane structure and function. Prog. Biophys. Mol. Biol. 43: 71-93, 1984.
67. Lal, M. Radiolytic oxidation of cysteine in aerated and oxygen saturated solution. Radiat. Phys. Chem. 19: 427-434, 1982.
68. Robinson, J. D. Interaction between protein sulphydryl groups and lipid double bonds in biological membranes. Nature, Vol. 212(58): 199-200, 1966.
69. Miller, G. G., and Raleigh, J. A. Action of some hydroxyl radical scavengers on radiation-induced haemolysis. Int. J. Radiat. Biol. 43: 411-419, 1983.
70. Hatefi, Y., and Hanstein, W. G. Solubilization of particulate proteins and non-electrolytes by chaotropic agents. Proc. Natl. Acad. Sci. USA 62: 1129-1136, 1969.

ROLE OF MEMBRANE LIPID COMPOSITION IN RADIATION-INDUCED DEATH OF MAMMALIAN CELLS

A. W. T. Konings

Department of Radiopathology
University of Groningen
Groningen, The Netherlands

ABSTRACT

There have been many speculations during the last 10 years concerning a possible role of membrane lipids in cellular radiation injury. This article addresses this question and discusses recent experiments related to this topic. If lipids are involved in the mechanism of radiation-induced cell death, then the most probable molecular reactions underlying this effect are related to lipid peroxidation.

In the first part of this paper, it is shown that the naturally occurring polyunsaturated fatty acyl (PUFA) chains of phospholipids, arachidonic acid (20:4), and docosahexanoic acid (22:6) are the most radiosensitive lipid moieties and are damaged by peroxidative reactions. Protection by vitamin E is very efficient. In the second part, experiments are discussed in which mammalian cells in culture are modified so that more PUFA is present in the membrane phospholipids. Also the antioxidant status of the cells is manipulated. The effect of these membrane modifications on radiation-induced reproductive death is reported. The third part of this article is concerned with the role of membrane lipids and protective systems in radiation-induced interphase death.

INTRODUCTION

Although DNA is considered to be a very important cellular target, there has always been a suspicion that some non-DNA structures might also play a role in the mechanism of cell killing by ionizing radiation. The most popular candidate proposed for this role has been the membranous structure of the cell (1). Especially Alper (2) has stressed the possibility that membrane lipids might be the "O" type target and DNA the "N" type target. From data extracted from the literature, Alper collected a "hierarchy" of oxygen enhancement ratios (OER's) ranging from 0.1 to 10. She postulated that a high OER corresponded with substantial "O" type damage in the cell, while a low OER was related to relatively more "N" type damage.

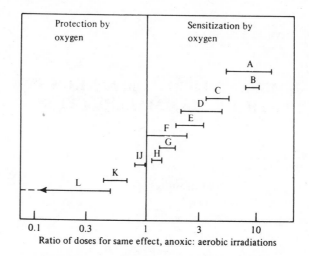

Fig. 1. Hierarchy of OER's for various classes of observations (A-L). Oxygen may protect or sensitize radiation-induced changes. A, release of lysosomal enzymes; L, bacteriophage irradiated in buffer. For other types of observations (B-K), see Alper, 1979, p. 211. (After ref. 2, Alper, 1979)

The two extreme situations (Figure 1) were data from experiments with bacteriophages (OER 0.1-0.5) and lysosomes (enzyme release, OER 5-15).

For a proper functioning of the cells, it is important that the membranes are well preserved. This is true not only for metabolic and regulatory functions within the cell, but also for contacts of cells with each other and with extracellular material. This article reviews part of the research performed in our laboratories that is related to the possible involvement of membrane lipids in radiation-induced cell death.

At least two different mechanisms of cell death (3) can be distinguished: reproductive death as determined by the loss of clonogenic ability, and interphase death as assayed by the uptake of an exclusion dye such as trypan blue (TB) or the loss of intracellular potassium. Both types of cell death are considered here and exemplified for normal and fatty acid-substituted mouse fibroblasts in culture.

LIPID PEROXIDATION

Vulnerability of Different Fatty Acids; Role of Structural Organization

The structures of biological membranes are maintained essentially by complex interactions of lipids and proteins. Model studies on the interaction of the primary free radicals with fatty acids have shown (see, e.g., ref. 4) that the hydrated electrons are generally unreactive ($k = 10^6$ dm^3 mol^{-1} s^{-1}). However, hydroxyl radicals react at rate constants that are between 10^8 and 10^{10} dm^3 mol^{-1} s^{-1}. The hydroxyl radicals can nonspecifically abstract hydrogen atoms from all the C-H bonds. The resulting radicals can react with oxygen ($k = 3\text{-}10 \times 10^8$ dm^3 mol^{-1} s^{-1}) to form peroxyl radicals ($LO_2\cdot$), which in turn can react with the α-methylenic hydrogen atoms of the unsaturated moieties.

Initiation:	$LH + OH\cdot \rightarrow L\cdot + H_2O$	(reaction 1)
Propagation:	$L\cdot + O_2 \rightarrow LO_2\cdot$	(reaction 2)
	$LO_2\cdot + *LH \rightarrow LO_2H + *L\cdot$	(reaction 3)
	$*L\cdot + O_2 \rightarrow *LO_2\cdot$	(reaction 4)
Termination:	$LO_2\cdot + LO_2\cdot \rightarrow (LO_2)_2$	(reaction 5)
	$(LO_2)_2 \rightarrow 2LO\cdot + O_2$	(reaction 6)
	$LO\cdot + LO\cdot \rightarrow (LO)_2$	(reaction 7)

Strong indications have been obtained that in ordered lipid structures like micelles and liposomes, reaction 3 from the propagation step determines the relative radiosensitivity of different fatty acids. The radiosensitivity (lipid peroxidation) of fatty acyl chains in phospholipids, structured as bilayers in liposomes, increases when the methylene-interrupted double-bond system is extended (5): docosahexanoic (22:6) > arachidonic (20:4) > linolenic (18:3) > linoleic (18:2) whereas in single-component micelles, the order of reactivity is reversed (6).

Figure 2 illustrates the differing radiosensitivities of fatty acyl chains in phospholipid membranes. The liposomes prepared were multilamellar vesicles (MLV). After exposure to different doses of X rays, the liposomal suspension was subjected to lipid extraction. The phospholipids were hydrolyzed, and the methyl esters of the fatty acids were analyzed with the aid of gas-liquid chromatography. The liposomes used were prepared (5) from phospholipids extracted from rat liver. The approximate phospholipid composition was as follows: phosphatidylcholine 65%, phosphatidylethanolamine 20%, sphingomyelin 5%, phosphatidylinositol plus phosphatidylserine 5%, and diphosphatidylglycerol 5%.

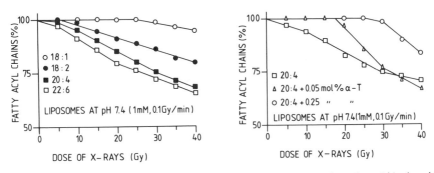

Fig. 2. Radiation-induced lipid peroxidation in liposomes and protection of arachidonic acid (20:4) by α-tocopherol. In left panel, relative radiosensitivity of different phospholipid acyl chains is given. Disappearance of intact acyl groups after X irradiation was measured (ref. 5) with aid of gas-liquid chromatography. (After ref. 13, Konings, 1986)

The average fatty acid composition of the phospholipids was 16:0 (24%), 18:0 (18%), 18:1 (8%), 18:2 (17%), 20:4 (22%), and 22:6 (11%). As can be seen in Figure 2 (left panel), the polyunsaturated fatty acyl (PUFA) chains 22:6 and 20:4 are more vulnerable to radiation damage than are 18:2 and 18:1, respectively. Up to a dose of 40 Gy, no loss of the saturated fatty acids was observed. When α-tocopherol (α-T, vitamin E) is incorporated in the bilayers of the MLV's, damage of the PUFA's may be totally prevented up to a certain dose, depending on the amount of α-T incorporated. This is illustrated in the right panel of Figure 2. As reported earlier (7,8), the rate of lipid peroxidation after the protection period is faster than the rate observed in the liposomes without the antioxidant present. It may be that during the period that intact α-T is still present, lipid hydroperoxides or other products have accumulated, providing substrates for many initiation reactions when the antioxidant is exhausted.

It appears that the responsiveness of the peroxidative mechanism is dependent on rather subtle changes in molecular organization. Alterations in close packing of micelles by added electrolytes could clearly change the chain propagation step (9), and addition of linoleic acid (18:2) to docosahexanoic acid (22:6) could greatly enhance the radiosensitivity of the latter fatty acids (6). The structure of fatty acid micelles is strongly dependent on pH; this pH dependency was also found for lipid peroxidation (6).

Apart from the fact that the extent of radiation damage to membranes is strongly dependent on the organization and cooperation of the different molecular entities in the membrane, it is true that radiation may alter membrane structure because of peroxidative damage (illustrated in Figure 3). In the present study, liposomes prepared from phospholipids extracted from rat liver were irradiated with doses up to 120 Gy. Lipid peroxidation was indicated by the production of conjugated dienes (absorption at 233 nm). Structural changes in the membrane were assayed by the measurement of fluorescence polarization using the probe diphenylhexatriene (DPH). Two different concentrations of α-T were incorporated in the bilayer of the liposomes during preparation. The graphs in the left and right panels of Figure 3 are almost identical, which means that the dose-dependent peroxidation of lipids is directly correlated with the dose-dependent structural changes in the membrane. Prevention of lipid peroxidation by α-T corresponds with the prevention of fluidity changes. DPH is generally localized in the hydrophobic midzone of the bilayer. The fluidity as indicated by DPH fluorescence polarization (P) (a low P value is related to a high degree of fluidity) is usually high when the PUFA's are still intact. After the peroxidative radical damage, fragmentation of fatty acid tails and cross-linking of peroxidized fatty acids may result in the observed enhanced membrane rigidity at those lipophylic regions in the membrane where DPH likes to be concentrated.

Dose Rate Dependency

It has been shown by several authors (e.g., refs. 8,10) that radiation-induced peroxidation of unsaturated fatty acids is inversely dose rate dependent. When

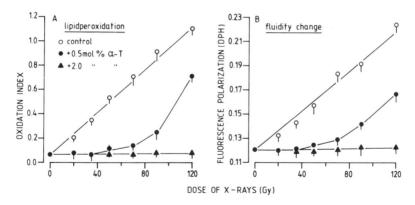

Fig. 3. Relation between lipid peroxidation and structural alterations in liposomes after X irradiation. Protection by α-tocopherol. Lipid peroxidation is measured as yield of conjugated dienes (A_{234} nm). Fluidity changes are indicated by changes in fluorescence polarization (P value) of diphenylhexatriene (DPH). Liposomes were MLV's prepared as published in ref. 5.

relatively high dose rates are applied, many peroxyl radicals will be formed in time (reaction 2). In that case, relatively more peroxyl radicals will react with each other (reaction 5) instead of reacting with an intact neighboring unsaturated fatty acid (reaction 3). This results in rather short chain reactions at high dose rates. Longer chain reactions will take place and more damage will occur at lower dose rates. The inverse dose-rate effect in liposomes is illustrated in Figure 4. Especially at the very low dose rates, the peroxidative yield may be very high. Dose-rate effects may conveniently be used to probe possible membrane damage after irradiation. It has been shown (11) that the loss of hemoglobin (Hb) and potassium ions (K+) from erythrocytes after an X-ray treatment is inversely dose rate dependent (shown in Figure 5, panels A and B, respectively). More damage was observed at lower dose rates. Especially at very low dose rates (e.g., 0.0013 against 0.27 Gy/min, see ref. 11), the difference in radiosensitivity (Hb loss) is impressive.

The uptake of TB in lymphocytes is generally (3) taken as a measure of interphase death. As can be seen in Figure 5, panel C, more cells die at a certain dose when the dose rate is lower (12). This is in contrast to the well-known observation that

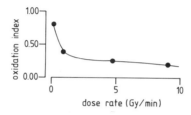

Fig. 4. Inverse dose-rate effects after X irradiation of liposomes. Liposomes were MLV's prepared as published in ref. 5.

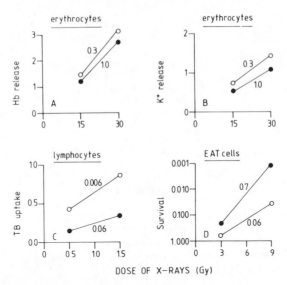

Fig. 5. Dose-rate effect after irradiation of different cell systems.

Panel A: Inverse dose-rate effect after X irradiation on hemoglobin (Hb) release in bovine erythrocytes. Liberation of Hb of control cells is set at 0. Numbers at vertical axis indicate how much more release is observed above control value. If a spectrophotometric absorbance reading (A540 nm) for lost hemoglobin was 0.10, then number 1, 2, or 3 means an extra loss because of radiation of 0.10, 0.20, and 0.30 absorbency units. Dose rates applied were 0.3 Gy/min (o) and 1.0 Gy/min (●).

Panel B: Inverse dose-rate effect after X irradiation on K+ release in suspensions of bovine erythrocytes. Dose rates applied were same as in panel A.

Panel C: Inverse dose-rate effect after X irradiation on trypan blue (TB) uptake in lymphocytes. Numbers at vertical axis indicate fraction of cells that took up dye after radiation exposure. Dose rates applied were 0.006 (o) and 0.06 (●) Gy/min.

Panel D: Dose-rate effects on clonogenic ability of Ehrlich ascites tumor (EAT) cells. Fractional survival is indicated on vertical axis. Dose rates applied were 0.06 (o) and 0.7 (●) Gy/min.

the loss of clonogenic ability (reproductive death) is less at the lower dose rates (Figure 5D). If lipid peroxidation were the most important molecular mechanism explaining reproductive cell death, one would not expect a higher cell killing at higher dose rates. A more extensive discussion on the use of dose rate as a probe to identify membrane damage in mammalian cells is found elsewhere (13).

ROLE OF MEMBRANE LIPIDS IN RADIATION-INDUCED REPRODUCTIVE DEATH

Because it is known (5) that arachidonic acid (20:4) and docosahexanoic acid (22:6) are the most radiosensitive constituents of the membrane phospholipids, it was decided to enrich the PUFA content of cells in culture. For this purpose, mouse fibroblast LM cells were grown in suspension in a serum-free, protein-free medium (14). Fatty acid-supplemented cells were obtained by culturing exponentially growing cells in the medium supplemented with 100 μM 5,8,11,14-eicosatetraenoic acid (arachidonic acid 20:4) complexed to 25 μM fatty acid-free bovine serum albumin (BSA). Normal cells were grown in the presence of 25 μM fatty acid-free BSA. The cells were cultured in the "PUFA"-medium for 24 hours before they were harvested, washed, and used in the radiation experiments.

As can be seen in Table 1, the PUFA content in the phospholipids has been drastically changed in the modified cells. The plasma membranes and the nuclear membranes contain about eight times more PUFA. The main PUFA species in the modified cells were 20:4 and 22:4, covering 18% and 13%, respectively, of the total fatty acid content (15).

Modified and normal fibroblasts were X-irradiated in air with doses up to 10 Gy at 0°, 10°, 22°, and 37°C. No differences in radiosensitivity could be observed between the PUFA and the normal fibroblasts (14). This result could be confirmed by two other groups of investigators. In 1983 a report by George et al. (16) showed data of a human lymphoid cell line. In 1986, Goulet et al. (17) reported similar results with V79 cells. It was decided to deplete the cells of reduced glutathione (GSH) and also in separate experiments to supplement the cells with α-T in order to assess the possible role of antioxidants in protecting the PUFA in the membrane during and after irradiation. The glutathione content was reduced by incubating the cells with diethylmaleate (DEM) for 1 hour at 37°C (50 nmol DEM/10^6 cells). Thereafter the DEM was washed out. Vitamin E cells were obtained by culturing the cells for at least 4 days in medium containing 230 μM DL-α-tocopherol complexed to 25 μM BSA. Cells were washed before use.

Both cell lines contained about 26 nmol GSH per mg protein. After the (nontoxic) DEM treatment, the GSH content decreased to less than 5% of the untreated cells. Cells grown in the presence of vitamin E contained about 200 ng vitamin E per 10^6 cells, whereas normal cells contained less than 0.5 ng per 10^6 cells. The fatty acid composition of vitamin E-supplemented cells (normal and PUFA cells) was

Table 1. Polyunsaturated Fatty Acyl Chains and Degree of Fluidity in Subfractions of Mouse Fibroblasts

Cell fractions	Normal cells		Modified cells	
	% PUFA[1]	ᴾDPH[2]	% PUFA	ᴾDPH
Plasma membranes	6.2 ± 0.8	0.247 ± 0.006	44.3 ± 6.8	0.223 ± 0.005
Nuclear membranes	5.1 ± 1.3	0.207 ± 0.022	42.4 ± 9.3	0.182 ± 0.006
Mitochondria	5.5 ± 1.1	0.200 ± 0.007	37.9 ± 8.1	0.184 ± 0.010
Microsomes	9.0 ± 2.4	0.242 ± 0.005	41.7 ± 9.7	0.202 ± 0.008

[1]Polyunsaturated fatty acyl (PUFA) chains of phospholipid fraction
[2]Fluidity measured as degree of fluorescence polarization (ᴾ) of diphenylhexatriene (DPH). Data given as means ± SD of four independent experiments.

comparable to the control cells (18). Although drastic differences in the PUFA content and in the anti-oxidant status of the fibroblasts could be obtained, no differences in radiosensitivity were observed (Figure 6) when the cells were X-irradiated under either oxic or anoxic conditions (19). So neither the OER nor the absolute radiosensitivity of the cells seems to be dependent on the amount of radiosensitive fatty acids in the membrane. When the OER for cell survival was assayed at different oxygen concentrations during the radiation treatment, no difference was observed for the PUFA and the normal cells (20) (illustrated in Figure 7).

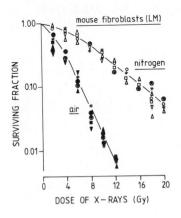

Fig. 6. Survival of normal and "PUFA" mouse fibroblast LM cells, containing different concentrations of reduced glutathione (GSH) and vitamin E (α-T). Closed symbols are derived from radiation experiments under oxic conditions; open symbols relate to anoxic conditions. o●, normal cells; □■, GSH-depleted normal cells; *, α-T supplemented normal cells; △▲, PUFA cells; ▽▼, GSH-depleted PUFA cells; ⊛, α-T supplemented PUFA cells.

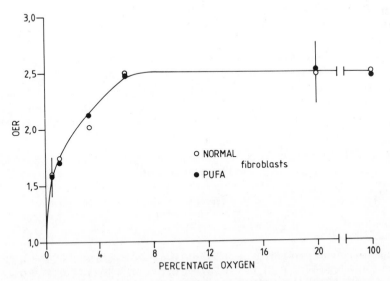

Fig. 7. OER of mouse fibroblast LM cells X irradiated at different oxygen conditions. o, normal cells; ●, PUFA cells. OER was calculated at 10% survival level. Bars represent SD. Absence of bars indicates a single experiment. (After ref. 20, Wolters et al., 1987)

Table 2. Molecular Substitution in Different Cellular Targets; Effect on Radiation and Heat Sensitivities[1]

Cellular target	Substituent	Sensitization of:	
		Radiat. eff.	Heat eff.
Membranes	PUFA[2]	No (ref.14)	Yes (ref. 21)
DNA	BrdUrd[3]	Yes (ref. 22)	No (ref. 22)

[1]Survival of mammalian cells measured as clonogenic after irradiation or hyperthermia
[2]Polyunsaturated fatty acyl (PUFA) chains in phospholipid
[3]5'-bromodeoxyuridine (BrdUrd), a thymidine analogue

If the membranes (and especially the lipids) were important targets for radiation-induced reproductive death, one would expect a modification of radiosensitivity after membrane substitution just as found for thermosensitivity (21). This was not the case. However, when DNA is the modified target, a change of radiosensitivity is apparent (22) (illustrated in Table 2).

ROLE OF MEMBRANE LIPIDS IN RADIATION-INDUCED INTERPHASE DEATH

Interphase death takes place in nonproliferating cells like lymphocytes after sublethal doses of radiation and in proliferating cells after supralethal doses (3). Membrane damage may play a decisive role in these cases (13,23), and enhanced possibilities for radiation-induced peroxidation in cellular membranes may result in higher sensitivity for interphase death of the cells.

It was decided to investigate possible membrane dependence of interphase death after high-dose X irradiation of mouse fibroblast LM cells. The same experimental approach was chosen as discussed earlier with respect to loss of clonogenic ability: the effect of GSH depletion and α-T suppletion on the radiosensitivity of normal and PUFA-substituted cells. This time, TB uptake and K+ loss were chosen as indicators for cell death. A full report on these experiments has been published recently (24).

As can be seen in Figure 8, PUFA-substituted cells lose more K+ after X irradiation than do normal cells. A clear protection of vitamin E suppletion is observed in both cell lines. Depletion of GSH by DEM caused a dramatic radiosensitization in PUFA cells. No effect of DEM treatment was found in normal cells. This finding indicates that the GSH level is not critical in protecting "normal membranes" against radiation damage but is of special importance in protecting the polyunsaturated fatty acids against lipid peroxidation. Vitamin E protection is of a more general

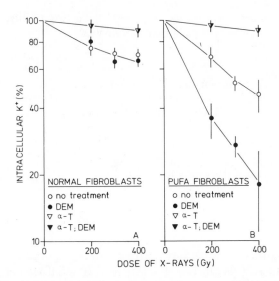

Fig. 8. Dose dependency of intracellular potassium content in normal (A) and PUFA (B) mouse fibroblasts. Effect of vitamin E supplementation (α-T), GSH-depletion (DEM), and combination (α-T; DEM). Data are mean of at least four experiments. Bars represent SD. (After ref. 24, Wolters and Konings, 1985)

nature, and it fully protects against all oxygen-dependent damage in both cell lines. The same radioprotecting effects of GSH and vitamin E in PUFA cells were found when TB uptake was assayed instead of K+ (24). Applied differences in the antioxidant status of the cells had no effect on the anaerobic radiosensitivity. The result of this fact is that the OER's of membrane damage differ considerably, depending on the amount of oxidizable substrate in the membranes (PUFA) and the level of antioxidants (GSH, α-T) (illustrated in Table 3). OER's vary from 1 to 17, depending on the quality of the cellular membranes and endogenous redox protectors.

Table 3. Radiation Sensitivity (Do) and Oxygen Enhancement Ratio for Interphase Death in Modified Mouse Fibroblasts

| Fibroblast LM cells | Do (kGy) for K^+ loss: | | |
	In air	In nitrogen	OER
Normal	1	4	4
PUFA-substituted	0.5	4	8
PUFA + α-T[1]	4	4	1
PUFA-GSH[2]	0.23	4	17

Do, radiation sensitivity; OER, oxygen enhancement ratio; K^+ loss, interphase death
[1]Vitamin E (α-tocopherol, α-T) cells were cultured for 4 days in medium containing 230 μM DL-α-tocopherol, which resulted in an α-T content of 200 ng per 10^6 cells. Normal cells contained less than 0.5 ng per 10^6 cells.
[2]PUFA-GSH cells were fibroblasts substituted with polyunsaturated fatty acids (PUFA) and depleted of GSH ($<$ 5%) by a treatment with diethylmaleate (DEM). For methods, see ref. 24.

The results in this section show that when TB uptake and K+ loss of cells are accepted as measures for interphase death, this type of radiation-induced cell death seems membrane dependent. When lipid peroxidation plays a predominant role in the mechanism of cell killing, a high OER is found. To a certain extent, this observation is in line with the suggestions of Alper (2) illustrated in Figure 1; however, there is no evidence that DNA damage is involved when a low OER is found. Differences in OER also can be found in pure lipid systems when the dose rate is altered (5).

COMPARISON OF REPRODUCTIVE DEATH AND INTERPHASE DEATH OF CELLS AFTER EXPOSURE TO IONIZING RADIATION

The main results of the experiments discussed in the two previous sections are summarized in Figure 9. In the dose range relevant for reproductive cell death (0-20 Gy), almost no loss of K+ could be observed. Clonogenic ability after irradiation was not influenced by substituting radiosensitive fatty acids in the membrane, not even after depletion of glutathione. Above 20 Gy, K+ loss becomes apparent and is markedly influenced by the lipid composition of the membranes (PUFA) and the cellular antioxidant status (GSH, α-T). The observations reported and discussed in this article strongly suggest that deoxyribonucleic acid has to be considered as the cellular target involved in radiation-induced reproductive death and that cellular membranes are important targets for radiation-induced interphase death (Table 4).

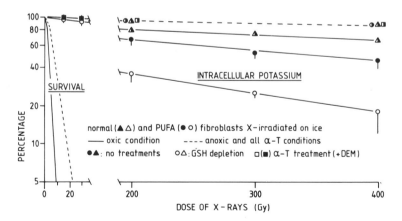

Fig. 9. Comparison of effect of X irradiation on clonogenic ability (survival) and potassium content in normal and PUFA cells. No modification (by PUFA substitution and altered antioxidant status) of radiation effect is observed when survival is taken as end point of radiation damage (reproductive death), but clear effects are observed when K+ content is measured.

Table 4. Cellular Targets Determining Mode of Cell Killing in Mouse Fibroblasts

			Modifying effect by:		
Type of cell death	Dose range (Gy)	Detection method	PUFA subst.	α-T, GSH depl.	Target involved
Reproductive	0-20	Clonog.ab.[1]	No	No	DNA
Interphase	20-400	K+/TB[2]	Yes	Yes	Membrane

[1]Clonogenic ability after plating on soft agar
[2]Loss of potassium or uptake of dye trypan blue (TB)

However, the practical meaning of membrane involvement in high-radiation-dose interphase death of proliferating cells is limited. No experiments have been performed as yet with lipid-substituted membranes of lymphocytes or other nonproliferating cells that undergo interphase death at low radiation doses. The observed inverse dose-rate effect in bovine lymphocytes (Figure 5) may be taken as an indication of membrane involvement in low-dose-radiation interphase death but can also be explained otherwise. From the experiments reported thus far, one cannot definitively conclude that no lipid peroxidation is going on in the membranes of the cells during and after exposure to low doses (0-20 Gy) of X rays. Although lipid peroxides or other products have never been detected and no preferential decrease of fatty acids in membrane phospholipids was observed, this lack of detection may be a matter of sensitivity. However, when preferential damage of radiosensitive fatty acids has taken place, it may be expected that with the use of radioactive fatty-acid incorporation studies, one would detect a preferential increase of fatty-acid turnover rate for PUFA's after low dose irradiation. This was not found. Fonck et al. (25,26) published some studies on turnover rates of palmitic acid (16:0) and arachidonic acid (20:4) in membrane phospholipids after X irradiation. These were *in vivo* (mice) and *in vitro* radiation studies with lymphosarcoma cells. After a dose of 5 Gy, an immediate increase in label incorporation started, which lasted about 50 minutes. However, there was no difference in turnover when the radioresistant (16:0) or the radiosensitive (20:4) fatty acid was monitored.

Data in the literature (27-30) show that mouse survival after ionizing irradiation, in the dose range causing bone marrow syndrome, is influenced by the amount of vitamin E in food and tissues of the animals. The reason for this protective effect may be indirect (31) and may not be taken as proof of membrane involvement in cell killing.

This paper has discussed the possible role of cellular membranes in radiation-induced cell killing. However, it is important to realize that other membrane-related end points also may be relevant in the case of radiation injury. Recently Ojeda and co-workers (32) reported experiments in which the presence of surface immunoglobin (S-Ig) of mouse lymphocytes were assayed after X irradiation. They observed an inverse dose-rate effect with respect to radiation-induced loss of S-IgG expression. Derangements of the plasma membrane of the cell at low nonlethal doses of radiation are well documented (33), and they may be of consequence for cell-cell contacts in immunology and other physiological important phenomena.

ACKNOWLEDGMENTS

The author is very grateful to all present and past co-workers who participated in this work. Financial support for these investigations was obtained by the Koningin Wilhelmina Fonds, Netherlands Cancer Foundation, and Interuniversitair Instituut voor Radiopathologie en Stralingsbeschermig, IRS.

REFERENCES

1. Cole, A. Radiation effects on DNA and membranes. In: "Radiation Research." Reviews and Summaries, Proceedings of the Seventh International Congress of Radiation Research. J. J. Broerse, G. W. Barendsen, H. B. Kal, and A. J. van der Kogel, eds. Martinus Nijhof Publishers, Amsterdam, 1983, pp. 225-230.

2. Alper, T., ed. "Cellular Radiobiology." Cambridge University Press, Cambridge/London, 1979.

3. Okada, S. Radiation-induced death. In: "Radiation Biochemistry," Vol. I. K. Z. Altman, G. B. Gerber, and S. Okada, eds. Academic Press, New York/London, 1970, pp. 247-260.

4. Butler, J., Land, E. J., and Swallow, A. J. Chemical mechanisms of the effects of high energy radiation on biological systems. Radiat. Phys. Chem. 24: 273-282, 1984.

5. Konings, A. W. T., Damen, J., and Trieling, W. B. Protection of liposomal lipids against radiation-induced oxidative damage. Int. J. Radiat. Biol. 35: 343-350, 1979.

6. Mooibroek, J., Trieling, W. B., and Konings, A. W. T. Comparison of the radiosensitivity of unsaturated fatty acids, structured as micelles or liposomes, under different experimental conditions. Int. J. Radiat. Biol. 42: 601-609, 1982.

7. Konings, A. W. T. Radiation protection of membranes by α-tocopherol. Int. J. Radiat. Biol. 38: 119, 1980.

8. Konings, A. W. T. Lipid peroxidation in liposomes. Preparation of liposomes. In: "Liposome Technology," Vol. I. G. Gregoriades, ed. CRC Press, London/New York, 1984, pp. 139-162.

9. Raleigh, J. A., and Kremers, W. Promotion of radiation peroxidation in models of lipid membranes by caesium and rubidium counter ions: Micellar linoleic and linolenic acids. Int. J. Radiat. Biol. 34: 439-447, 1978.

10. Raleigh, J. A., Kremers, W., and Gaboury, B. Dose-rate and oxygen effects in models of lipid membranes: Linoleic acid. Int. J. Radiat. Biol. 31: 203-213, 1977.

11. Konings, A. W. T. Radiation-induced efflux of potassium ions and haemoglobin in bovine erythrocytes at low doses and low dose-rates. Int. J. Radiat. Biol. 40: 441-444, 1981.

12. Konings, A. W. T. Dose-rate effects on lymphocyte survival. J. Radiat. Res. 22: 282-285, 1981.

13. Konings, A. W. T. Role of biologically available antioxidant compounds in the regulation of cellular radiosensitivity. In: "Oxygen and Sulfur Radicals in Chemistry and Medicine." A. Breccia, M. A. J. Rodgers, and G. Semerano, eds. Edizioni Scientifiche "Lo Scarabeo", Bologna, 1986, pp. 257-269.

14. Wolters, H., and Konings, A. W. T. Radiation effects on membranes. III. The effect of X-irradiation on survival of mammalian cells substituted by polyunsaturated fatty acids. Radiat. Res. 92: 474-482, 1982.

15. Wolters, H., Kelholt, D., and Konings, A. W. T. Effect of membrane fatty acid substitution and temperature on repair of sublethal damage in mammalian cells. Radiat. Res. 102: 206-212, 1985.

16. George, A. M., Lunec, J., and Cramp, W. A. Effect of membrane fatty acid changes on the radiation sensitivity of human lymphoid cells. Int. J. Radiat. Biol. 43: 363-378, 1983.

17. Goulet, D. L., Fisher, G. J., Pageau, R., and Van Lier, J. E. Effect of membrane fatty acid composition on radiosensitivity of V79 Chinese hamster cells. Biochim. Biophys. Acta 875: 414-417, 1986.

18. Konings, A. W. T. Mechanisms of ozone toxicity in cultured cells. I. Reduced clonogenic ability of polyunsaturated fatty acid-supplemented fibroblasts. Effect of vitamin E. J. Toxicol. Environ. Health 18: 491-497, 1986.

19. Wolters, H., and Konings, A. W. T. Radiosensitivity of normal and polyunsaturated fatty acid supplemented fibroblasts after depletion of glutathione. Int. J. Radiat. Biol. 46: 161-168, 1984.

20. Wolters, H., Van Tilburg, C. A. M., and Konings, A. W. T. Radiation induced lipid peroxidation: Influence of oxygen concentration and membrane lipid composition. Int. J. Radiat. Biol. 51: 619-629, 1987.

21. Konings, A. W. T., and Ruifrok, A. C. C. Role of membrane lipids and membrane fluidity in thermosensitivity and thermotolerance of mammalian cells. Radiat. Res. 102: 86-98, 1985.

22. Raaphorst, G. P., Vadasz, J. A., and Azzam, E. I. Thermal sensitivity and radiosensitization in V79 cells after BrdUrd or IdUrd incorporation. Radiat. Res. 98: 167-175, 1984.

23. Konings, A. W. T. The involvement of polyunsaturated fatty acyl chains of membrane phospholipids in radiation-induced cell death of mammalian cells. In: "Oxygen Radicals in Chemistry and Biology." W. Bors, M. Saran, and D. Tait, eds. Walter de Gruyter & Co., Berlin/New York, 1984, pp. 593-602.

24. Wolters, H., and Konings, A. W. T. Membrane radiosensitivity of fatty acid supplemented fibroblasts as assayed by the loss of intracellular potassium. Int. J. Radiat. Biol. 48: 963-973, 1985.

25. Fonck, K., Scherphof, G. L., and Konings, A. W. T. Control of fatty acid incorporation in membrane phospholipids; X-ray-induced changes in fatty acid uptake by tumor cells. Biochim. Biophys. Acta 692: 406-414, 1982.

26. Fonck, K., Scherphof, G. L., and Konings, A. W. T. The effect of X-irradiation on membrane lipids of lymphosarcoma cells *in vivo* and *in vitro*. J. Radiat. Res. 23: 371-384, 1982.

27. Konings, A. W. T., and Drijver, E. B. Radiation effects on membranes. I. Vitamin E deficiency and lipid peroxidation. Radiat. Res. 80: 494, 1979.

28. Malick, M. A., Roy, R. M., and Sternberg, J. Effect of vitamin E on post-irradiation death in mice. Experientia 34: 1216, 1978.

29. Sakamoto K., and Sakka, M. Reduced effect of irradiation on normal and malignant cells irradiated *in vivo* in mice pretreated with vitamin E. Brit. J. Radiol. 46: 538, 1973.

30. Srinivasan, V., and Weiss, J. F. Vitamins and radiation injury: Vitamin E, vitamin B6, and liver microsomal drug-metabolizing system. Radiat. Res. 87: 385, 1981.

31. Bichay, T. J. E., and Roy, R. M. Modification of survival and hematopoiesis in mice by tocopherol injection following irradiation. Strahlenther. & Onkologie 162: 391-399, 1986.

32. Ojeda, F., Moraga, D., Guarda, M. I., and Folch, H. Dose rate dependence of radiation induced IgG membrane receptor alteration. Z. Naturforsch. 39c: 1021-1022, 1984.

33. Koteles, G. J., Kubasova, T., Somosy, Z., and Horvath, L. Derangement of cellular plasma membranes due to non-lethal radiation doses. In: "Biological Effects of Low-level Radiation." IAEA-SM-266/37. International Atomic Energy Agency, Vienna, 1983, pp. 115-128.

INFLUENCE OF MEMBRANE LIPID COMPOSITION AND ORGANIZATION ON RADIO- AND THERMOSENSITIVITY OF BACTERIA

M. B. Yatvin[1], O. Yukawa[2], and W. H. Dennis[3]

Departments of [1]Human Oncology, [1]Radiology, and [3]Physiology
University of Wisconsin
Madison, Wisconsin

[2]Division of Biology
National Institute of Radiological Sciences
Anagawa, Chiba, Japan

ABSTRACT

Numerous investigators have studied and reported on the effects of ionizing radiation on cellular membranes. The effects reported range from inactivation of membrane-bound enzymes to increased ion permeability. This review concerns radiation effects on bacterial membranes and related structures such as "DNA-membrane complexes." In addition, the relationship of composition and organization of the *Escherichia coli* membrane to hyperthermic cell killing is considered. In particular, evidence for the role played by fatty acids on the movement of proteins to the outer membrane and the relevance of such protein redistribution on cell survival in heat-stressed cells are discussed.

Bacteria, especially auxotrophs of *Escherichia coli*, provide powerful tools for study of the role of the membrane in radiation sensitivity (1-5) and hyperthermic sensitivity (6-9). The non-auxotrophic strains most studied by radiobiologists have been the radiation-resistant strain, $B_{/r}$, and the radiation-sensitive strain, B_{s-1}. The unsaturated fatty acid-requiring auxotroph, K1060, has provided a cell whose membrane composition can be controlled in a predictable manner. Simply by varying the growth temperature and the unsaturated fatty acid provided during growth, one can control the biophysical properties of the membrane. Such modified cells are important tools for researchers interested in studying the influence of membrane composition on cell killing by radiation and heat.

The role of the membrane in radiation injury has intrigued researchers for many years. Alper and Gillies (10) in 1958 implied that damage to a target other than

DNA was involved in radiation injury to cells. With this background, Alper and her co-workers (2,3,11-14) and others, including our laboratory (1,4,5,15,16), have attempted to elucidate the role of the membrane in radiation killing of cells.

In our laboratory, we found that immediately following irradiation, the membranes of *E. coli* are "damaged" by exposure to gamma radiation. This was inferred from the fact that spheroplasts made immediately after irradiation were more easily disrupted when exposed to an osmotic gradient than were unirradiated cells (15). In contrast, compositional changes (phospholipids) following irradiation occurred more slowly, within 1½-3 hr (16). In bacteria (*E. coli* K1060) with alterable composition, those cells whose transition temperatures are elevated display a differential sensitivity to irradiation at 0°C compared to bacteria with lower transition temperatures. Cell membrane organization appears to be critical in this phenomenon (1,4,5,17).

In an extension of Alper's hypothesis, we and others (13,14,18,19) have studied "DNA-membrane complexes" isolated from cells by different methods. The "DNA-membrane complex" isolated in this laboratory has characteristics that are related to radiation sensitivity (19). Both the amount of DNA and the ability of the DNA to withstand enzymatic degradation are positively associated with the relative cell survival of *E. coli* $B_{/r}$ and B_{s-1}. One approach taken by researchers interested in studying the importance of DNA in the response of cells to radiation has been to alter the DNA, particularly with analogues of thymidine halogenated in the 5 position (20,21). We have chosen to modify the lipid, especially in ways that will enable us to affect membrane lipid organization.

Figure 1 depicts the "anatomy" of *E. coli* with regard to its membrane structure. Shown are an outer membrane, a peptidoglycan layer, the periplasmic space, the inner membrane to which the DNA is attached, and the cytoplasm.

The radiation-resistant strain of *E. coli* $B_{/r}$ and the radiosensitive strain B_{s-1} (Figure 2) provide a system for investigating how membrane organization affects radiation survival. The survival after radiation is not dependent on growth stage in B_{s-1}. For $B_{/r}$, the surviving fraction may be slightly increased in stationary phase for a given dose of irradiation, but the increase is not marked (22). Earlier, Ginsberg and Webster (23) reported similar results.

The membrane fatty acid compositions in both strains are similar in mid-log cells and in stationary-phase cells. Figure 3 presents the fatty acid composition of $B_{/r}$ for both growth phases. Especially noteworthy are the changes in the 16:1 and 17:0 and in the 18:1 and 19:0 fatty acids. In progressing from mid-log to the stationary phase, the delta cyclopropane fatty acid contents (17:0, 19:0) show the biggest increases with concomitant decreases in the 16:1 and 18:1 fatty acids, which are the precursors for the 17:0 and 19:0 fatty acids. The membrane fatty acid contents of B_{s-1} in mid-log and in stationary-phase cells follow the same pattern as $B_{/r}$ and are not significantly different statistically (data not shown).

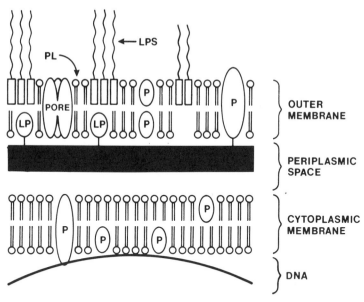

Fig. 1. Scheme of principal elements of *E. coli* membranes and adjacent spaces. Solid black bar in periplasmic space represents the peptidoglycan layer.

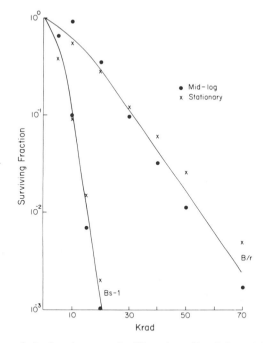

Fig. 2. Radiation survival of stationary and midlog-phase *E. coli* B$_{/r}$ and B$_{s-1}$ at room temperature as a function of irradiation dose. (Reproduced with permission from Academic Press 1986: Radiat. Res. 106:78-88.)

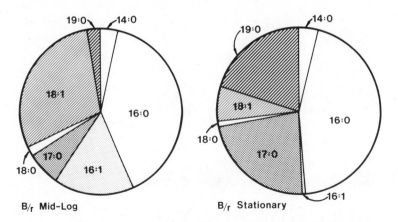

Fig. 3. Fatty acid composition expressed as percentage of total membrane lipids of *E. coli* B$_{/r}$ in mid-log and stationary phase, presented as pie charts. Noteworthy is increase in 17:0 and in 19:0 on growth from mid-log to stationary phase with compensating reduction in 16:1 and 18:1. Similar results were obtained with *E. coli* B$_{s-1}$.

The membrane microviscosity, as estimated by fluorescent polarization techniques using diphenylhexatriene, is greater by about 8% at 37°C in both strains in stationary-phase cells. These increases occur coincidentally with the formation of increased 17:0 and 19:0 cyclopropane fatty acids (22). The membrane microviscosity of B$_{/r}$ cells is consistently slightly greater than that of B$_{s-1}$ cells.

Changes in osmotic fragility of spheroplasts from *E coli* B$_{/r}$ can be detected within minutes after irradiation. This was determined by exposing aliquots of spheroplasts obtained from an unirradiated, irradiated, and irradiated-reincubated cells to solutions of decreasing osmolarity and determining the osmolarity at which 50% lysis occurs. Initially, spheroplasts from irradiated B$_{/r}$ cells showed greater lysis than did spheroplasts from B$_{s-1}$ cells. Spheroplasts from irradiated cells of both strains were more fragile than their unirradiated control (15). When the cells were reincubated at 37°C for 20 min after irradiation but prior to spheroplast formation, B$_{/r}$ showed evidence of "repair" by decrease of osmotic fragility. In contrast, B$_{s-1}$ spheroplasts were even more fragile after reincubation than immediately postirradiation (15).

Sublethal irradiation of *Micrococcus radiodurans* with gamma rays resulted in the release of cell surface exonuclease (24) and in the loss of some cell wall polysaccharides. These included N-acylated glucosamine and a polymer of glucose, which seem to come from a lipid-rich region of the cell wall (25). Particularly strong support for an ionizing radiation-induced alteration in membrane structure that influences cell survival is forthcoming from the work of Gillies et al. (26). Penicillin enhanced lethal damage to *E. coli* B$_{/r}$ after irradiation under aerobic conditions but not under anoxic conditions or following UV irradiation (260 nm). If penicillin treatment after aerobic irradiation was delayed 2 hours, no enhancement in cell killing was noted. In subsequent work, these authors (27) demonstrated that interaction of oxygen with an X-ray lesion, either in the membrane or bacterial cell envelope, plays the primary role in oxygen radiosensitization.

$E.$ $coli$ $B_{/r}$ and B_{s-1} also differ in how their phospholipid compositions change following irradiation (16). There are strain- and dose-dependent increases in the percent and total amount of cardiolipin (CL). Over the dose ranges studied of gamma irradiation (0-540 Gy), the percent of CL increased continuously in $B_{/r}$, attaining a level of 14% at 540 Gy. In contrast, in B_{s-1}, CL content plateaued at 9% after 90 Gy. Thus, at lower doses, the effect on B_{s-1} was markedly greater. While the changes in phospholipid distribution correlate with radiosensitivity, the plateau in CL percent after 90 Gy has yet to be explained for B_{s-1}. It is possible that all damage in B_{s-1} has been accomplished by 90 Gy and that a similar plateau at higher levels might be achieved in $B_{/r}$ at doses of greater than 540 Gy. Nevertheless, it may be that the absolute percent CL levels achieved may reflect an index of tolerance to damage in the two strains (16).

We have extended these observations at different radiation doses and with better analytical techniques to more carefully evaluate phospholipid changes other than CL (Ghoulipour-Khalili and Yatvin, unpublished). Table 1 shows the phospholipid composition of $E.$ $coli$ $B_{/r}$ after exposure to 400 or 540 Gy. The changes in phospholipid composition were followed in cells labeled prior to irradiation with either ^{14}C-serine or ^{14}C-glycerol-3-PO_4. The most dramatic changes noted are the fourfold increase in phosphatidylserine (PS) and the two- to threefold increase in CL following 540 Gy. The phosphatidylglycerol (PG) content is reduced about 40%, and phosphatidylethanolamine (PE) showed the least change, being reduced 4-6%. The increase in PS is about the same as the decrease in PE; the increase in CL is about the same as the decrease in PG. With 400 Gy, the alteration in phospholipid composition is approximately intermediate between the control and the 540-Gy values.

After bacteria are lysed with sarcosyl prior to centrifugation, a cell fraction from $E.$ $coli$ can be sedimented during alkaline sucrose gradient analysis of DNA. This

Table 1. Phospholipid Distribution (%) in $E.$ $coli$ $B_{/r}$ 180 Minutes After Irradiation in Prelabeled Cells

	^{14}C-Serine				^{14}C-Glycerol-3-PO_4			
	PS	PE	PG	CL	PS	PE	PG	CL
0 time	.65	76.5	18.0	3.4	.6	76.3	19.0	3.2
0 Gy + 180 min	.65	78.5	18.5	2.9	.4	77.5	18.5	3.2
400 Gy + 180 min	1.9	74.0	15.0	7.0	2.0	74.0	16.0	6.5
540 Gy + 180 min	2.4	74.0	10.8	9.6	2.1	74.4	11.3	11.0

Ghoulipour-Khalili and Yatvin, unpublished

sediment contains a "DNA-membrane complex." The proportion of total DNA found in the complex is strongly influenced by the isolation procedure and the solutions used. When phosphate-buffered saline is used, the percent of DNA from $B_{/r}$ is 4.4% and B_{s-1} is 1.8%; with tris-HCl buffer, $B_{/r}$ is 14.1% and B_{s-1} 4.6%. The tris buffer protects against alkaline-induced DNA lability. Regardless of the buffer used, the fraction of DNA is always 2½- to threefold greater for $B_{/r}$ than for B_{s-1}. Irradiation (180 Gy) reduces the amount of DNA associated with "DNA-membrane complex." Reincubation before lysing and gradient centrifugation returns the values to control level.

Differential sensitivity of cells to killing by ionizing radiation may thus be mediated to a large degree by the "packaging" of cellular DNA. For example, the number and kind of DNA-membrane attachment sites, as well as the composition of the cell membrane, contribute to the nature of the "package," and these stereoscopic factors could be a critical determinant in a cell's response to a radiation insult. Burrell et al. (18) have suggested that the extraordinary radioresistance of *Micrococcus radiodurans* is mediated by such a mechanism.

During postirradiation reincubation, 35%-50% of the DNA radioactive label is lost. However, the fraction of DNA associated with the "complex" in *E. coli* $B_{/r}$ after reincubation is greater than the fraction immediately after irradiation. Thus, in the face of overall degradation of DNA, the "DNA-membrane" complex appears to be reforming (Table 2, last column).

The trend toward increased survival after irradiation of stationary-phase $B_{/r}$, as seen in Figure 2, may be associated with the small but consistent increase in the percent DNA associated with the membrane. The marked difference in survival of $B_{/r}$ and B_{s-1} may also be related to the differences in the attachment of DNA to the membrane (19). The DNA associated with the $B_{/r}$ "complex" is more resistant to the action of DNAase than is the DNA in the B_{s-1} complex. A three- to fourfold

Table 2. Evidence for Repair of "DNA Membrane Complex"

		Total Counts	% Control	Pellet[1] Count	Pellet % Total Control	% Control Pellet
Mid-log:	Control[2]	61,250	100	8,372	14	100
	+ 180 Gy	62,300	102	1,748	3	21
	Reincubated 40 min	44,460	73	5,306	12	86
Stationary:	Control[3]	8,700	100	1,584	18	100
	+ 180 Gy	8,400	97	590	7	39
	Reincubated 40 min	5,800	67	1,002	17	94

[1] Putative "DNA-Membrane Complex"
[2] 3H-thymidine added when subculture started and incubated 1 hr
[3] 3H-thymidine added after 3 hr growth of subculture and incubated 1 hr
From Yatvin, unpublished

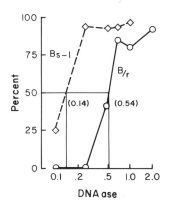

Fig. 4. Percentage of DNA released after exposing complex to several concentrations of DNAase. Complex from *E. coli* $B_{/r}$ releases DNA at lower DNAase concentrations than those required for equivalent release from B_{s-1} derived complex.

higher concentration of DNAase is required to degrade equal amounts of ^{3}H-thymidine radioactivity from the $B_{/r}$ "complex DNA" (Figure 4). However, since three to four times more DNA was associated with the $B_{/r}$ "complex" than with the B_{s-1} complex, it is conceivable that in the former, the level of DNAase was insufficient at the lower concentration.

The membrane composition of the unsaturated fatty acid-requiring auxotroph of *E. coli*, K1060, may be altered in a predictive manner by varying the growth temperature of the unsaturated fatty acid present during growth. The percent of DNA in the "complex" in unirradiated cells held at 4° or less just prior to sarcosyl lysing is greater than when the cells were held for 40° prior to sarcosyl lysing. This occurs independently of the transition temperature (T_t) of the membrane lipids obtained from cells supplemented with linolenic (18:3), oleic (18:1), palmiteladic (trans-16:1), or elaidic acid (trans-18:1). Further, more DNA was lost from the "complex" when irradiation of the cells occurred at lower temperatures regardless of whether the membrane was below the T_t of the membrane. Thus, the T_t of membrane lipids is not the only critical factor in the response of the "DNA membrane complex" in K1060 cells. However, it is notable that, when cells are irradiated at various temperatures and reincubated at 37°C following irradiation, "repair" of the "complex" was markedly reduced in those cells irradiated at lower temperatures (19).

Membrane fluidity influenced the survival following irradiation of bacteria whose membranes are substituted with various pairs of cis- and trans-unsaturated fatty acids. With cells grown in oleic acid, survival after irradiation at ice bath temperature (below T_t) is depressed with respect to cells exposed above the T_t of their membrane lipids either in the presence of oxygen or nitrogen (28). In contrast, while linolenic-supplemented cells show the expected oxygen effect, no temperature effect was observed for these cells when they were irradiated under ice bath conditions (above T_t) in the presence of either oxygen or nitrogen (28).

When irradiated above their transition temperature, the survival of cells supplemented with elaidic acid (trans-18:1) was similar to survival of cells grown in cis-18:1. In contrast, the survival of trans-fatty acid-substituted cells is lower than that of cis-fatty acid-substituted cells after irradiation at ice bath temperatures. Furthermore, unlike the cis-substituted cells, cells grown on elaidic acid and irradiated at temperatures between 21° and ice bath temperature had a greater decrease in survival as the temperature was reduced (28). Figure 5 presents the striking influence of the increase in membrane microviscosity associated with temperature below the T_t on the radiation killing of elaidic acid supplemented *E. coli*. The levels of survival obtained at the low temperature (approximately 4°C) and the low dose (270 Gy) are similar to those at the higher temperature (21°-37°C) and the higher dose (900 Gy). The reasons for the greater killing of *E. coli* cells whose membrane lipids are below the T_t is unresolved. George et al. (4) have found that under conditions where bacteria are more sensitive to radiation (below T_t), there was no accompanying increase in residual lesions in the DNA (Figure 6).

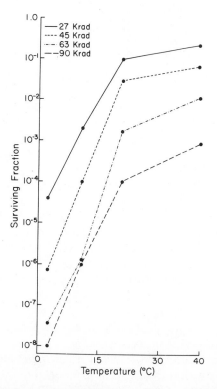

Fig. 5. *E. coli* K1060 grown with elaidic acid (18:1-trans) and irradiated at temperatures from 4° to 40°C. For each irradiation dose, surviving fraction is plotted against temperature during irradiation. Marked increase in cell killing occurred at temperatures below 21°C.

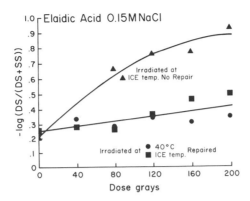

Fig. 6. Plot of -log DS/(DS + SS) against dose showing extensive and equivalent repair in elaidic grown bacteria irradiated at ice temperature and 40°C in saline followed by a postirradiation incubation time of 30 min at 37°C. (Drawing reproduced from ref. 4.)

The greater killing effect of irradiating trans-fatty acid grown cells in the gel state may be because these acids cause a greater modification of the membrane proteins. Shechter et al. (29), using freeze-fracture electron microscopy, have shown that as the temperature is lowered below T_t, lipid and protein particles aggregate into separate domains, with this lateral segregation being more striking in trans-grown cells. The protein-attached DNA may represent a different and more sensitive target when the proteins are aggregated. This cannot be the whole explanation because proteins in linolenic acid-grown cells also aggregate at ice bath temperature (29), although to a lesser extent. The greater rigidity of aggregated proteins of oleic acid- and trans-acid-supplemented cells below the T_t may account for the increased cell killings, perhaps by influencing either energy transfer in the "target" or the physical state of the "target."

The radiosensitivity of cells reportedly changes when they are treated with compounds that perturb cytoplasmic membranes (local anesthetics, analgesics, and tranquilizers). Excellent reviews were recently written on this subject (30,31). Such compounds only sensitized bacteria when they were irradiated under hypoxic conditions. Interestingly, the enhancement of radiation killing when E. coli K1060 cells were irradiated below the transition temperature of their membrane lipids was blocked by procaine (17). Thus, membrane "fluidity" is a critical factor in radiation lethality.

Numerous investigators have postulated a correlation between loss of reproductive capacity in cells and lipid peroxidation; however, a variety of studies in cells have failed to extend conclusions based on experiments using lipid model systems (1,32,33). The correlations predicted by such *in vitro* systems are not observed when cell survival is compared to lipid peroxidation, to sulfhydryl content, or to damage of membrane-bound enzymes.

Recently we investigated the correlation between malonyldialdehyde (MDA) production and survival of K1060 cells grown in medium supplemented with either oleic or linolenic acid. MDA is essentially unchanged from 0.12 nmol MDA formed/ mg protein (no incubation) to 0.08 and 0.18 at room temperature for oleic-grown cells irradiated at 250 and 500 Gy, respectively, while a surviving fraction decreased 100-fold. MDA increased from a control level of 0.16 to 5.6 and to 8.0 for linolenic-grown cells. The largest radiation-induced changes in MDA were observed in linolenic acid-grown cells with identical survival curves. The survival curves of these cells showed a much greater resistance to irradiation at ice bath temperature than oleic acid-grown cells.

The role of the membrane, particularly membrane lipids, also exerts a critical influence on cellular hyperthermic sensitivity. When membrane composition of the K1060 auxotroph is altered, one observes a variable response to heat that is dependent on its fatty acid composition at the time of heat insult. Cells show an increased sensitivity to heat as the unsaturated fatty acid is changed from oleic acid (which contains one unsaturated bond and is 18 carbons in length) to linolenic acid (which contains three unsaturated bonds and is 18 carbons in length) (6). Hyperthermic sensitivity can be further increased in bacterial cells by the addition of membrane-active agents such as local anesthetics (6).

Likewise, supplementation with fatty acids containing the same degree of unsaturation but of differing chain length also produces cells that display differing sensitivity to heat. For example, bacteria supplemented with palmitoleic acid with a 16 carbon chain length with one unsaturated bond are more sensitive to hyperthermia than are eicosenoic acid-supplemented cells. The latter fatty acid has one unsaturated bond but is 20 carbons long (34).

Stationary-phase cells are substantially more resistant to hyperthermic exposure than are mid-log-phase cells (Figure 7). There is little difference in the survival after hyperthermia between the two strains. This is in sharp contrast to the lack of growth-phase effect following irradiation in these strains. Nevertheless, the studies with $B_{/r}$ and B_{s-1} cells suggest that while the membrane is critical in both heat and radiation killing, different sites of injury must be involved.

The exposure of cells to hyperthermia leads to the formation of heat-stress proteins. This universal response leads one to conclude that these proteins play a role in cell survival after an insult. In E. coli, the translocation of nascent proteins to the outer membrane is observed (35). This translocation occurs within 10 seconds of hyperthermic exposure, and the translocated proteins are not heat-stress proteins (36). Heat-stress proteins occur in significant amounts only after several minutes have elapsed, and they are confined to the cytoplasm.

The rate and amount of protein translocation are dependent on the unsaturated fatty acid used during growth. Cells grown in 11-eicosenoic acid (20:1) show greater translocation at 3°-4°C above growth temperature compared to cells grown with

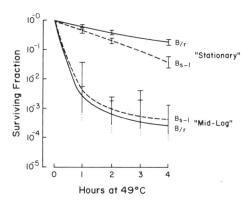

Fig. 7. Survival of stationary and mid-log-phase *E. coli.* $B_{/r}$ and B_{s-1} at 49°C as a function of time of heating.

palmitoleic acid (16:1). This is true both with total outer-membrane proteins on SDS-PAGE gels and with the non-signal sequence containing 2-galactosidase (34).

In the case of the nascent proteins, one may infer that the membrane disorder associated with hyperthermia leads to the formation of "bridges" between the inner and outer membranes, which facilitate rapid protein translocation (35). Once the proteins are on the outer surface, they may serve a protective function by stiffening the membrane. Evidence for such stiffening is the increase in membrane microviscosity of the outer membrane prepared from cells exposed to hyperthermia, especially at elevated temperatures. In the case of the heat-stress proteins, they remain in the cytoplasm and perhaps play a protective role in maintaining the DNA structure of the nucleoid.

ACKNOWLEDGMENT

Supported in part by NIH grant CA24872. O. Yukawa is supported by the Japanese Science and Technology Agency.

REFERENCES

1. Yatvin, M. B. Evidence that survival of gamma-irradiated *E. coli* is influenced by membrane fluidity. Int. J Radiat. Biol. 30: 571-575, 1976.
2. Redpath, J. L., and Patterson, L. K. The effect of membrane fatty acid composition on the radiosensitivity of *E. coli* K1060. Radiat. Res. 75: 443-447, 1978.
3. Suzuki, S., and Akamatsu, Y. Involvement of membrane lipids in radiation damage to potassium ion permeability of *Escherichia coli*. Int. J. Radiat. Biol. 33: 185-190, 1978.

4. George, A. M., Cramp, W. A., and Yatvin, M. B. The influence of membrane fluidity on radiation-induced changes in the DNA of *E. coli* K1060. Int. J. Radiat. Biol. 38: 427-438, 1980.

5. Alper, T., Cramp. W. A., George, A. M., Lunec, J., and Yatvin, M. B. Membrane fluidity and the radiosensitivity of *E. coli* K1060. Int. J. Radiat. Biol. 40: 211-215, 1981.

6. Yatvin, M. B. The influence of membrane lipid composition and procaine on hyperthermic death of cells. Int. J. Radiat. Biol. 32: 513-521, 1977.

7. Dennis, W. H., and Yatvin, M. B. Correlation of hyperthermic sensitivity and membrane microviscosity in *E. coli* K1060. Int. J. Radiat. Biol. 39: 265-271, 1981.

8. Yatvin, M. B., Gipp, J. J., Rusy, B. F., and Dennis, W. H. Correlation of bacterial hyperthermic survival with anaesthetic potency. Int. J. Radiat. Biol. 42: 141-149, 1982.

9. Simone, G., Gipp. J. J., Dennis, W. H., and Yatvin, M. B. Lowered pH eliminates the enhanced hyperthermic killing of *E. coli* induced by procaine and exposure to N_2 gassing. Int. J. Radiat. Biol. 44: 87-95, 1983.

10. Alper, T., and Gillies, N. E. Restoration of *E. coli* strain B after irradiation: Its dependence on sub-optimal growth conditions. J. Gen. Microbiol. 18: 461-472, 1958.

11. Alper, T. "Cellular Radiobiology." Cambridge University Press, New Rochelle, New York, 1979.

12. Alper, T. Lethal mutations and cell death. Phys. Med. Biol. 8: 365-385, 1963.

13. Watkins, D. K. High oxygen effect for the release of enzymes from isolated mammalian lysosomes after treatment with ionizing radiation. Adv. Biol. Med. Phys. 13: 289-305, 1970.

14. Cramp, W. A., Watkins, D. K., and Collins, J. Effects of ionizing radiation on bacterial DNA membrane complexes. Nature [New Biol.] 235: 76-77, 1972.

15. Yatvin, M. B., Wood, P. G., and Brown, S. M. "Repair" of plasma membrane injury and single strand breaks in gamma-irradiated *Escherichia coli* $B_{/r}$ and B_{s-1}. Biochim. Biophys. Acta 287: 390-403, 1972.

16. Jacobson, A. F., and Yatvin, M. B. Changes in phospholipid composition of *Escherichia coli* following gamma and UV-irradiation. Radiat. Res. 66: 247-266, 1976.

17. Yatvin, M. B., Schmitz, B. J., and Dennis, W. H. Radiation killing of *E. coli* K1060: Role of membrane fluidity, hypothermia and local anesthetics. Int. J. Radiat. Biol. 37: 513-519, 1980.

18. Burrell, A. D., Feldschreiber, P., and Dean, C. J. DNA-membrane association and the repair of double breaks in X-irradiated *Micrococcus radiodurans*. Biochim. Biophys. Acta 247: 38-53, 1971.

19. Yatvin, M. B., Gipp, J. J., Werts, E. D., Tamba, M., and Simone, G. Membrane aspects of radiation biology. In: "Oxygen Radicals in Chemistry and Biology." W. Bors, M. Saran, and D. Tait, eds. Walter de Gruyter and Co., Berlin, 1984, pp. 133-143.

20. Heidelberger, C., Griesbach, L., Montag, B. J., Mooren, D., Cruz. O., Schnitzer, R. J., and Grunberg, E. Studies on fluorinated pyrimidines. II. Effects on transplanted tumors. Cancer Res. 18: 305-317, 1958.

21. Djordjevic, B., and Szybalski, W. Genetics of human cell lines. III. Incorporation of 5-bromo and 5-iododeoxyuridine into deoxyribonucleic acid of human cells and its effects on radiation sensitivity. J. Exp. Med. 12: 509-531, 1960.

22. Yatvin, M. B., Gipp, J. J., Klessig, D. R., and Dennis, W. H. Hyperthermic sensitivity and growth stage in *Escherichia coli*. Radiat. Res. 160: 78-88, 1986.

23. Ginsberg, D. M., and Webster, H. K. Chemical protection against single strand breaks in DNA of gamma-irradiated *E. coli*. Radiat. Res. 39: 421-435, 1969.

24. Gentner, N. E., and Mitchel, R. E. J. Ionizing radiation-induced release of a cell surface nuclease from *Micrococcus radiodurans*. Radiat. Res. 61: 204-215, 1975.

25. Mitchel, R. E. J. Ionizing radiation damage in *Micrococcus radiodurans* cell wall: Release of polysaccharide. Radiat. Res. 66: 158-169, 1976.

26. Gillies, N. E., Obioha, F. I., and Ratnajothi, N. H. An oxygen dependent x-ray lesion in *Escherichia coli* strain $B_{/r}$ detected by penicillin. Int. J. Radiat. Biol. 36: 587-594, 1979.

27. Gillies, N. E., and Ratnajothi, N. H. Introduction of resistance to x-rays in *E. coli* by toluene. Int. J. Radiat. Biol 38: 193-198, 1980.

28. Yatvin, M. B., Gipp, J. J., and Dennis, W. H. Influence of unsaturated fatty acids, membrane fluidity and oxygenation on survival of an *E. coli* fatty acid auxotroph following gamma-irradiation. Int. J. Radiat. Biol. 35: 539-548, 1979.

29. Shechter, E., Letellier, L., and Gulik-Krywicki, L. E. Relationship between structure and function in cytoplasmic membrane vesicles isolated from an *Escherichia coli* fatty-acid auxotroph. Eur. J. Biochem. 49: 61-76, 1974.

30. Yonei, S., Malsutani, C. Y., Todo, T., and Niwa, O. Modification of radiosensitivity of bacterial and mammalian cells by membrane-specific drugs. In: "Modification of Radiosensitivity in Cancer Treatment." Academic Press Japan, Inc., 1984, pp. 291-311.

31. Shenoy, M. A., and Singh, B. B. Non-nitro radiation sensitizers. Int. J. Radiat. Biol. 48: 315-326, 1985.

32. Edwards, J. C., Chapman, D., and Cramp, W. A. Radiation studies of *Acholeplasma laidlawii*: The role of membrane composition. Int. J. Radiat. Biol. 44: 405-412, 1983.

33. Konings, A. W. T., Gipp, J. J., and Yatvin, M. B. Radio- and thermosensitivity of *E. coli* K1060 after thiol depletion by diethylmaleate. Radiat. Environ. Biophys. 23: 245-253, 1984.

34. Yatvin, M. B. Influence of membrane lipid composition on translocation of nascent proteins in heated *Escherichia coli*. Biochim. Biophys. Acta 901: 147-156, 1987.

35. Yatvin, M. B., Smith, K. M., and Siegel, F. L. Translocation of nascent non-signal sequence proteins in heated *Escherichia coli*. J. Biol. Chem. 261: 8070-8075, 1986.

36. Yatvin, M. B., Clark, A. W., and Siegel, F. L. Major heat-stress proteins do not translocate: Implications for cell survival. Int. J. Radiat. Biol., in press.

ROLE OF MEMBRANES, FREE RADICALS, AND COPPER IN RADIATION-INDUCED CHANGES IN QUATERNARY STRUCTURE OF DNA: SOME CLINICAL IMPLICATIONS

W. A. Cramp[1], A. M. George[1], J. C. Edwards[1], S. A. Sabovljev[1],
G. Harris[2], L. E. Hart[3], H. Lambert[4], and M. B. Yatvin[5]

[1]MRC, Radiobiology Unit
Hammersmith Hospital
London, United Kingdom

[2]Kennedy Institute
London, United Kingdom

[3]Department of Histopathology and Rheumatology
Hammersmith Hospital
London, United Kingdom

[4]Radiotherapy Department
Hammersmith Hospital, United Kingdom

[5]Department of Human Oncology
University Hospitals
Madison, Wisconsin

ABSTRACT

On challenge with 2 M NaCl, the nuclei of human lymphocytes yield an aggregate of DNA-protein material. The density of the material is less when isolated from irradiated cells than when isolated from nonirradiated cells. The density of this material [designated histone-free DNA (HF-DNA)] from irradiated cells returns to that from nonirradiated cells if the irradiated cells are allowed time at 37°C in nutrient conditions.

Lymphocytes from elderly donors and from patients with some connective tissue diseases were significantly less able to restore the density of HF-DNA after irradiation. In addition, lymphocyte HF-DNA from patients who have exhibited hypersensitivity to radiotherapy exhibits slower repair characteristics than does lymphocyte HF-DNA from the average normal subject. It may be possible from prospective studies to identify potentially sensitive patients, especially where radio-therapy regimens unavoidably include lymphocytic populations in the field of irradiation, and to modify such regimens accordingly.

The radiobiological properties of this HF-DNA from human lymphocytes suggest that radiation-induced alterations in the relatively weak cellular bonds responsible for the structure of chromatin are prime reasons for the consequent cellular malfunction. Where the proliferation of human lymphocytes, from which HF-DNA has been derived, has been measured by concanavalin A stimulation, oxygen enhancement ratios, relative biological effectiveness of densely ionizing radiation, and chemical modification have been of the same order as density changes in HF-DNA.

Also, we have found that the density of HF-DNA is heavily dependent on Cu (copper) content. This has led us to propose that cell killing or malfunction at the nuclear level caused by ionizing radiation is a result of the conversion of Cu^{II} → Cu^{I} and specific $\cdot OH$ attack on DNA or proteins at a Cu site.

It is conceivable that these events interact and that the new molecular species produced predominate in the presence of oxygen. There will be a variability in sensitivity between cell types dependent on intrinsic chromatin levels of Cu^{II} and Cu^{I} and, of course, capacities to restore original configurations of macromolecular complexes.

INTRODUCTION

The majority of work measuring the effects of ionizing radiation on membranes has been done with the purpose of relating radiation-induced damage to the membranes with cellular malfunction. This approach was stimulated by Alper, who as long ago as 1958 (1) implied that damage to a target other than DNA was involved, which interacted with damaged DNA with consequent lethal effects. A summary of Alper's work, where she clearly reasons that a non-DNA target more sensitive in the presence of oxygen must be considered, can be found in her text *Cellular Radiobiology* (2).

In the last few years, the structure and location of DNA-associated biological macromolecules within the cell have become better defined (3,4). It is clear that many of the relatively simple "model" systems used in radiation studies (see review, reference 5), which consist of membrane and other cellular ingredients, are very different from the extremely complex intermingling of macromolecules found within cells. For example, the role of the radiation-sensitive unsaturated centers in the fatty acid constituents of membranes appears to be (6-8) of negligible importance in modifying the lethal effects of radiation when the fatty acid is in the intact cell.

In addition, the interpretation of the results from much of the work where "model" membranes, cell extracts, or partially disrupted cells are used must be viewed with caution when very large "supralethal" doses are used to bring about measurable changes. When sufficient information is available, "back" extrapolation from these high-dose results often indicates a threshold dose where no effect in the membrane structure is measurable but where marked loss in biological function occurs.

The hypotheses of Alper (2) have been a guide to experimental work in our laboratories for the past 2 decades. We have attempted to correlate the extent of lesions and interaction of lesions in DNA and membrane complexes with cell death. On the whole, we have been unsuccessful in achieving close correlation, probably because the crucial lesion(s) remains unidentified.

As mentioned earlier, changes in the levels of unsaturated fatty acid constituents of membranes in either bacteria (6,7) or mammalian cells (8) have failed to alter radiation sensitivity. However, where such modifications allow a change in the fluidity from the "liquid-crystal" to "gel" state of the cellular membranes (as cooling the cells to ice temperature at the time of irradiation), a change in radiation sensitivity can be achieved (9,10). Thus we conclude that there is in bacteria some physical interaction between the gelled membrane and a closely neighboring vital target.

The lesions in DNA that have been extensively investigated are single- and double-backbone breaks in the twin-stranded DNA molecule. However, in general there is not a close relationship between such lesions and cell survival. In particular, oxygen enhancement ratios for strand breaks are usually much higher than for cell killing; sensitizing and protective agents do not have equivalent effects on breaks and lethality; and the relative biological effectiveness of high-linear-energy-transfer radiation, such as neutrons, is not positively correlated with cell death. Also, radiation-sensitive cells derived from ataxia telangiectasia patients show the same extent of induction of strand breaks (single and double) and the same rapid rate of repair as normally tolerant cells (11). Work with supercoiled phage also shows inactivation before strand breaks are measurable (12).

The main drawback in interpreting the measurement of lesions in a specific macromolecule is that interactions between macromolecules are disregarded. Thus, although breaks in DNA may indicate the energy deposited in this macromolecule, the loss of some more subtle but essential intermolecular structure dependent on weaker non-covalent bonding is not measured, nor is its reestablishment.

We have recently investigated radiation-induced changes in the interaction between DNA and protein by density measurements on histone-free-DNA-protein (HF-DNA) complexes released from nuclei isolated from human peripheral blood mononuclear cells. Such changes, including repair characteristics of the system, correlate reasonably closely with conventional radiobiological end points.

METHODS

Isolation of Lymphocytes From Whole Blood

Venous blood samples were drawn by syringe and mixed with heparin (sodium, 20 μg/ml) and stored at room temperature until lymphocyte separation on the same day. Whole blood was mixed with GIBCO RPMI 1640 + (L)-glutamine (GIBCO, Grand Island, NY) to give a diluted suspension (10:14, v/v). Lymphocytes were isolated by centrifugation on a discontinuous Histopaque 1077 (Sigma Chemical Co., St. Louis, MO) gradient using standard methods (Sigma Technical Bulletin No. 1077) and washed twice in RPMI 1640 + (L)-glutamine. The washed lymphocytes were finally resuspended in RPMI 1640 + (L)-glutamine and 5% fetal calf serum (GIBCO) at a concentration of 1 x 10^6 cells/ml and either stored overnight at 4°C or used immediately. All procedures were under sterile conditions in a Class II microbiological safety cabinet.

Assay of Density of HF-DNA

Lymphocyte suspensions (2 x 10^6/ml in RPMI 1640) were irradiated in air or under N_2 gassing at a dose rate of approximately 2 Gy/min at 0°C either with 240 kV X rays or cyclotron-produced 7-MeV neutrons. When chemical modifiers were used, they were added 30 min before irradiation. Immediately after irradiation, the cells were pelleted by centrifugation and resuspended in 0.2 ml RPMI 1640. Then 2.0 ml of ice-cold Non-idet reagent [Non-idet P-40 (0.25% w/v) + sucrose (0.25 M) + $CaCl_2$ (3.3 mM)] was added. After 10 min at 0°C, the nuclei released from the lysed cells were pelleted by centrifugation (500 x g, 5 min, 0°C). The nuclear pellets were agitated to resuspend in residual liquid, and an ice-cold lysis mixture [NaCl (2 M) + EDTA-Na_2 (10 mM) + Tris-HCl (pH 8.0, 10 mM)] was added to each sample. The resulting material, designated HF-DNA (13), was applied using a plastic disposable pipette to the top of a cold, 5%-20% (w/v) sucrose (Sigma Grade I) gradient containing NaCl (2 M), EDTA-Na_2 (10 mM), Tris-HCl (pH 8.0, 10 mM), and Hoechst Dye 33258 (1 μg/ml) in a 25-ml polycarbonate tube. Centrifugation followed at 48,000 x g for 15 min at 4°C. This procedure is diagrammed in Figure 1. Location of this HF-DNA material was determined by illumination with long-wavelength UV light (365 nm) in a dark room. The HF-DNA appears as a bluish band, 1 to 2 mm wide, the position of which can be defined by measuring its distance from the bottom of the tube. In most experiments, other aliquots of the irradiated cells were incubated at 37°C for various times before preparation of the nuclei. Nonirradiated controls, held at ice temperature or incubated at 37°C, were processed as described and run together (in separate tubes) with irradiated material.

The conditions for centrifugation were optimal. Further centrifugation at 48,000 x g did not move the bands significantly. Nuclei were stable for up to 2 hours at ice temperatures before HF-DNA isolation.

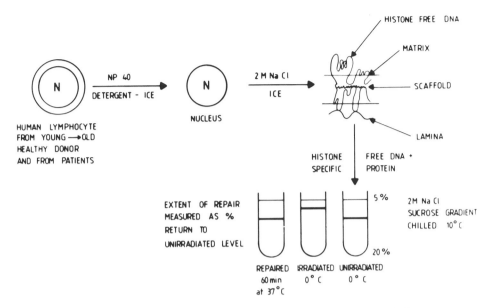

Fig. 1. Diagram of isolation of nuclei from peripheral blood lymphocytes and subsequent release of HF-DNA containing all DNA and structural proteins

Proliferation Studies

Concanavalan A (Con A, 6 μg/ml, Sigma) was added to triplicate 3-ml cultures (0.5 x 10⁶ cells/ml) in RPMI 1640 with 5% fetal calf serum and 2-mercaptoethanol (0.5 mM) in "Linbro" tissue culture plates. Proliferation was assessed on day 7 by counting cells using a hemocytometer.

Modification Procedures: Application, Concentrations, and Results

The position of the banded HF-DNA could be altered without radiation by the following procedures:

(a) Addition of calcium (5 mM CaCl₂) to the sucrose-2M NaCl gradient, which caused a small but measurable increase in density (band nearer to bottom of tube than in gradient without calcium).

(b) The absence of calcium from the Non-idet enucleation procedure and from the gradients gave rise, on 2-M NaCl lysis, to HF-DNA material of much decreased density. The material remained in the surfaces of the gradient and did not move under the standard centrifugation conditions.

(c) The addition of a free-SH compound, dithiothreitol (0.1 mM), to the lymphocytes in RPMI 1640 (30 min at 37°C) or the isolated HF-DNA at ice temperatures caused a marked decrease in density.

(d) Incubation of the whole lymphocytes with Cu^{II} (0.5 mM) for 30 min in RPMI, before separation and washing, dramatically increased the density of HF-DNA.

(e) Addition of sodium diethyl dithiocarbamate (SDDC, 0.1 mM) (a powerful chelator of copper) directly to the HF-DNA complex caused a large decrease in density.

(f) Addition of the radiation sensitizer misonidazole (4 mM) to lymphocytes suspended in RPMI 1640 for 30 min decreased the density of HF-DNA before irradiation. In addition, the presence of this compound at the time of irradiation under anoxic conditions prevented repair over a 60-min period.

(g) The ADPRT inhibitor 3-acetamido-benzamide, which has been shown to be a potent postirradiation sensitizer (14), also prevented postirradiation repair to nonirradiated densities of HF-DNA (15).

DNA-Protein Binding

Lymphocytes that had been triggered into DNA-synthetic activity (day 4 after Con A stimulation) were labeled overnight with ^{3}H-thymidine and ^{14}C-leucine or ^{3}H-leucine and ^{14}C-thymidine (all at 50 μCi/ml). Unlabeled thymidine and leucine were added as dilutants. The activities and concentrations of these radiolabeled compounds were adjusted to give counts of approximately 5,000-10,000 dpm for whole nuclei from 0.5 x 10^6 cells.

The labeled cells were processed to release nuclei (as above) and the nuclei (0.5 x 10^6 per aliquot) collected on Millipore Membrane "Durapore" HVLP filters (Millipore, Bedford, MA) made of polyvinylidene fluoride designed to retain large-molecular-weight species (MW >10^6) but to which DNA, RNA, and protein do not adhere. Half of the nuclei were treated with ice-cold 2-M NaCl on the filter (5 x 2 ml) to release HF-DNA. After extensive washing, the filters containing whole nuclei and HF-DNA were added to scintillant (Ultrafluor, National Diagnostics Ltd, Somerville, NJ) and counted.

RESULTS

Density of HF-DNA

The distribution of initial densities of HF-DNA material from a variety of lymphocyte sources has been reported (16,17). In general, the location of the HF-DNA band does not predict either the effect of radiation on Con A-stimulated proliferation or the capabilities of the cells to restore the density of the HF-DNA to nonirradiated levels.

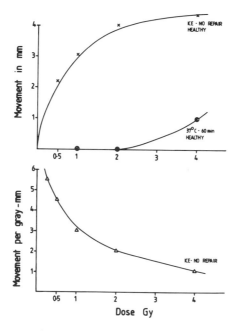

Fig. 2. Dose-effect curves (upper) indicating that HF-DNA material becomes less dense with dose. "Movement" means distance farther from bottom of tube than nonirradiated sample. Effect of dose (lower) expressed as movement per Gy. Lymphocytes were obtained from a normal healthy female, age 29 years.

Figure 2 shows the effect of a range of doses on decreasing the density (movement upward from original nonirradiated position) and the effectiveness of incubation at 37°C in RPMI 1640 for a set period of 60 min. The lower graph in this figure shows (as expected) that the effect per Gy is greater at the lower doses. An effect is easily measurable at 0.25 Gy. These values were obtained using lymphocytes from a healthy 29-year-old donor.

Figure 3 shows the results obtained when lymphocytes from a patient with active systemic lupus erythematosus were used. The repair in a 60-min period was much reduced. The two donors used for the results yielded HF-DNA of initially the same density. We have not examined all lymphocyte samples as extensively as shown in Figures 2 and 3. In summary, from healthy donors we have detected approximately 5% with diminished repair capacities. This figure is higher for the older end of this group. Mononuclear cells from patients with connective tissue disease (CTD) show a significantly higher lack of repair capability.

Proliferation

Figure 4 shows (upper) the type of result obtained when lymphocytes are stimulated into division with Con A irradiated and nonirradiated and (lower) a diagram of the position of HF-DNA in centrifuge tubes. Percentage proliferation was taken

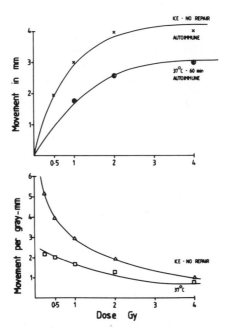

Fig. 3. Same as in Figure 2, but lymphocytes were obtained from a female donor, 33 years old, suffering from systemic lupus erythematosus.

as 100 in nonirradiated conditions and after irradiation was dependent on the relative number of cells at day 7. There was a positive correlation between the increased sensitivity measured in this way and the lack of repair (return to nonirradiated density) of HF-DNA in all samples examined. For example, Figure 5 clearly shows that the lymphocytes from a geriatric group were, on the average, more sensitive to radiation in proliferation studies. These hypersensitive individuals, who weighted these results to the mean values shown, had very poor capacities to restore the density of HF-DNA to nonirradiated values. However, there were a few unexplained exceptions in this group and in a larger group of patients with CTD.

A cytogenetic study of cells from the group of old donors showed that lymphocytes that were sensitive, as shown by proliferation and HF-DNA repair, had a higher background of chromosome aberrations (rings and dicentrics) and were more sensitive to aberration induction after irradiation than were the insensitive donors from the same group.

Modification Procedures for HF-DNA

Results have been noted in the Methods section. In summary, the density of HF-DNA material is dependent on the calcium and copper contents of nuclei derived from lymphocytes.

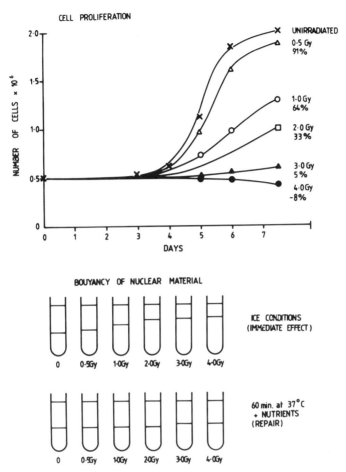

Fig. 4. Typical example of proliferation of Con A-stimulated lymphocytes, irradiated and nonirradiated (upper). Note that the nonirradiated lymphocytes increased fourfold. If increase in numbers was less than factor of 2, results were disregarded. It is possible to have less than starting number of lymphocytes at day 7 after higher doses. Diagram of position of HF-DNA bands (lower) from same sample of lymphocytes as used to obtain results in upper figure.

In preliminary studies, the oxygen enhancement ratio for the system approximates to 3, and, using cyclotron-produced neutrons, the relative biological effectiveness of these particles producing "knock-on" protons is significantly greater than 1. Also, using chemicals that modify whole-cell sensitivity, dramatic effects have been observed at the HF-DNA level.

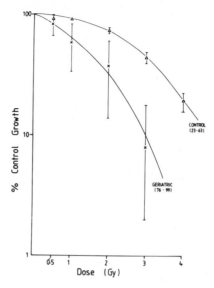

Fig. 5. Proliferation curves derived from two age groups. Points are means; lines, standard errors. Within "geriatric" group, 9/27 were markedly more sensitive and showed concomitant lack of restoration of density of HF-DNA and a higher-than-normal background of chromosomal aberrations.

DNA-Protein Binding

With lymphocytes from three donors (one healthy and two radiotherapy patients who had shown late complications after conventional radiation regimens), we attempted to measure postirradiation losses of protein attached to DNA by a dual radiolabeling technique as described. No significant loss in counts of either labeled DNA or labeled protein material was measured (0-4 Gy), and the ratio of DNA to protein remained constant. Approximately 5% of total nuclear protein remained associated with the DNA at all doses. No loss of nuclear DNA was observed in the filtration process described. In earlier experiments with other donors where radiolabeled HF-DNA was extracted from the sucrose gradients (a more difficult process), the same results were obtained.

DISCUSSION

We have found that probably because of the nonreplicative state of the lymphocytes and mononuclear cells used in these studies, the DNA banding, measured by the gradient technique, yields narrow, easily located bands. We have attempted the same studies with asynchronous populations of murine lymphoma cells and a human transformed lymphoid line, and we believe that some modification of the described techniques is required before good results can be obtained.

It must be stated that a major problem when using human lymphocytes is the difficulty in obtaining samples from the same donor on a sufficient number of occasions to establish reproducibility.

The central end point that has been described in this paper is the sedimentation property of all the DNA from the nucleus associated with a tenaciously bonded amount of protein. Sedimentation properties are altered after irradiation whereas the total DNA-protein content apparently remains constant. We interpret these results to mean that DNA-DNA and DNA-protein or protein-protein interactions are radiation sensitive. The interactions are different from donor to donor as shown by differences in initial dose response; in some selective instances, the cells are poor in restoring the original configuration of these interactions. Membranes will be important to these interactions if DNA or protein is in close juxtaposition with, or possibly embedded in, the nuclear membrane. On this basis we have proposed a model on which some radiobiological phenomena can be explained (18).

With the appearance of the hypothesis of Laemmli and his colleagues (3,4) that copper plays a major role in the three-dimensional structure of DNA, bringing together DNA and protein by coordinate bonding around a copper atom, we recently defined (19) in some detail the "direct" and "indirect" radiation effects on such structures. In brief, the "direct" effect of radiation may be the conversion of Cu^{II} to Cu^{I} by interaction with the aqueous electron (e^-aq) or the peroxy free radical $-\dot{O}_2$:

$$Cu^{II} + e^-aq \rightarrow Cu^{I}$$

$$Cu^{II} + \dot{O}_2 \rightarrow Cu^{I} + O_2$$

An accompanying "indirect" effect in the opposite direction $Cu^{I} \rightarrow Cu^{II}$ may be brought about by the attack on Cu^{I} by an H_2O_2 molecule generated in the location of Cu^{I} with the result that a very site-specific $\cdot OH$ radical is produced:

$$Cu^{I} + H_2O_2 \rightarrow Cu^{II} + \cdot OH + {}^-OH$$

This radical can then attack either DNA or protein neighboring a copper atom link between these two macromolecules. Although we have observed the return of HF-DNA material to nonirradiated densities, we assume that the relationship of DNA and protein is different; that is, misrepair may result. In consequence, such vital roles as transcription, replication, or condensation at division may be upset, causing death or malfunction of the cells.

The clinical implications that arise from our work with lymphocytes can be divided into three areas. In recent years, small subsets of patients with various forms of intractable CTD have been treated with either total lymphoid irradiation (TLI) (20-22) or total-body irradiation (TBI) (23). However, because of an excessively high incidence of radiation-induced toxicity in these patients, the role of radiation in the general management of severe CTD has been widely discredited. To date, there has been no attempt to establish whether an assessment of lymphocyte radiosensitivity *in vitro* might reflect *in situ* sensitivity to either TLI or TBI. If it is shown that *in vitro* assays such as ours can predict thresholds for toxicity in persons about to receive therapeutic irradiation, clinicians would be provided with a useful instrument for individualizing radiation doses and thereby minimizing dangerous side effects. It is in this area that further refinements of our assays might find their best clinical application.

The role of copper in the etiology of immune diseases has been discussed (24), and there may be a direct correlation between radiation sensitivity and disease status.

We have examined peripheral blood lymphocytes from patients who exhibited severe complications after large-field radiotherapy to the pelvic area. In the large majority of cases (n = 16/18), these patients (retrospectively investigated) showed a higher than normal lymphocyte radiation sensitivity. By prospective examination it may be feasible to predict those patients who may be unable to tolerate the conventional fractionated regimen.

Although we cannot see marked radiation-induced changes until 0.25 Gy, our dose-effect curves suggest a very small threshold effect, and it is possible that small but important permanent changes occur at low doses. A very similar dose-effect curve has been reported in a study by Waldren et al. (25), where the potential mutagenic effect of ionizing radiation on human chromosome 11 incorporated into rodent DNA was measured. In long-lived lymphocytes exposed to "self" metabolites and extraneous environmental hazards, a further insult could potentially be dangerously mutagenic. A survey of persons in occupations where "low" radiation doses are now allowable might reveal hypersensitive individuals whose health should be safeguarded by removal to other areas.

ACKNOWLEDGMENTS

L. W. Hart is supported by a research fellowship from the Medical Research Council of Canada. Supported in part by National Cancer Institute Grant CA24872 to M. B. Y.

REFERENCES

1. Alper, T., and Gillies, N. E. Restoration of *E. coli* Strain B after irradiation: Its dependence on suboptimal growth conditions. J. Gen. Microbiol. 18: 461-472, 1958.
2. Alper, T. "Cellular Radiobiology." Cambridge University Press, New Rochelle, New York, 1979.
3. Lewis, C. E., and Laemmli, U. K. Higher order metaphase chromosome structure: Evidence for metalloprotein interactions. Cell. 29: 171-181, 1982.

4. Lewis, C. E., Lebkowski, J. S., Daly, A. K., and Laemmli, U. K. Interphase nuclear matrix and metaphase scaffolding structures. J. Cell. Sci. Suppl. 1: 103-122, 1984.

5. Edwards, J. C., Cramp, W. A., Chapman, D., and Yatvin, M. B. The effects of ionizing radiation on biomembrane structure and function. Prog. Biophys. Mol. Biol. 43: 71-93, 1984.

6. Yatvin, M. B. Evidence that survival of γ-irradiated *E. coli* is influenced by membrane fluidity. Int. J. Radiat. Biol. 30: 571-575, 1976.

7. Edwards, J. C., Chapman, D., and Cramp, W. A. Radiation studies of *Scholeplasma laidlawii:* The role of membrane composition. Int. J. Radiat. Biol. 44: 405-412, 1983.

8. George, A. M., Lunec, J., and Cramp, W. A. Effect of membrane fatty acid changes on the radiation sensitivity of LDV human lymphoid cells. Int. J. Radiat. Biol. 43: 363-378, 1983.

9. Yatvin, M. B., Gipp, J. J., and Dennis, W. Influence of unsaturated fatty acids, membrane fluidity and oxygenation on the survival of an *E. coli* auxotroph. Int. J. Radiat. Biol. 35: 539-548, 1979.

10. Alper, T., Cramp, W. A., George, A. M., Lunec, J., and Yatvin, M. B. Membrane fluidity and the radiosensitivity of *E. coli* K1060. Int. J. Radiat. Biol. 40: 211-215, 1981.

11. Lehmann, A. Defects in DNA repair and carcinogenesis. Inaugural meeting of British Oncological Society, 17, Abstracts, 1986.

12. Kohnlein, W., and Schafers, F. The effect of ionizing radiation and bleomycin on transfecting ability of bacillus subtilis phage DNA. In: "Radiation Biology and Chemistry." H. E. Edwards, ed. Elsevier, New York, 1978, pp. 345-357.

13. Cook, P. R., and Brazell, I. S. The superhelical density of nuclear DNA from human cells. Eur. J. Biochem. 74: 527-531, 1977.

14. George, A. M., Lunec, J., Cramp, W. A., Brennan, S., Lewis, P. D., and Whish, W. J. D. The effects of benzamide ADP-ribosyl transferase inhibitors on cell survival and DNA strand-break repair in irradiated mammalian cells. Int. J. Radiat. Biol. 49: 783-798, 1986.

15. Durkacz, B. W., Irwin, J., and Shall, S. The effect of inhibition of (ADP-ribose)n biosynthesis on DNA repair assayed by the nucleoid technique. Eur. J. Biochem. 121: 65-69, 1981.

16. Harris, G., Cramp, W. A., Edwards, J. C., George, A. M., Sabovljev, S. A., Hart, L. E., Hughes, G. R. V., Denman, A. M., and Yatvin, M. B. Radiosensitivity of peripheral blood lymphocytes in auto-immune disease. Int. J. Radiat. Biol. 47: 689-699, 1985.

17. Harris, G., Holmes, A., Sabovljev, S. A., Cramp, W. A., Hedges, M., Hornsey, S., Hornsey, J. M., and Bennett, G. C. J. Sensitivity to X-radiation of peripheral blood lymphocytes from aging donors. Int. J. Radiat. Biol. 50: 685-694, 1986.

18. Yatvin, M. B., Cramp, W. A., Edwards, J. C., George, A. M., and Chapman, D. The effects of radiation on biological membranes. Journal of Nuclear Instruments and Method in Physics Research A255, 306-316, 1987.

19. George, A. M., Sabovljev, S. A., Hart, L. E., Cramp, W. A., Harris, G., and Hornsey, S. DNA quaternary structure in the radiation sensitivity of human

lymphocytes: A proposed role of copper. 13th L. H. Gray Meeting. Brit. J. Cancer 55 (Suppl. VIII): 141-144, 1987.

20. Trentham, D. E., Weinblatt, M. E., and Austen, K. F. Total lymphoid irradiation in tertiary care for rheumatoid arthritis. Ann. Intern. Med. 102: 544-545, 1985.

21. Hart, L. E. Total lymphoid irradiation to treat rheumatoid arthritis? Can. Med. Assoc. J. 134: 218-219, 1986.

22. Ben-Chetrit, E., Gross, D. J., Braverman, A., Weshler, Z., Fuks, Z., Slavin, S., and Eliakim, M. Total lymphoid irradiation in refractory systemic lupus erythematosus. Ann. Intern. Med. 105: 58-60, 1986.

23. Morgan, S. H., Bernstein, R. M., Coppen, J., Halnan, K. E., and Hughes, G. R. V. Total body irradiation and the course of polymyositis. Arthritis Rheum. 28: 831-835, 1985.

24. Davis, P. Penicillamine metal chelates and their possible importance in rheumatoid arthritis: A brief review. Clinical and Investigative Medicine 7: 41-44, 1984

25. Waldren, C., Correll, L., Sognier, M. A., and Puck, T. T. Measurement of low levels of X-ray mutagenesis in relation to human disease. Proc. Natl. Acad. Sci. USA 83: 4839-4843, 1986.

ROLE OF LIPID PEROXIDATION IN RADIATION DAMAGE AND REPAIR OF CELL MEMBRANES IN THE THERMOPHILIC BACTERIUM *THERMUS THERMOPHILUS*

S. Suzuki

Radiology Unit
Institute of Medical Science
University of Tokyo
Tokyo 108, Japan

ABSTRACT

The thermophilic bacterium *Thermus thermophilus* HB-8, which does not contain unsaturated fatty acids, lost its ability to take up extracellular K^+ in a temperature-dependent manner after its exposure to either UV radiation or cobalt-60 gamma rays. The dose response of HB-8 cells in terms of K^+ permeability was not significantly different from that of an unsaturated fatty acid-requiring auxotroph of *E. coli* UFA[ts] grown with linoleate. These results indicate that a radiation-induced change in K^+ permeability of cell membranes is not attributable to lipid peroxidation.

INTRODUCTION

It has been suggested that a radiation-induced change in the K^+ permeability of cells is associated with either impairment of SH groups of the proteins or peroxidation of the lipids in the membrane. Many lines of evidence support this concept, but they are not completely decisive. It is doubtful that lipid peroxidation is responsible for the damage, because the fatty acid composition or phospholipid of bacterial and mammalian cell membranes does not change significantly before and after irradiation (1-3). To clarify this problem, the thermophilic bacterium *Thermus thermophilus* HB-8 was used, which does not contain unsaturated fatty acids in the membrane. Instead, it has iso- and anteiso-branched-chain fatty acids to maintain the fluidity of membranes (4):

$$\text{iso: } (CH_3)_2CH(CH_2)_nCOOH, \text{ anteiso: } C_2H_5CH(CH_3)(CH_2)_mCOOH$$

If lipid peroxidation were the main damage to membranes, then little or no damage should occur in HB-8 cells. On the contrary, if damage occurs in HB-8 cells, lipid peroxidation is not essential for damage to cell membranes.

MATERIALS AND METHODS

Thermus thermophilus HB-8 and *E. coli* UFA[ts] were grown with shaking in media at the optimum temperatures of 75° and 42°C, respectively. Cultured cells were kept for the indicated time at 0°C unless otherwise stated. Cells in growth medium were irradiated with cobalt-60 gamma rays or ultraviolet light (UV) at 0°, 20° or 37°C. After irradiation, the cells were kept at 0°C for 20 hr unless otherwise stated. Then the cells were incubated at 20°C for 2 hr, washed twice with 0.15 M choline chloride, and suspended in 0.01% LiCl, and their K$^+$ content was measured directly with a flame photometer. The ratio of the net K$^+$ content of irradiated cells to that of nonirradiated ones was used as a marker of radiation damage to membranes.

Cell survival was determined from the number of colonies formed on agar plates.

RESULTS

HB-8 cells were irradiated with cobalt-60 gamma rays at 0°, 20°, or 37°C. As shown in Figure 1, the ratio of the K$^+$ content of irradiated cells to that of nonirradiated ones decreased with increase in dose, although the decrease was reduced

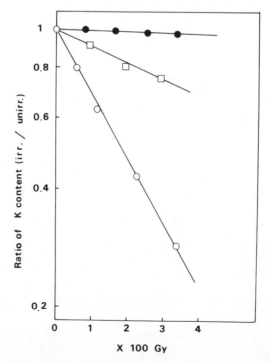

Fig. 1. Dose-response curves of relative K$^+$ content of HB-8 cells irradiated with ^{60}Co *y*-rays at 0° (o), 20° (□), or 37°C (●)

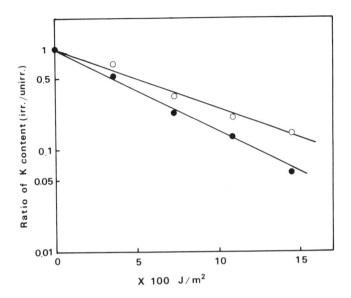

Fig. 2.   Dose-response curves of relative K$^+$ content of HB-8 (o) and *E. coli* UFA^{ts} with 18:2 (●) assayed 20 hr after UV exposure

at high temperature. These results indicate that radiation damage to membranes can occur without lipid peroxidation since the membrane lipids of HB-8 cells contain no unsaturated fatty acyl groups. Since the structure of membrane constituents of thermophilic bacteria is not essentially different from that of bacteria apart from the absence of unsaturated fatty acid, lipid peroxidation is probably not essential for radiation damage of the membranes of cells that do have polyunsaturated fatty acids in their membranes.

Radiation affects directly or indirectly more or less every structure of cells. If damage to membranes of other cells containing polyunsaturated fatty acids were due to lipid peroxidation, the membrane of HB-8 cells should be much more radioresistant than that of other cells. To examine this possibility, the K$^+$ contents of HB-8 without polyunsaturated fatty acids and of *E. coli* UFA^{ts} supplemented with linoleate (18:2) were compared after UV exposure (Figure 2). The radiosensitivities of these cells were not extremely different. Therefore, it is unlikely that lipid peroxidation is responsible for the membrane damage.

Cells exposed to UV at 0°C were held at 0° or 37°C for 80 min and further at 0°C for 20 hr. After these treatments, their ability to take up K$^+$ was determined. As shown in Figure 3, cells incubated at 0°C were more sensitive than at 37°C, whereas surviving fractions of cells were not different at both temperatures. This difference in K$^+$ uptake can be explained by repair of membrane damage at the higher temperature. The result also implies that radiation-induced change in K$^+$ permeability of cell membranes was not positively correlated with cell death.

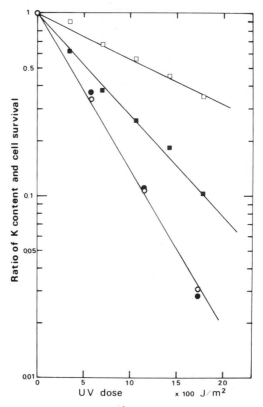

Fig. 3. Dose-response curves of relative K⁺ contents (□,■) and surviving fractions (○,●) of HB-8 cells. Cells were incubated at 0° (■,●) or 37°C (□,○) for 80 min after UV exposure and then held at 0°C for 20 hr.

DISCUSSION

Parts of this work were previously reported (5,6). One of the criticisms against our work is that membrane damage in HB cells does not necessarily mean little involvement of lipid peroxidation in the membrane damage in other cells that have polyunsaturated fatty acids. But the present work showed, at least for UV, that the membrane damage of HB-8 cells was not extremely different from that of *E. coli* UFA^ts grown with linoleate (Figure 2). It is therefore unlikely that UV-induced lipid peroxidation is mainly responsible for membrane damage in cells with polyunsaturated centers.

ACKNOWLEDGMENT

I am grateful to Drs. Y. Akamatsu and M. Oshima for their help with this work.

REFERENCES

1. Suzuki, S., and Akamatsu, Y. Involvement of membrane lipids in radiation damage to potassium ion permeability of *Escherichia coli.* Int. J. Radiat. Biol. 33: 185-190, 1978.
2. Suzuki, S., and Akamatsu, Y. Increase of radiation damage to potassium ion permeability in *E. coli* cells with decrease in membrane fluidity. Int. J. Radiat. Biol. 37: 475-482, 1980.
3. Suzuki, S., Doi, O., and Akamatsu, Y. Alteration in radiation-induced change in K^+ permeability of mouse fibroblast LM cells by modification of their membranes with unsaturated fatty acids. J. Radiat. Res. 23: 129-140, 1982.
4. Oshima, M., and Miyagawa, A. Comparative studies on the fatty acid composition of moderately and extremely thermophilic bacteria. Lipids 9: 476-480, 1974.
5. Suzuki, S., Oshima, M., and Akamatsu, Y. Radiation damage to membranes of the thermophilic bacterium, *Thermus thermophilus* HB-8: Membrane damage without concomitant lipid peroxidation. Radiat. Res. 91: 564-572, 1982.
6. Suzuki, S. Membrane damage and its repair in the thermophilic bacterium, *Thermus thermophilus* HB-8 exposed to U.V. radiation and ^{60}Co γ-rays. Int. J. Radiat. Biol. 47: 291-297, 1985.

MACROMOLECULE EFFECTS ON PROSTACYCLIN PRODUCTION AND MEASUREMENT

R. M. Price[1], D. M. Gersten[2], and P. W. Ramwell[3]

[1]Departments of Microbiology and Biophysical Sciences
State University of New York at Buffalo
Buffalo, New York 14214

[2]Department of Pathology
[3]Department of Physiology
Georgetown University Medical Center
Washington, DC 20007

ABSTRACT

Incubation of human umbilical artery tissue in a crystalloid solution elicits the production of prostacyclin (PGI_2). However, the inclusion of macromolecules in the incubation solution reduces the amount of PGI_2 produced by the umbilical artery. This effect is independent of charge since serum proteins, polyglutamic acid, polylysine, or dextran all reduce PGI_2 release by 50% at the same concentration of macromolecule (4.4 mg/ml). However, this effect is dependent on molecular size. Serum proteins with molecular weights (MW) below 60,000 are less effective than serum proteins above 60,000 MW (excluding albumin), and dextrans of 10,000 and 40,000 MW are not effective whereas dextrans of 70,000 and 500,000 MW are effective in reducing PGI_2 release. The effective macromolecules act at the exterior surface of the cell and do not shunt arachidonic acid into other catabolic pathways. Although the presence of these macromolecules affects the radioimmunoassay measurement of 6-keto-$PGF_{1\alpha}$, this effect can be quantitatively accounted for and does not explain the observed reduction in PGI_2 release. These studies appear to demonstrate a connection between receptor-independent cell surface events (adhesion of macromolecules) and the initiation of intracellular enzymatic activity (PGI_2 production). We hypothesize that macromolecules associate with the cell surface and retard the lateral mobility of externally projecting cell surface components. This would decrease the formation of cell membrane complexes responsible for arachidonic acid release. The observed reduction in PGI_2 release would result, then, from decreased precursor arachidonic acid release. Thus, the presence of macromolecules at the cell surface decreases the cellular response (e.g., prostaglandin production) to membrane perturbations and represents the *in vivo* situation. This may be an important consideration clinically (e.g., the choice between colloid or crystalloid solution for infusion) when increased prostaglandin release would be deleterious to the patient.

INTRODUCTION

Macromolecules are ubiquitous in the external environment of mammalian cells. This ubiquity prompted us to ask whether the presence of macromolecules, such as serum proteins, polyamino acids, or clinically used dextrans, can affect cell processes. We are particularly interested in those processes that lead to membrane damage and impaired cell function.

One way that cells respond to perturbation of their membranes is by producing prostaglandins. Although the role of prostaglandin release has yet to be determined for many cell types, the magnitude of such release seems to occur in proportion to the stimulus. Thus, prostaglandin release may be used as an index of membrane function. We have measured the *in vitro* release of prostacyclin (PGI$_2$) from human vascular tissue by radioimmunoassay (RIA) and have used these measurements as indicators of endothelial cell membrane function. When macromolecules are present in the incubation solution, less PGI$_2$ is produced by the vascular tissue than when incubated in a crystalloid solution. This effect is dependent on the size of the macromolecule but not on the charge or enzymatic activity. Further, the presence of macromolecules is a confounding factor in RIA due to the nonspecific binding of small ligands, e.g., prostaglandins, by macromolecules. However, this effect can be quantitatively accounted for in the RIA.

MATERIALS AND METHODS

Human umbilical artery is used as the vascular tissue source of PGI$_2$. Artery segments are incubated in Kreb's solution (buffer) or buffer plus the appropriate concentration of macromolecule (1). After incubation, the artery segments are removed and air-dried to constant weight. Aliquots of the incubation solution are assayed for released PGI$_2$ by competitive RIA of 6-keto-PGF$_{1\alpha}$, the spontaneous hydrolysis product of PGI$_2$. Two aliquots are used as duplicate samples; a third aliquot does not receive antibody and acts as a nonspecific (i.e., macromolecule) binding control (1,2). Other techniques, such as autoradiography and thin layer chromatography, have been described in detail (1).

RESULTS

When human umbilical artery segments are incubated in a crystalloid buffer, PGI$_2$ is released at a relatively constant rate. Addition of human serum to the incubation medium reduces PGI$_2$ release by 50%, at a serum protein concentration of 4.4 mg/ml or greater. We envision three mechanisms by which serum might reduce endogenous PGI_2 release from vascular tissue: binding of the released ligand and subsequent exclusion from RIA measurement, diversion of arachidonic acid to some other metabolic product, or reduction of arachidonic acid release and subsequent metabolism to PGI_2.

Serum binding of the immunoreactive ligand 6-keto-$PGF_{1\alpha}$ is assayed by incubation of radiolabeled 6-keto-$PGF_{1\alpha}$ with whole human serum. Bound ligand is separated from unbound ligand by gel filtration. Serum binds only 3% of the available ligand. Therefore, we conclude that the observed reduction in PGI_2 release from human vascular tissue incubated in the presence of human serum is not accounted for by serum binding of the ligand and subsequent exclusion from the RIA.

To test whether significant diversion of arachidonic acid to alternate metabolic pathways occurs, the incubation solutions are extracted and examined by thin layer chromatography. Both buffer and serum-containing incubation solutions show one spot each chromatographically identical to 6-keto-$PGF_{1\alpha}$. Serum-containing incubation solutions also show a spot at the solvent front; however, this spot is present in extracts of whole human serum that had not been incubated with vascular tissue and that did not contain detectable 6-keto-$PGF_{1\alpha}$ (by RIA). Therefore, we conclude that the observed reduction in PGI_2 release from human vascular tissue incubated in the presence of human serum is probably due to an actual reduction in release.

Incubation of vascular tissue in solutions containing protein fractions of human serum (on an equal mass basis, 4.4 mg/ml) does not reveal a significant enrichment in inhibitory activity associated with any one protein fraction. To model some general properties of proteins, artery segments are incubated with polyanionic macromolecules (polyglutamic acid, M_r 50,000-100,000), polycationic macromolecules (polylysine, M_r 70,000-150,000), the two in combination, or neutral macromolecules (dextran, M_r 70,000), all on an equal mass basis with human serum (4.4 mg/ml). All types of macromolecular incubation solution reduced PGI_2 release from vascular tissue to 50% of that released in buffer. Therefore, we conclude that macromolecular charge does not play a significant role in the mechanism of the reduction of PGI_2 release from vascular tissue.

To determine the effect of macromolecular size in this system, defined M_r dextrans (10,500, 40,000, 70,000, 500,000), on an equal mass basis (4.4 mg/ml), are incubated with artery segments. Only the dextrans with M_r of 70,000 and 500,000 inhibit PGI_2 release (by 50% and 35%, respectively). Therefore, a minimum macromolecular size is necessary to modulate PGI_2 release. Autoradiography of radiolabeled macromolecules incubated with artery segments does not indicate macromolecular entry into the cells on the time scale of the experiments.

Arachidonic acid added to the incubation solutions acts as both a precursor to prostaglandin production and a membrane-perturbing agent (3). Measurable PGI_2

release after the addition of arachidonic acid to the incubation solutions is increased in all solutions tested; however, release in solutions containing serum or dextran (M_r 70,000) is significantly less than that in solutions not containing macromolecules. Since arachidonic acid acts by at least two mechanisms to stimulate prostaglandin release, we cannot discern whether the macromolecules act to reduce PGI_2 release at a point before or after precursor arachidonate release. However, we have presented a model that attempts to explain these observations in terms of an interaction between the macromolecules and the cell surface (1).

Although nonspecific binding of PGI_2, or its hydrolysis product, by macromolecules can account for only 3% (at most) of the measurable reduction in PGI_2 release, this degree of binding could be significant when one tries to measure small amounts of prostaglandin in biological solutions (which contain significant amounts of macromolecules). We account for such nonspecific binding in the RIA for 6-keto-$PGF_{1\alpha}$ by using an additional tube, for each sample, to which antibody is not added. Since unbound ligand (PG) is absorbed by incubation with charcoal, the degree to which the ligand is bound by macromolecules is readily measured. We simply subtract the amount of ligand bound nonspecifically (as shown by the count rate) from the count rate in tubes containing antibody (specifically bound ligand plus nonspecifically bound ligand). The resulting count rate thus reflects specifically bound ligand. This technique is valid for RIA tubes containing up to 17 mg/ml of serum protein. Depending on the size of the sample aliquot used in the RIA (from 10 to 50 μl), this concentration of serum protein in an RIA tube can represent a tenfold to twofold concentration, respectively, of whole human serum (results submitted for publication).

DISCUSSION

Our results indicate that macromolecules, acting externally to the cell, can influence intracellular enzyme activities such as prostaglandin production. Since measurement of prostaglandin production may be a useful index of membrane perturbation, it is significant that macromolecules such as dextran reduce prostaglandin release. Such macromolecules are available for clinical use [e.g., colloid solutions such as Macrodex (dextran, M_r 70,000), Hespan, or Plasmanate] and may be useful adjuncts in moderating membrane damage. However, our results also indicate that additional macromolecules above a fairly low concentration (4.4 mg/ml) do not confer additional inhibition of prostaglandin release in our in vitro system. Therefore, we conclude that there is normally sufficient serum protein, which adheres to cell surfaces, to modulate membrane damage in vivo.

In the present context, it is also significant that macromolecular agents, such as certain polysaccharides (4,5), are radioprotective in vivo. Our results suggest that one must consider the possibility that these agents may be protective by modulating radiation-induced cell membrane damage. Our results also suggest that such a mechanism may show a dependence on macromolecular size.

ACKNOWLEDGMENTS

This research was supported in part by a grant from the American Heart Association, Nation's Capital Affiliate, and grant no. HL17516 from the Public Health Service. R.M.P. is a recipient of a University at Buffalo Presidential Postdoctoral Fellowship.

REFERENCES

1. Price, R. M., Gersten, D. M., and Ramwell, P. W. Macromolecules mediate prostacyclin release from human umbilical artery. Biochim. Biophys. Acta 836: 246-254, 1985.
2. Price, R. M., Gersten, D. M., and Ramwell, P. W. Serum inhibition of vascular prostacyclin release measurable by RIA. Book of Abstracts, 184th ACS National Meeting, Kansas City, Missouri, MEDI 47, 1982.
3. Sedar, A. W., Silver, M. J., Ingerman, C. M., Nissenbaum, M., and Smith, J. B. Model system for the study of initial damage to arterial endothelial cells *in situ* by scanning electron microscopy. Scanning Electron Microsc. 3: 235-241, 1980.
4. Behling, U. H. The radioprotective effect of bacterial endotoxin. In: "Beneficial Effects of Endotoxin." A. Nowotny, ed. Plenum Press, New York, 1983, pp. 127-148.
5. Maisin, J. R., Kondi-Tamba, A., and Mattelin, G. Polysaccharides induce radioprotection of murine hemopoetic stem cells and increase the $LD_{50/30}$ days. Rad. Res. 105: 276-281, 1986.

INOSITOL LIPID TURNOVER, CA^{2+} MOBILIZATION, AND INCREASED CA^{2+} INFLUX IN HEAT-STRESSED CELLS

M. A. Stevenson[1,2], E. K. Farnum[1,3], G. M. Hahn[4], and S. K. Calderwood[1]

[1]Joint Center for Radiation Therapy
Dana Farber Cancer Institute
Harvard Medical School
Boston, Massachusetts 02115

[2]Department of Medicine
Massachussetts General Hospital
Boston, Massachusetts

[3]George Washington University Medical School
Washington, DC

[4]Cancer Biology Research Laboratory
Stanford University
Stanford, California

ABSTRACT

We have investigated membrane permeability to Ca^{2+} and Ca^{2+} levels in cells subjected to heat stress. Heat caused a large and rapid rise in intracellular free calcium (Ca$^{2+}_i$) within 2 min at 45°C. Effects on transmembrane Ca^{2+} influx were more gradual, requiring 30 min at 45°C before increases were observed. The early rise in Ca$^{2+}_i$ appeared to be due to release from internal Ca^{2+} stores. Changes in Ca^{2+} homeostasis were correlated with alterations in membrane phosphoinositide (PI) turnover. The early rise in Ca$^{2+}_i$ was preceded by rapid (30 sec) release of inositol triphosphate, a PI catabolite that mobilizes Ca^{2+} from internal pools. Increased Ca^{2+} influx was closely correlated with membrane accumulation of phosphatidic acid, a lipid catabolite of PI with Ca^{2+} ionophoretic properties. The results indicate that heat, acting at the membrane, precipitates a cascade of interacting lipid and ionic changes that may be involved in the mediation of thermal injury.

INTRODUCTION

Heat generates a wide spectrum of cellular and molecular changes in tissues. These include mitogenesis, cytotoxicity, and induction of heat-shock gene transcription (1,2). However, there is no clear hypothesis to explain the interactions between these heat-induced events. We have studied the effect of heat stress on intracellular free calcium (Ca^{2+}_i) regulation through phosphoinositide (PI) metabolism (3). The rationale for investigating Ca^{2+}_i is that ionized calcium is both a regulator of normal metabolism and a mediator of cell death (4). Similarly, heat has both pleiotropic and toxic properties; Ca^{2+}_i thus seems a strong candidate as a mediator of the effects of thermal injury.

MATERIALS AND METHODS

Experiments were carried out on HA-1 Chinese hamster fibroblasts, BALB/c 3T3 mouse fibroblasts, and PC12 rat pheochromocytoma cells growing in monolayer culture. The major aim of the project was to characterize the response of the cellular inositol lipid/Ca^{2+} system to heat stress. We also wished to compare the effects of heat with those of physiological agonists of the PI/Ca^{2+} system. We chose to examine mitogenic, α_1 adrenergic, and cholinergic agonists, each of which is known to be effective in inducing PI turnover and Ca^{2+} mobilization (5,6). We examined serum-induced mitogenic stimulation in BALB/c 3T3 and HA-1 cells because we showed previously (7) that this treatment induces IP₃ release in these cells. Similarly, HA-1 cells respond to treatment with the α_1 adrenergic agonist phenylephrine (7). For PC12 cells (a line of neurosecretary cells), we used the muscarinic agonist carbachol, which causes IP₃ release in this cell line (6).

Ca^{2+}_i and Ca^{2+} influx were measured as described previously (8). Inositol phosphates were assayed in acid extracts of cells after separation on 1-cm Dowex-1 columns eluted with an ammonium formate/formic acid gradient (7,8). For separation of lipids, cells were dissolved in $CHCl_3$:MeOH:12 N HCl (2:1:0.12); 2 M KCl was added and the mixture swirled gently. Separated lipids were stored dry at -80°C. PA and PI were separated on silica gel-60 thin-layer chromatography plates with a $CHCl_3$:MeOH:C_2H_5COOH (13:5:2) solvent system. PIP and PIP₂ (9) were resolved with $CHCl_3$:MeOH:H_2O:NH_4OH (40:48:5:10). Lipids were identified using standards (Sigma, St. Louis, MO).

RESULTS AND DISCUSSION

The effect of heat (45°C) on cellular IP₃ concentration is shown in Figure 1. Heat induced rapid IP₃ release in HA-1, 3T3, and PC-12 cells. IP₃ release induced by heat stress was of a similar order to that generated by physiological agonists. In HA-1 cells, the α_1 adrenergic agonist phenylephrine and mitogenic stimulation with serum both caused a level of IP₃ release that was similar to that induced by heat stress. We have shown previously that heat and serum induce additive IP₃ production, probably indicating that heat acts at the post-receptor stage of phospholipase C activity (7). In BALB/c 3T3 cells, heat and serum induced similar degrees of IP₃ release (Fig. 1), which were again additive (7). In PC-12 cells, heat was a less effective inducer of IP₃ release than was the cholinergic agonist carbachol

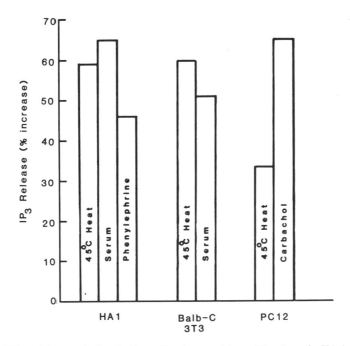

Fig. 1. Effect of heat and phospholipase C-active agonists on IP$_3$ release in HA-1, 3T3, and PC12 cells. HA-1 cells were treated for 5 min at 45°C, in 10% FCS or in 10^{-4} M phenylephrine. 3T3 cells were given 5 min at 45°C in 10% FCS. PC12 cells were incubated for 5 min at 45°C in 10^{-5} M carbachol. HA-1 and 3T3 cells were incubated for 48 hr in medium + 0.2% FCS before treatments in order to accumulate them in a mitogen-sensitive G$_0$ phase. PC12 cells were in normal growth medium. Datum points are means of five determinations.

(Fig. 1). The reason for this is not clear, but heat- and agonist-induced IP$_3$ release probably involve different mechanisms (7). Many cell types appear to contain discrete pools of PIP$_2$ (10), which may differ in size and responsiveness to treatments.

An increase in IP$_3$ preceded a rapid rise in Ca$^{2+}_i$ (Fig. 2). IP$_3$ levels were increased by at least 1 min. Ca$^{2+}_i$, as indicated by fluorescent dye Quin-2, was increased

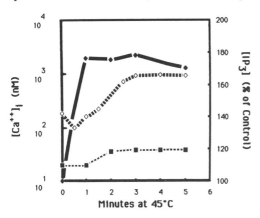

Fig. 2. Changes in IP$_3$ and Ca$^{2+}_i$ during 45°C heat shock: (◆) IP$_3$; (◇) Ca$^{2+}_i$ when Ca$^{2+}_e$ = 0.47 mM; (■) Ca$^{2+}_i$ when Ca$^{2+}_e$ < 1.0 μM. IP$_3$ datum points are means of five to seven estimations. Ca$^{2+}_i$ profiles are aggregates of 10-12 experiments.

from control levels of 180 nM by 2 min, reaching levels of 450-1100 nM by 3 min at 45°C. A rise in Ca^{2+}_i was also observed in conditions of extremely low Ca^{2+}_e (calcium-free medium + 10 mM EGTA). This indicates that the heat-induced Ca^{2+}_i increase is due at least partially to mobilization of internal Ca^{2+} stores; such an effect is consistent with the Ca^{2+}-mobilizing properties of IP_3 (3).

The effect of continuous heating at 45°C on $^{45}Ca^{2+}$ influx into HA-1 cells is shown in Fig. 3. A major increase in permeability to Ca^{2+} commenced at 30 min of heating. Increased permeability to Ca^{2+} coincided with a rise in PA concentration in the membranes of heated cells. PA is a metabolite of the PI pathway shown to increase the Ca^{2+} permeability of artificial and mammalian membranes (ref. 8; Fig. 4). We have shown (8) that $^{45}Ca^{2+}$ influx in HA-1 cells is strongly correlated with PA content.

CONCLUSIONS

Thus it seems likely that heat, acting on the PI pathway, causes a rapid increase in IP_3 concentration, leading to Ca^{2+} mobilization from an internal site and subsequent buildup of PA, which leads to enhanced Ca^{2+} influx (Fig. 4). IP_3 release is probably due to stimulation of phospholipase C; PA accumulation may be due to a block in its conversion to PI (ref. 8; Fig. 4).

The large and rapid rise in Ca^{2+}_i seen after heat stress may mediate some of the effects of heat. Depleting the internal Ca^{2+} stores of cells protects them from injury by heat (11), and treatment with Ca^{2+} ionophore A23187 enhances thermal

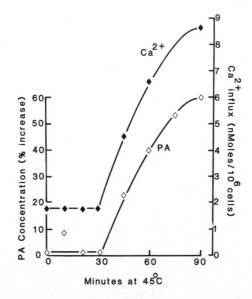

Fig. 3. PA concentration (◊) and $^{45}Ca^{2+}$ influx (◆) at 45°C. Datum points are means of triplicate assays. Experiments were repeated three times.

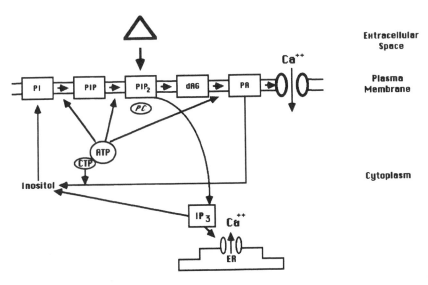

Fig. 4. Effect of heat on inositol lipid turnover and Ca^{2+} homeostasis. Changes in Ca^{2+} homeostasis appear to be initiated by heat-induced cleavage of phosphatidylinositol-4,5-bisphosphate (PIP$_2$) by phospholipase C (PC). Generation of compounds (IP$_3$, PA) that increase the permeability of membranes to Ca^{2+} then ensues. Products of PIP$_2$ cleavage, IP$_3$, and diacylglycerol (dAG), are metabolized in a bifurcating pathway to yield, respectively, PA in the membrane and free inositol in the cytosol (3). These compounds are then resynthesized into phosphatidylinositol (in membrane), and latter compound is converted by successive phosphorylations into PIP and PIP$_2$.

cell killing (12). Therefore, it seems that heat-induced perturbation of inositol lipid metabolism and Ca^{2+} homeostasis may play an important role in the mediation of thermal injury.

REFERENCES

1. Leeper, D. B. Molecular and cellular mechanisms of hyperthermia alone or in combination with other modalities. In: "Hyperthermic Oncology," Vol. 2. J. Overgaard, ed. Taylor & Francis, London, 1985, pp. 9-40.
2. Schlesinger, M. J., Ashburner, M., and Tissieres, A., eds. "Heat Shock." Cold Spring Harbor Laboratory, Cold Spring Harbor, NY, 1982.
3. Berridge, M. J., and Irvine, R. F. Inositol triphosphate, a novel second messenger in cellular signal transduction. Nature (Lond.) 312: 315-321, 1984.
4. Rasmussen, H., and Barrett, P. Q. Calcium messenger system: An integrated review. Physiol. Rev. 64: 938-976, 1984.
5. Whitman, M., Fleischman, L., Chahwala, S. B., Cantley, L., and Rosoff, P. Phosphoinositides, mitogenesis, and oncogenesis. In: "Phosphoinositides and Receptor Mechanisms." J. W. Putney, Jr., ed. Alan Liss, New York, 1986, pp. 197-218.

6. Vicentini, L. M., Ambrosini, A., Di Virgilio, F., Pozzan, T., and Meldolesi, J. Muscarinic receptor-induced phosphoinositide hydrolysis at resting cytosolic Ca^{2+} concentration in PC12 cells. J. Cell Biol. 100:1330-1333, 1985.
7. Calderwood, S. K., Stevenson, M. A., and Hahn, G. M. Heat stress stimulates inositol triphosphate release and phosphorylation of phosphoinositides in CHO and BALB/c 3T3 cells. J. Cell Physiol., in press.
8. Stevenson, M. A., Calderwood, S. K., and Hahn, G. M. Rapid increases in inositol trisphosphate and intracellular Ca^{2+} after heat shock. Biochem. Biophys. Res. Commun. 137: 826-833, 1986.
9. Irvine, R. F., Letcher, A. J., and Dawson, R. M. C. PIP phosphodiesterase and phosphomonoesterases in rat brain. Biochem. J. 218: 177-185, 1984.
10. Cockroft, S. Phosphoinositides and neutrophil action. In: "Phosphoinositides and Receptor Mechanisms." J. W. Putney, Jr., ed. Alan Liss, New York, 1986, pp. 287-310.
11. Lamarche, S., Chretien, P., and Landry, J. Inhibition of heat shock response and synthesis of glucose regulated proteins in Ca^{2+}-deprived rat hepatoma cells. Biochem. Biophys. Res. Commun. 131: 868-876, 1985.
12. Malhotra, A., Kruuv, J., and Lepock, J. Sensitization of rat hepatocytes to hyperthermia by calcium. J. Cell Physiol. 128: 279-285, 1986.

ROLE OF ANTIOXIDANT NUTRIENTS IN PREVENTING RETINAL DAMAGE DURING HYPERBARIC OXYGEN TREATMENT

A. L. Hollis[1], R. A. Henderson[2], and W. L. Stone[1]

[1]Departments of Pediatrics and Biomedical Sciences
Meharry Medical College
Nashville, Tennessee 37208

[2]Armed Forces Institute of Pathology
Washington, DC 20306-6000

ABSTRACT

Tissue damage due to hyperbaric oxygen treatment or radiation treatment may be mediated by free radicals and lipid peroxidation by-products. We have found that the retinas of animals deficient in antioxidant nutrients are very susceptible to tissue damage induced by hyperbaric oxygen treatment. Hyperbaric oxygen treatment was found to adversely affect the electroretinograms of rats fed either a diet deficient in both vitamin E and selenium (the basal or B diet) or a diet deficient in vitamin E alone (B + Se diet). Vitamin E and selenium are micronutrients thought to play essential roles in preventing *in vivo* lipid peroxidation. After 4 weeks of hyperbaric oxygen treatment (3.0 ATA of 100% oxygen, 1.5 hr per day, 5 days/week), rats fed the B diet deficient in vitamin E and selenium showed decreased ($p < 0.05$) a-wave amplitudes ($85 \pm 10 \mu V$, N = 11) compared with a-wave amplitudes ($150 \pm 10 \mu V$, N = 21) for rats fed an identical diet but not treated with hyperbaric oxygen. Rats fed a basal diet supplemented with both vitamin E and selenium (the B + E + Se diet) or with vitamin E alone (the B + E diet) showed constant a- and b-wave amplitudes that did not decrease after 15 weeks of hyperbaric oxygen treatment. After 15 weeks of hyperbaric oxygen treatment, rats fed the B + Se diet deficient in vitamin E alone showed decreased ($p < 0.01$) a-wave ($61 \pm 9 \mu V$, N = 4) and b-wave ($253 \pm 23 \mu V$, N = 4) amplitudes compared with a-wave ($115 \pm 7 \mu V$) and b-wave amplitudes ($450 \pm 35 \mu V$) for rats fed the same diet but not treated with hyperbaric oxygen. Dietary antioxidants appeared to provide protection from hyperbaric oxygen damage to the retina. Quantitative histopathology studies are now under way to determine which cell types in the retina of antioxidant-deficient rats have been damaged by hyperbaric oxygen treatment.

The livers of rats fed the B diet had significantly higher levels of malondialdehyde (MDA) than rats fed the B + E + Se diet. However, hyperbaric oxygen treatment (4 wk) did not increase the levels of liver MDA. The polyunsaturated fatty acid contents of livers from rats fed the B or B + E + Se diets were very similar and not affected by hyperbaric oxygen treatment (4 wk).

INTRODUCTION

In this review we focus on the biochemical and physiological factors that contribute to the extreme sensitivity of the visual system to oxidative stress and free-radical damage. We discuss the role of nutritionally modulated antioxidant mechanisms in protecting the visual system from oxidative stress. We also summarize our work showing that vitamin E and selenium can protect retinal tissues from damage caused by hyperbaric oxygen treatment. Hyperbaric oxygen therapy is increasingly being used to treat a variety of clinical disorders and to enhance wound healing (1). The beneficial effects of hyperbaric oxygen therapy are limited by the toxic effects of oxygen on lung, retina, and other organs (2-4). Under normal therapeutic protocols, the toxic effects of hyperbaric oxygen are minimal, but they may be increased in patients with impaired antioxidant mechanisms and/or in patients with pro-oxidant stressors. For example, hyperbaric oxygen is used to enhance wound healing in patients with radiation necrosis, usually many months after radiation therapy has been terminated. On the other hand, ionizing radiation concomitant with hyperbaric oxygen therapy could be synergistic sources of oxidative stress due to the increased production of free radicals.

NUTRITIONALLY MODULATED ANTIOXIDANT MECHANISMS

Antioxidant mechanisms play a central role in minimizing the production and concentration of free radicals and their chemical by-products (5). Many antioxidant mechanisms are dependent on the nutritional bioavailability of micronutrients. It is clearly important to understand the relationships between hyperbaric oxygen toxicity, radiation treatment, and nutritionally modulated antioxidant mechanisms. Animal models are a useful means of exploring these interrelationships and in determining which organ systems may be most susceptible to oxidative damage. Tissue levels of vitamin E are dependent on the dietary intake and the absorption of vitamin E. Vitamin E is thought to quench the lipid peroxy radicals produced during lipid peroxidation. Glutathione peroxidase activity in most tissues is dependent on the dietary intake of selenium. Selenium is a cofactor for glutathione peroxidase, which can reduce lipid hydroperoxides (or hydrogen peroxide) to the corresponding lipid alcohol (or dihydrogen oxide). Rat retinas normally have very high levels of vitamin E (6,7) and significant glutathione peroxidase activity (8). Nutritional deficiencies of vitamin E and selenium dramatically decrease the levels of retinal vitamin E (6,7) and glutathione peroxidase (8), respectively. Other important antioxidant nutrients are vitamin C, beta-carotene (a singlet oxygen quencher), sulfur amino acids (precursors of glutathione), and zinc (a component of superoxide dismutase).

Animals deficient in antioxidant nutrients show evidence of increased *in vivo* lipid peroxidation. For example, intact rats deficient in vitamin E or selenium undergo increased *in vivo* lipid peroxidation as measured by exhalation of small-chain hydrocarbons (9), which are chemical by-products of lipid peroxidation. Deficiencies of vitamin E and selenium act synergistically in increasing *in vivo* lipid peroxidation (7,9). Organisms deficient in antioxidant nutrients and subjected to ionizing radiation or hyperbaric oxygen may be particularly susceptible to tissue damage caused by free radicals and *in vivo* lipid peroxidation. The extent to which a specific organ system suffers from oxidative damage depends on many factors: type of oxidative stress (hyperbaric or radiation), levels of nutritional modulated antioxidant mechanisms, blood flow, and oxygen concentration.

SUSCEPTIBILITY OF RETINA TO LIPID PEROXIDATION DAMAGE AND HYPERBARIC OXYGEN TOXICITY

A significant body of data (6-8,10,11) suggests that the retina is very sensitive to oxidative damage caused by lipid peroxides or by deficiencies of nutritionally modulated antioxidant mechanisms. This is because the retina has (a) one of the highest contents of polyunsaturated fatty acid of any tissue in the body (about 30% 22:6n3), (b) an extremely high oxygen consumption (about seven times more per gram of tissue than the brain) and therefore an extremely high flux of oxygen from the choroidal capillaries across the photoreceptor disc membranes (rod outer segment area) to the rod inner segments, (c) a high concentration of retinal (the aldehyde form of vitamin A), which can induce photosensitized oxidation reactions (6), and (d) a highly membranous structure in which oxygen is very soluble.

Hyperbaric oxygen toxicity is very dependent on blood flow. Blood supply to the retina, via the ophthalmic artery, is well oxygenated under normal circumstances. Under hyperbaric oxygen conditions, plasma can carry a significant amount of dissolved oxygen, and the retina would receive a large supply of hyperoxygenated plasma from the ophthalmic artery. The liver, in contrast to the retina, receives the majority of its blood supply from the portal vein and therefore would not be oxidatively stressed since the portal vein carries deoxygenated blood that has already passed through the capillary beds of the intestine.

QUANTITATION OF RETINAL DAMAGE BY LIPID PEROXIDATION AND BY HYPERBARIC OXYGEN TOXICITY

We have recently found (unpublished results) that hyperbaric oxygen treatment adversely affects the electroretinograms (ERG's) of rats fed a diet deficient in both vitamin E and selenium (the basal or B diet) or a diet deficient in vitamin E alone (B + Se diet). After only 4 weeks of hyperbaric oxygen treatment (3.0 ATA of 100% oxygen, 1.5 hr per day, 5 days/week), rats fed the B diet deficient in vitamin

E and selenium showed decreased ($p < 0.05$) a-wave amplitudes (85 ± 10 μV, $N = 11$) compared with a-wave amplitudes (150 ± 10 μV, $N = 21$) for rats fed an identical diet but not treated with hyperbaric oxygen. Rats fed a basal diet supplemented with both vitamin E and selenium (the B + E + Se diet) or with vitamin E alone (the B + E diet) showed constant a-wave and b-wave amplitudes that did not decrease after 15 weeks of hyperbaric oxygen treatment. After 15 weeks of hyperbaric oxygen treatment, rats fed the B + Se diet deficient in vitamin E alone showed decreased ($p < 0.01$) a-wave (61 ± 9 μV, $N = 4$) and b-wave (253 ± 23 μV, $N = 4$) amplitudes compared with a-wave (115 ± 7 μV) and b-wave amplitudes (450 ± 35 μV) for rats fed the same diet but not treated with hyperbaric oxygen. Both these dietary antioxidants thus appear to provide protection from hyperbaric oxygen damage to the retina, but vitamin E protects significantly more effectively in this model.

Quantitative histopathology studies are now under way to determine which cell types in the retina of antioxidant-deficient rats have been damaged by hyperbaric oxygen treatment. The a-wave is thought to be generated from rod outer-segment cells and the b-wave from the bipolar cells. We expect that both these cell types will show damage in the vitamin E-deficient and selenium-deficient rats treated with hyperbaric oxygen.

ACKNOWLEDGMENTS

The authors thank the Air Force Systems Command, the Air Force Office of Scientific Research, and Universal Energy Systems for support (Subcontract No. S-760-OMG-037 to W.L.S). Additional support was provided by a Minority Biomedical Support Grant from the National Institutes of Health (S06RR008037-13 to W.L.S).

REFERENCES

1. Meyers, R. A., and Schnitzer, B. M. Hyperbaric oxygen use. Postgraduate Med. 76: 83-95, 1984.
2. Clark, J. M., and Fisher, A. B. Oxygen toxicity and extension of tolerance in oxygen therapy. In: "Hyperbaric Oxygen Therapy." J. C. Davis and T. K. Hunt, eds. Undersea Medical Society, Bethesda, MD, 1977, pp. 61-77.
3. Small, A. New perspectives on hyperoxic pulmonary toxicity: A review. Undersea Biomed. Res. 11: 1-24, 1984.
4. Jamieson, D., Chance, B., Cadenas, E., and Boveris, A. The relation of free radical production to hyperoxia. Ann. Rev. Physiol. 48: 703-719, 1986.
5. Fridovich, I., and Freeman, B. Antioxidant defenses in the lung. Ann. Rev. Physiol. 48: 693-702, 1986.
6. Stone, W. L., Katz, M. L., Lurie, M., Marmor, M. F., and Dratz, E. A. Effects of dietary vitamin E and selenium on light damage to the rat retina. Photochem. Photobiol. 29: 725-730, 1979.

7. Katz, M. L., Stone, W. L., and Dratz, E. A. Fluorescent pigment accumulation in retinal pigment epithelium of antioxidant-deficient rats. Invest. Ophthalmol. 17: 1049-1058, 1978.

8. Stone, W. L., and Dratz, E. A. Selenium and non-selenium glutathione peroxidase activities in selected ocular and non-ocular rat tissues. Exp. Eye Res. 35: 405-412, 1982.

9. Hafeman, D. G., and Hoekstra, W. G. Lipid peroxidation *in vivo* during vitamin E and selenium deficiency in the rat as monitored by ethane evolution. J. Nutr. 107: 666-672, 1977.

10. Armstrong, D., Hiramitsu, T., Gutterridge, J., and Nilsson, S. E. Studies on experimentally induced retinal degeneration. 1. Effects of lipid peroxides on electroretinographic activity in the albino rabbit. Exp. Eye Res. 35: 157-171, 1982.

11. Katz, M. L., Parker, K. R., Handelman, G. J., Bramel, T. L., and Dratz, E. A. Effects of antioxidant nutrient deficiency on the retina and retinal pigment epithelium of albino rats: A light and electron microscopic study. Exp. Eye Res. 34: 339-369, 1982.

EFFECTS OF PHENYLHYDRAZINE ON HUMAN RED CELL MEMBRANE AMINOPHOSPHOLIPIDS: A FLUORESCAMINE STUDY

Z. Chen[1] and A. Stern[2]

[1]Cancer Center
University of Rochester
Rochester, New York 14642

[2]Department of Pharmacology
New York University Medical Center
New York, New York 10016

ABSTRACT

Phenylhydrazine was used as a peroxidative probe for studying human red cell membrane aminophospholipid peroxidation. A fluorescamine labeling method for aminophospholipids (1) was modified for direct measurement of the labeled aminophospholipids by fluorescence spectroscopy following separation of the phospholipid derivatives by thin layer chromatography. Phenylhydrazine-treated red cells had a significant decrease of phosphatidylethanolamine (PE) labeling by fluorescamine. This finding is consistent with our previous results in which we observed a significant degradation of PE associated with alterations in red cell membrane surface lipid packing (2). Further experiments indicated that the decreased labeling of PE was due to degradation of PE to lysophosphatidylethanolamine (LPE) and other oxidized lipids (CL). In seven persons studied, four degraded PE to LPE and three degraded PE to oxidized lipids other than LPE. These results indicate that individual variation may exist in the degradation of PE under oxidative conditions.

INTRODUCTION

Phenylhydrazine has been used as a probe for studying membrane damage in red cells (3). During the autoxidation of phenylhydrazine, both reduced products of oxygen and reactive free-radical intermediates are generated. Phenylhydrazine has recently been shown to cause red cell membrane spectrin (4) and PE degradation (2).

97

In this study, we used fluorescamine, a fluorescent compound that readily reacts with free amino groups on aminophospholipids (1). We present relative fluorescence measurements of fluorescamine labeling of aminophospholipids and their derivatives in red cells treated with phenylhydrazine. We also present evidence that individual variation may exist in the degradation of red cell PE when red cells are exposed to phenylhydrazine.

MATERIALS AND METHODS

Treatment of Erythrocytes With Phenylhydrazine

Blood was drawn fresh daily from healthy donors. Red cells were washed three times with 0.9% NaCl and resuspended to 50% hematocrit (Hct) with KRG buffer (120 mM NaCl, 4.7 mM KCl, 2.5 mM $CaCl_2$, 1.2 mM KH_2PO_4, 1.2 mM $MgSO_4$, and 50 mM glycylglycine, pH 7.4). One mM phenylhydrazine was added last to each experimental sample (final Hct 25%) and incubated with shaking in sealed flasks at 37°C for 1 hour. After incubation, the red cells were washed twice with 0.9% NaCl and once with fluorescamine buffer (150 mM $NaHCO_3$ and 20 mM HEPES, pH 8.3).

Fluorescamine Labeling and Phospholipid Extraction

Fluorescamine labeling and phospholipid extraction was by a modification of the method of Rawyler et al. (1). Briefly, 20 μl (0.2 M) of fluorescamine stock solution (in DMSO) was added to each blood sample (0.75 ml) and shaken for 30 minutes at room temperature. Then 0.25 ml 1 M NH_4OH was added to all

Table 1. Effects of Phenylhydrazine on Reactivities of Human Red Cells Aminophospholipids With Fluorescamine

	Labeling Before Extraction		Labeling After Extraction	
	Control	Experimental	Control	Experimental
CL	8.1 ± 1.2	11.9 ± 1.9^1	11.0 ± 3.1	19.6 ± 2.9^2
PE	84.6 ± 2.1	77.2 ± 4.5^1	79.6 ± 3.3	67.6 ± 3.9^1
LPE	7.0 ± 2.2	10.8 ± 5.9	3.4 ± 0.7	6.9 ± 1.8^2
PS	1.8 ± 1.3	3.5 ± 0.9^1	3.0 ± 0.5	2.4 ± 0.5
LPS	1.7 ± 0.6	1.5 ± 0.2	3.0 ± 0.8	3.7 ± 1.4

CL, oxidized lipids; PE, phosphatidylethanolamine; LPE, lysophosphatidylethanolamine; PS, phosphatidylserine; LPS, lysophosphatidylserine.
Data are calculated as fluorescamine labeling of each individual phospholipid versus total fluorescamine labeling of five lipid components (% of total labeling). Values are mean ± SE of seven experiments.
Significant difference between control and experimental: [1]$P \leq 0.05$, [2]$P \leq 0.01$
Experimental samples were treated with 1 mM phenylhydrazine.

samples to stop the reaction and lyse the red cells. Chloroform/methanol (5:8) was added to each sample and shaken for 30 minutes at room temperature. The supernatants were then partitioned by adding chloroform/0.9% NaCl (3.7/1) and centrifuged at 400 x g for 5 minutes. The upper phase was removed and the supernatant washed twice with 0.9% NaCl. An extract of the lower phase was dried under N_2.

Labeling After Lipid Extraction

Labeling is the same as above except that fluorescamine was added before drying under N_2. Then 20 μl (0.2 M) of fluorescamine stock was added to each 4 ml of extract and vortexed for 15 seconds, and the reaction was stopped by the addition of 0.25 ml of 1 M NH_4OH. The supernatant was washed three times with 0.9% NaCl and the upper phase removed. An extract of the lower phase was dried under N_2.

Thin Layer Chromatography of Phospholipids

The dried samples were resuspended in 100 μl of chloroform/methanol (2:1) and loaded on 10 x 10 cm thin-layer chromatography plates (Silica gel HL, Analtech, Inc., Newark, NJ). The running solvent in the first dimension was chloroform/methanol/NH_4OH (28%)/H_2O (90:54:5.5:5.5) and in the second dimension was chloroform/methanol/H_2O (65:25:4). The plates were dried with an air blower, and the fluorescent spots were observed under visible fluorescent light.

Fluorescence Intensity Measurements

Fluorescent spots were scraped immediately and the powder extracted for 20 minutes with shaking at room temperature. The fluorescent intensities of the extracts were measured (excitation 390 nm, emission 460 nm) on a Perkin-Elmer 650-10S spectrofluorometer (Perkin-Elmer Corp., Norwalk, CT).

RESULTS

Table 1 shows that phenylhydrazine treatment caused a decrease in fluorescamine labeling of PE, which is consistent with data obtained when the phospholipids were analyzed by inorganic phosphorus determinations (2). Accompanying this change were increases in CL and LPE after extraction and an increase in phosphatidylserine (PS) before extraction. No significant change was seen in lysophosphatidylserine (LPS). In further analysis of the data, we observed that the PE degradation due to phenylhydrazine treatment proceeded by two different pathways, depending on the individual. In one case (four persons), LPE increased with no change in CL, while in the other case (three persons), CL increased with no change in LPE (Table 2).

Table 2. Effects of Phenylhydrazine on Reactivities of CL and LPE With Fluorescamine (Labeling After Lipid Extraction) in Human Red Cells

	Type 1		Type 2	
	Mean ± SE	Change[1]	Mean ± SE	Change[1]
CL:				
Control	35.1 ± 3.1		14.7 ± 7.5	
Experimental	35.1 ± 3.5	0.0%	35.4 ± 1.6	141.5%
LPE:				
Control	9.1 ± 1.9		5.8 ± 3.4	
Experimental	19.1 ± 4.6	109.9%	4.9 ± 2.6	-15.5%

Data are presented as relative fluorescence.
Type 1, data from four persons; type 2, data from three persons.

[1]Calculated as $\dfrac{\text{Mean of Expt.} - \text{Mean of Control}}{\text{Mean of Control}}$

DISCUSSION

Phenylhydrazine specifically degrades the PE content of the intact human red cell (2). This was confirmed by the observation that fluorescamine labeling of PE is decreased following treatment of red cells with phenylhydrazine. Cytoskeletal protein degradation can be induced in red cells treated with phenylhydrazine (4). Furthermore, major changes in lipid packing of the bilayer can occur in red cells treated with phenylhydrazine at concentrations that do not cause degradation of PE (2). In phenylhydrazine-induced membrane damage, both protein and lipid changes generally are observed.

Use of fluorescamine allowed us to differentiate two groups of individuals: those that degraded PE to LPE and those that degraded PE to CL. Each person showed a change in one or the other product of PE degradation. The mechanism explaining this observation is unknown.

Fluorescamine labeling of amino phospholipids is a sensitive and alternative method to inorganic phosphorus determinations for evaluating changes in membrane aminophospholipids labeled with fluorescamine. The modification we introduce, in which relative fluorescence of each derivative can be obtained, allows both greater sensitivity and simplicity.

ACKNOWLEDGMENT

Supported by a grant from the National Institutes of Health (ES03425).

REFERENCES

1. Rawyler, A., Roelofson, B., and op den Kamp, J. A. F. The use of fluorescamine in intact Friend erythroleukaemic cells. Biochim. Biophys. Acta 769: 330-336, 1984.
2. Arduini, A., Chen, Z., and Stern, A. Phenylhydrazine-induced changes in erythrocyte membrane surface lipid packing. Biochim. Biophys. Acta 862: 65-71, 1986.
3. Stern, A. Red cell oxidative damage. In: "Oxidative Stress." H. Sies, ed. Academic Press, London, 1985, pp. 331-349.
4. Arduini, A., and Stern, A. Spectrin degradation in intact red blood cells by phenylhydrazine. Biochem. Pharm. 34: 4283-4289, 1985.

ROLE OF OXYGEN RADICALS IN PEROXIDATION OF DOCOSAHEXAENOIC ACID BY RAT BRAIN HOMOGENATE *IN VITRO*

T. Shingu and N. Salem, Jr.

National Institute of Alcohol and Alcoholism
National Institutes of Health
Bethesda, Maryland 20892

ABSTRACT

The effect of free radicals, enzyme inhibitors, and free radical scavengers on docosahexaenoic acid (22:6w3) metabolism by rat brain homogenate was studied *in vitro*. Rat brain homogenate was incubated with [^{14}C]-22:6w3 at 37°C for 15 min, and hydroxylated (1-3) 22:6w3 compounds were separated by reversed-phase high-performance liquid chromatography. Radioactivity of the fractions was measured by liquid scintillation counting. Ferrous iron, Fenton's reagent, and hydrogen peroxide stimulated the formation of 22:6w3 metabolites. When a boiled homogenate was incubated, the stimulatory effect of Fenton's reagent on metabolite formation was decreased markedly and that of hydrogen peroxide disappeared completely. Both nordihydroguaiaretic acid (NDGA) and alpha-tocopherol inhibited the stimulatory effect of ferrous iron, Fenton's reagent, and hydrogen peroxide. Superoxide dismutase (SOD) and indomethacin did not alter the hydrogen peroxide-induced stimulation of metabolite formation; however, mannitol potentiated this stimulation. These observations indicate that the mechanism of stimulation by ferrous iron and Fenton's reagent is probably mediated by a lipid hydroperoxide. The data also suggest (a) the participation of lipoxygenase in the stimulation of polyunsaturated fatty acid peroxidation by ferrous iron and Fenton's reagent, and (b) the importance of alpha-tocopherol as an effective antioxidant in the membrane lipid bilayer.

INTRODUCTION

It is well known that during cerebral ischemia, polyunsaturated free fatty acids (PUFA's) such as arachidonic acid (20:4w6) and docosahexaenoic acid (22:6w3) are released from membrane phospholipids, and that accumulated PUFA's rapidly disappear upon reperfusion of the cerebral vasculature (4). In view of these facts, it has been hypothesized that the peroxidized metabolites of PUFA's may play a role in ischemic brain damage. To date, available evidence is inconclusive. In our investigation of the effect of pH on the metabolism of 22:6w3 by rat brain homogenate, we observed that a pH of 6.5 stimulated peroxidation. Low pH conditions result in a number of biochemical changes: (a) mobilization of ferrous iron, (b) enhanced solubility of ferrous iron, (c) formation of hydrogen peroxide, and (d) formation of superoxide anion. Ferrous iron is a strong initiator of lipid peroxidation *in vitro,* and the reaction of ferrous iron and hydrogen peroxide (Fenton's reagent) generate hydroxyl radicals that are harmful for living tissue. The oxygenated metabolites of 22:6w3 may play an important role in lipid peroxidation and tissue damage associated with ischemic insult. The effect of iron and hydrogen peroxide on 22:6w3 are presented below.

METHOD

Male Sprague-Dawley rats weighing 200-250 g were decapitated; the brains were removed and homogenized in 50 mM tris buffer at pH 7.4 or 6.5. Following a 10-minute preincubation at 37°C, 25 μM [^{14}C]-22:6w3 was incubated with 1 ml of a 10% homogenate (w/v) for 15 min at 37°C. In various experiments, the following compounds were added to the incubate: 100 μM ferrous chloride, 100 μM ferric chloride, 100 μM hydrogen peroxide, 10 μM indomethacin, 50 IU superoxide dismutase (SOD), 40 μM nordihydroguaiaretic acid (NDGA), 100 mM mannitol, and 100 μM alpha-tocopherol (final concentration). Formic acid was added in order to terminate the reaction and bring the pH to 3.5. The incubate was then extracted twice with equal volumes of chloroform, and the extracts were concentrated and analyzed by HPLC. The HPLC conditions were as follows: (a) 5 μM Zorbax ODS column of 4.6 mm x 25 cm, (b) a mobile phase of acetonitrile and 0.1 M ammonium acetate, pH 7.0, (c) the eluant was run through a diode array UV detector at both

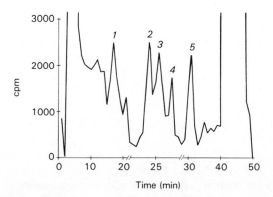

Fig. 1. Radiochromatogram of [^{14}C]-22:6w3 products after incubation with rat brain homogenate. See text for analysis conditions.

Table 1. Effect of pH on Docosahexaenoic Acid (22:6w3) Product Formation

Peak No.	pH 7.4		pH 6.7	
	Nonboiled	Boiled	Nonboiled	Boiled
1	1430(100)	3860(220)[1]	2290(300)[1]	3290(130)[2]
2	1870(220)	1820(40)[1]	2170(100)[1]	1600(70)[2]
3	2510(140)	2700(100)	3070(500)[1]	2440(60)
4	1730(120)	1910(60)[1]	1610(190)	1890(100)[2]
5	680(160)	3370(140)[1]	1210(210)[1]	3120(160)[2]

[1]Significantly different ($p < 0.05$) from nonboiled controls at pH 7.4
[2]Significantly different ($p < 0.05$) from nonboiled oontrols at pH 6.7

237 and 275 nm and then into a fraction collector, and (d) radioactivity of the collected fractions was measured by liquid scintillation counting.

RESULTS

A typical radiochromatogram is shown in Figure 1. The five major peaks were analyzed quantitatively.

Effect of pH

The formation of radioactive peaks was significantly enhanced at pH 6.7, compared to pH 7.4. However, the stimulatory effect of the lower pH was not observed for boiled homogenate (see Table 1).

Effect of Iron and Hydrogen Peroxide

The effect of ferrous iron, ferric iron, Fenton's reagent, and hydrogen peroxide is shown in Table 2. Ferrous iron, ferric iron, and Fenton's reagent stimulated the formation of 22:6w3 products. Ferric iron was a weak stimulating agent compared to the stimulatory effect of ferrous iron and Fenton's reagent. Hydrogen peroxide induced a mild increase in the metabolite peaks with the exception of peak 2 (see Table 2). Similar findings were observed at pH 6.7, although the stimulatory effect of ferric iron was more marked at pH 6.7 than at pH 7.4 (data not shown). In the boiled homogenate, Fenton's reagent was also stimulatory, although its effect was not as great as in the nonboiled homogenate. There was no detectable effect for hydrogen peroxide in the boiled homogenate at either pH condition.

Table 2. Effect of Iron, Fenton's Reagent, and Hydrogen Peroxide on Docosahexaeonic Acid Peroxidation[1]

Peak No.	Ferrous Iron	Fenton's Reagent	Hydrogen Peroxide
1	377[2]	369[2]	124[2]
2	211[2]	236[2]	112
3	207[2]	197[2]	114[2]
4	183[2]	167[2]	119[3]
5	457[2]	468[2]	118

[1]Values represent percent change relative to nontreated control (pH 7.4, n = 5).
[2]Significantly different (p <0.01) from control
[3]Significantly different (p <0.05) from control

Effect of Enzyme Inhibitors and Free Radical Scavengers

The effect of enzyme inhibitors and free radical scavengers on the stimulatory effects of ferrous iron, Fenton's reagent, and hydrogen peroxide are shown in Tables 3-5. Both NDGA and alpha-tocopherol inhibited ferrous iron, Fenton's reagent, and hydrogen peroxide-induced stimulation of the radioactive peaks. Mannitol potentiated the hydrogen peroxide-induced stimulation of metabolite formation, although both SOD and indomethacin were without effect.

DISCUSSION

It was observed that at pH values commonly observed during cerebral ischemic insult, 22:6w3 peroxidation was stimulated. This may be of significance due to

Table 3. Effect of Enzyme Inhibitors and Radical Scavengers on Ferrous Iron-Induced Docosahexaenoic Acid Peroxidation[1]

Peak No.	Indomethacin	NDGA	SOD	Mannitol	Alpha-Tocopherol
1	93	4[2]	88	104	71[3]
2	106	6[2]	107	121	72[3]
3	91	4[2]	102	97	73[3]
4	81	6[2]	86	79	51[3]
5	103	11[2]	94	112	72[3]

[1]Values represent percent change relative to ferrous iron-stimulated samples with no inhibitor present (pH 7.4, n = 5).
[2]Significantly different (p <0.001) from control
[3]Significantly different (p <0.01) from control
NDGA, nordihydroguaiaretic acid
SOD, superoxide dismutase

Table 4. Effect of Enzyme Inhibitors and Radical Scavengers on Fenton's Reagent-Induced Docosahexaenoic Acid Peroxidation[1]

Peak No.	Indomethacin	NDGA	SOD	Mannitol	Alpha-Tocopherol
1	101	4[2]	82	109	78[3]
2	92	6[2]	95	102	67[3]
3	88	5[2]	88	94	75[3]
4	74	6[2]	75	85	58[3]
5	106	9[2]	91	115	80[3]

[1]Values represent percent change relative to Fenton's reagent-treated controls (n = 5).
[2]Significantly different ($p < 0.001$) from control
[3]Significantly different ($p < 0.01$) from control

the hypothesized role of lipid peroxidation in cellular damage during ischemia (5). However, the primary factors leading to lipid peroxidation remained unclear. Furthermore, it has been proposed that oxygen free radicals are formed either during ischemia or after the reperfusion period (6), and that these oxygen free radicals may initiate lipid peroxidation. Under acidic conditions, ferrous iron is mobilized and superoxide anion stimulates the release of ferrous iron from ferritin (7). Our data indicate a strong stimulatory effect of ferrous iron and Fenton's reagent. The stimulatory effect of these agents was greatly diminished in boiled homogenate. Therefore, the possibility cannot be ruled out that the mechanism of stimulation is via such enzymes as lipoxygenase and monooxygenase. The observed inhibitory effect of NDGA supports this concept, although NDGA also inhibited the stimulatory effect of Fenton's reagent in boiled homogenate. This finding may indicate that the mechanism of the NDGA effect is not only through inhibition of lipoxygenase activity but also through a potent free-radical scavenging action. In view of this

Table 5. Effect of Enzyme Inhibitors and Radical Scavengers on Hydrogen Peroxide-Induced Docosahexaenoic Acid Peroxidation[1]

Peak No.	Indomethacin	NDGA	SOD	Mannitol	Alpha-Tocopherol
1	112	12[2]	108	149[2]	76
2	125	14[2]	111	158[3]	82
3	100	10[2]	121	143[2]	70[3]
4	105	8[2]	103	135[2]	60[2]
5	40	15[2]	41	50	31[2]

[1]Values represent percent change relative to hydrogen peroxide-treated controls (n = 5).
[2]Significantly different ($p < 0.001$) from control
[3]Significantly different ($p < 0.01$) from control

finding, the mechanism by which alpha-tocopherol works is important; it inhibited the stimulatory effect of ferrous iron, Fenton's reagent, and hydrogen peroxide. Recently Liebler et al. (8) reported that the antioxidant effect of alpha-tocopherol is due to its lipoxy-radical scavenging effect and not oxygen free-radical scavenging. NDGA decreases lipid peroxide levels that are absolutely necessary for the activation of both lipoxygenase and cyclooxygenase. Regulation of lipid peroxide levels is an important factor in the control of these different enzymatic pathways. The lack of effect in tbe presence of indomethacin suggests that cyclooxygenase may not play a role in the present results.

In conclusion, our data suggest that (a) lowering the pH to values somewhat lower than normal physiological levels leads to stimulation of polyunsaturate peroxidation, and (b) peroxidized PUFA compounds may play an important role in the propagation of lipid peroxidation. The observed stimulatory effect of lowered pH may be due, in part, to mobilized ferrous iron and oxygen free radicals. Such alterations have been observed in ischemic brain tissue (9); therefore, a role in the mechanism of ischemic brain damage must be considered.

REFERENCES

1. Yergey, J., Kim, H.-Y., and Salem, N., Jr. High performance liquid chromatography/thermospray mass spectrometry of eicosanoids and novel oxygenated metabolites of docosahexaenoic acid. Analyt. Chem. 58: 1344-1348, 1986.
2. Salem, N., Jr., Kim, H.-Y., and Yergey, J. Docosahexaenoic acid: Membrane function and metabolism. In: "Health Effects of Polyunsaturated Fatty Acids in Seafoods." A. P. Simopoulos, R. R. Kifer, and R. E. Martin, eds. Academic Press, New York, 1986, pp. 263-317.
3. Shingu, T., and Salem, N., Jr. Mechanism of docosahexaenoic acid oxidation in biological tissue. Trans. Am. Soc. Neurochem. 17: 193, 1986.
4. Rehncrona, S., Westerberg, E., et al. Brain cortical fatty acids and phospholipids during and following complete and severe incomplete ischemia. J. Neurochem. 38: 84-93, 1982.
5. Demopoulos, H. B. The basis of free radical pathology. Fed. Proc. 32: 1859-1861, 1973.
6. Siesjo, B. K. Cell damage in the brain: A speculative synthesis. J. Cereb. Blood Flow Metab. 1: 155-185, 1981.
7. Thomas, L. A., Morehouse, L. A., and Aust, S. D. Ferritin and superoxide dependent lipid peroxidation. J. Biol. Chem. 260: 3275-3280, 1985.
8. Liebler, D. C., Kling, D. S., and Reed, D. J. Antioxidant protection of phospholipid bilayers by α-tocopherol. Control of α-tocopherol status and lipid peroxidation by ascorbic acid and glutathione. J. Biol. Chem. 261: 12114-12119, 1986.
9. Nayini, N. R., White, B. C., et al. Post resuscitation iron delocalization and malondialdehyde production in the brain following prolonged cardiac arrest. J. Free Radical Biol. Med. 1: 111-116, 1985.

POSSIBLE INVOLVEMENT OF QUEUINE IN OXIDATIVE METABOLISM

L. Szabo and W. R. Farkas

University of Tennessee
College of Veterinary Medicine and The Program
  in Environmental Toxicology
Knoxville, Tennessee 37901

ABSTRACT

The possibility that the base queuine (Q) or tRNA's containing Q may play a role in controlling oxidative metabolism has been investigated. There was less thiobarbituric acid-reactive material in queuine-deficient (Q-) mouse liver and kidney than in (Q+) liver and kidney. (Q-) LM cells grown in culture had 53% less of the mitochondrial superoxide dismutase than did (Q+) cells. The enzymatic insertion of queuine into tRNA requires oxygen.

INTRODUCTION

The tRNA's for asparagine, aspartic acid, histidine, and tyrosine contain the hypermodified nucleoside queuine (Figure 1) in the first or wobble position of the anticodon (1-3). Higher mammals have a dietary requirement for queuine (4). The original transcripts of these tRNA's contain guanine in the wobble position, which is enzymatically excised and replaced by queuine (5,6). The biological function of queuine is not known. There are over 50 modified nucleosides in tRNA and, for the most part, their biological functions also remain to be elucidated. It has recently been suggested (7) that some of these modified nucleosides may be indicators of metabolic stress, including oxidative stress. E. coli mutants that cannot synthesize queuine grow as well as wild type. However, when grown under anaerobic conditions, wild type can use nitrate instead of oxygen, whereas the (Q-) mutant does so only poorly (8).

Based on the oxidative stress theory, we compared some of the enzymes involved in oxygen metabolism in (Q-) and (Q+) mouse liver and kidney. The levels of the thiobarbituric acid (TBA) reactive material in these tissues were also determined. We also reasoned that if queuine or the tRNA's that contain it are involved in oxidative metabolism, then the enzyme that inserts queuine into tRNA should be responsive to oxygen. The effect of anaerobiosis on the queuine insertion reaction was also studied.

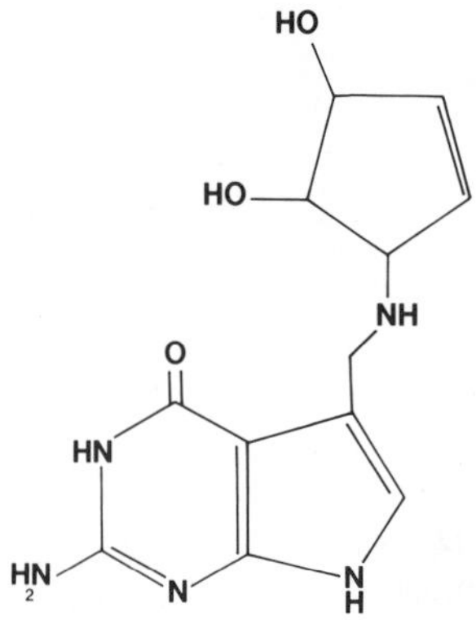

Fig. 1. Structure of queuine

METHODS

Germfree (GF) and conventional mice of the CD-1 strain (weight 28-32 g) were used in these studies. (Q-) GF mice were maintained on a chemically defined (CD) diet (9) that did not contain queuine for at least 6 weeks. The CD diet contains all essential macro- and micronutrients (including selenium and manganese) to support growth and reproduction (10). (Q+) mice were maintained as above, but queuine was added to the diet (2.5 μg/ml). Some GF mice were also fed a normal ingredient diet of Purina mouse chow.

(Q-) LM cells (ATCC CCL1.2, which is a murine fibroblast-like line) were cultured in Dulbeco's modified Eagle's medium (Gibco, Grand Island, NY), supplemented with 0.5% Bacto-peptone (Difco, Detroit, MI). The cells were checked for the content of queuine in their tRNA by RPC-5 chromatography (5). The medium for (Q+) LM cells contained 1 μM queuine. LM cells were purchased from the American type Culture Collection. There was no perceptible difference in the growth rate of (Q+) and (Q-) cells (personal communication, Dr. J. Katze). Queuine was obtained from bovine amniotic fluid (11). Queuine reduced with tritium (12) was a gift of Dr. J. Katze. The assay of the queuine insertion enzyme was carried out as described by Howes and Farkas (13). The assay when performed anaerobically was carried out in Thunberg tubes, the side arm of which contained the enzyme. The buffer, tRNA, potassium chloride, and ^{3}H guanine were in the bottom of the tube. The tubes were evacuated with a vacuum pump and then flushed with nitrogen or argon. This process was repeated three times. The control tubes were treated as above, but they were opened to the atmosphere before starting the reaction. The reaction was started by tipping the contents of the side arm into the reaction mixture. The guanine, queuine tRNA transglycosylase was purified from rabbit red blood cells (13).

Superoxide dismutase (SOD) activity was determined by the epinephrine-adrenochrome method of Misra and Fridovich (14) and Matkovics et al. (15). Catalase was determined by the spectrophotometric method of Beers and Sizer (16). Glutathione peroxidase was determined by a combination of the methods of Chiu et al. (17) and Sedlak and Lindsay (18). In this assay the amount of glutathione (GSH) remaining is determined with Elman reagent after the reaction of cumenehydroperoxide with GSH. The method of Placer et al. (19) was used to quantitate tissue TBA-reactive materials in mouse tissues. The TBA-reactive material, which is primarily malondialdehyde (20), indicates the degree of lipid peroxidation. The cytochrome P-450 and cytochrome b_5 levels in the microsomes were determined according to Greim (21). Protein was determined using the Bio-Rad (Richmond, CA) protein assay kit, as described by Bradford (22). The ^{3}H-guanine (specific activity 2.22 mCi/mmol) was from ICN (Costa Mesa, CA). The other reagents were purchased from Sigma (St. Louis, MO).

RESULTS

The effect of queuine was investigated using LM cells grown in tissue culture medium in the presence and absence of queuine. The data in Table 1 indicate that the absence of queuine in the medium decreased the amount of Mn- SOD by 53% but did not affect the level of Cu Zn- SOD. Catalase and glutathione peroxidase were not affected. In an experiment in which the CD diet was fed to mice, it was observed that the level of SOD was higher in mice fed the CD diet than in mice fed the normal ingredient diet regardless of the presence or absence of queuine.

The level of TBA-reactive material was, for the most part, lower in the livers and kidneys of GF mice than in conventional mice (Table 2). The addition of queuine to the CD diet resulted in an increase of TBA-reactive material in liver and kidney. The amount of cytochrome P-450 in mice fed the CD diet was 40% less than in mice fed normal ingredient diets, and it did not increase when queuine was added to the diet (data not shown). There was no significant difference in the cytochrome b_5 level in the four groups, and there was no detectable difference in the catalase activity. The amount of glutathione peroxidase activity was increased

Table 1. LM Cells Grown in Absence of Quenine Have 53% Less Mitochondrial Superoxide Dismutase

	Enzyme Activity Units/mg	
Enzyme	Q+	Q-
Copper, Zinc SOD	3.11 ± 0.49	3.00 ± 0.45
Manganese SOD	0.55 ± 0.16	0.25 ± 0.08
Catalase	5.87 ± 1.33	4.83 ± 1.81
Glutathione Peroxidase	1.52 ± 0.14	1.39 ± 0.11

Table 2. Queuine-Deficient Mice Have Less Malondialdehyde[1]

	(Q-)	(Q+)	GF Normal Diet	Conventional Normal Diet
Liver	19.5 ±2.1	25.3 ±1.6	26.5 ±3.3	55.8 ±4.4
Kidney	18.9 ±1.6	24.5 ±4.8	34.7 ±0.8	44.9 ±2.6

[1]nmoles malondialdehyde per g (wet wt)

(Q-), germfree mice were maintained on a chemically defined diet that did not contain queuine.
(Q+), same as above but queuine was added to diet. GF, germfree mice fed a diet of ordinary mouse chow. Conventional mice are non-germfree mice fed a diet of ordinary mouse chow.

in the livers and in the kidneys of mice that were fed queuine. The activity of this enzyme was also higher in mice fed CD diets than in those fed mouse chow.

If queuine or (Q+) tRNA plays a role in oxidative metabolism, one might expect that the enzymatic insertion of queuine into tRNA would be affected by oxygen. The data in Table 3 show three separate experiments in which guanine, queuine tRNA transglycosylase was assayed under nitrogen or under aerobic conditions. The data in Table 3 also show an experiment in which the enzyme was assayed under argon rather than nitrogen. We attribute the slight activity under nitrogen to be due to small amounts of oxygen in the nitrogen. Argon is not contaminated with low levels of oxygen, and the enzyme is completely inactive under argon.

DISCUSSION

The level of TBA-reactive material was, for the most part, lower in GF mice than in conventional mice. The amount of TBA-reactive material was increased

Table 3. Queuine Insertion Enzyme Requires Oxygen

Experiment No.	CPM Incorporated Into tRNA		
	Aerobic	Anaerobic	% Loss of Activity
1	1231	182	85
2	637	72	89
3	710	122	83
4[1]	206	0	100
5[2]	268	38	86

In experiments 1, 2, and 3, reaction tube was evacuated with a vacuum pump and then flushed with nitrogen three times before starting reaction.
[1]In experiment 4, argon was used instead of nitrogen.
[2]Tritiated queuine labeled by reducing queuine in presence of tritium gas was used instead of queuine.

in the livers and kidneys of mice fed queuine, indicating increased oxidative activity due to queuine. The level of TBA-reactive material was the lowest in the CD-diet mice that had not been fed queuine. This suggests the presence of more OH· radicals in queuine-treated mice. The mitochondrial SOD activity was increased in LM cells grown in the presence of queuine, which might be due to increased oxygen metabolism in (Q+) cells. As a result of SOD activity, H_2O_2 is produced and in subsequent reactions decomposed by the enzymes peroxidase and catalase.

Despite the fact that the Q insertion into tRNA is not a redox reaction, the reaction requires oxygen, suggesting that the queuine insertion into tRNA may modulate oxygen metabolism. The increased activity of some enzymes involved in oxygen metabolism and the higher level of TBA-reactive materials in mice fed queuine suggest that Q-containing tRNA's are involved with biological oxidations.

ACKNOWLEDGMENTS

We are grateful to Bill Connover and Kim Abney for their diligence with the germfree mouse colony and to Tom Stanawitz for purifying the queuine insertion enzyme. This work was supported in part by NIH grant HD14062.

REFERENCES

1. Kasai, H., Ohashi, Z., Harada, F., Nishimura, S., Oppenheimer, N. J., Crain, P. F., Liehr, J. G., Von Minden, D. L., and McCloskey, J. A. Biochem. 14: 4198-4208, 1975.
2. Kasai, H., Nakanishi, K., MacFarlane, R. D., Tongerson, D. F., Ohashi, Z., McCloskey, J. A., Gross, H. J., and Nishimura, S. J. Am. Chem. Soc. 98: 5044-5046, 1976.
3. Katze, J. R. Nucleic Acids Res. 5: 2513-2522, 1978.
4. Farkas, W. R. J. Biol. Chem. 255: 6832-6835, 1980.
5. Katze, J. R., and Farkas, W. R. Proc. Nat. Acad. Sci. (U.S.A.) 76: 3271-3275, 1979.
6. Okada, N. S., Okada, N., Ohgi, T., Goto, T., and Nishimura, S. Biochem. 19: 395-400, 1980.
7. Bochner, B. R., Lee, P. C., Wilson, S. W., Cutler, C. W., and Ames, B. N. Cell 37: 225-232, 1984.
8. Janel, G., Michelsen, U., Nishimura, S., and Kersten, H. EMBO Journal 3: 1603-1608, 1984.
9. Reyniers, J. P., Pleasants, J. R., Wostmann, B. S., Katze, J. R., and Farkas, W. R. J. Biol. Chem. 256: 11591-11594, 1981.
10. Pleasants, J. R., Reddy, B. S., and Wostman, B. S. J. Nutr. 100: 498-508, 1970.
11. Reyniers, J. P., and Farkas, W. R. Anal. Biochem. 130: 427-430, 1983.
12. Katze, J. R., Gunduz, U., Smith, D. L., Cheng, C. S., and McCloskey, J. A. Biochem. 23: 1171-1176, 1984.

13. Howes, M. K., and Farkas, W. R. J. Biol. Chem. 253: 9082-9087, 1978.

14. Misra, H. P., and Fridovich, I. J. Biol. Chem. 247: 3170-3175, 1972.

15. Matkovics, B., Novak, R., Hoang Duc Hanh, Szabo, L., Vanga, Sz. I., and Zolesna, G. Comp. Biochem. Physiol. 56B: 31-34, 1977.

16. Beers, R. F., Jr., and Sizer, I. W. J. Biol. Chem. 195: 133-140, 1952.

17. Chiu, D. T. Y., Stults, F. H., and Tappal, A. L. Biochem. et Biophys. Acta 445: 558-566, 1976.

18. Sedlack, I., and Lindsay, R. H. Analyt. Biochem. 25: 192-205, 1968.

19. Placer, Z. A., Chusham, L., and Johnson, B. C. Analyt. Biochem. 16: 359-364, 1966.

20. Patton, S., and Kurtz, G. J. Dairy Sci. 34: 669-674, 1951.

21. Greim, H. Naunyn-Schmiedebergs Arch. Exp. Path. Pharmak. 266: 261-275, 1970.

22. Bradford, M. M. Anal. Biochem. 72: 248-254, 1976.

CYTOCHROME P-450 INTERACTION WITH ARACHIDONIC ACID HYDROPEROXIDES: ROLE IN LIPID PEROXIDATION

R. H. Weiss, J. L. Arnold, and R. W. Estabrook

Department of Biochemistry
University of Texas Health Science Center at Dallas
Dallas, Texas 75235-9038

ABSTRACT

The addition of the methyl ester of 15-hydroperoxyeicosa-5,8,11,13-tetraenoic acid, an arachidonic acid hydroperoxide, to rat liver microsomes results in the uptake of molecular oxygen and the formation of thiobarbituric acid reactive products, indicative of lipid peroxidation. Addition of the hydroperoxide to boiled microsomes does not lead to lipid peroxidation. Metyrapone and imidazole anti-mycotic agents, inhibitors of cytochrome P-450-dependent reactions, inhibit the initial rates of lipid peroxidation. The antioxidant butylated hydroxyanisole inhibits lipid peroxidation at a sub-stoichiometric concentration relative to the hydro-peroxide, suggesting that the reaction is an autocatalytic free-radical process. The results indicate that microsomal cytochrome P-450 may play a role in the trans-formation of lipid hydroperoxides (such as those produced by the irradiation of membranes) to products of lipid peroxidation.

INTRODUCTION

Radiation-induced membrane damage is believed to be due to the peroxidation of unsaturated membrane lipids. Lipid hydroperoxides may be produced by the irradiation of membranes (1) and may be intermediates in the formation of lipid peroxidation products, such as thiobarbituric acid (TBA) reactive products (2). Svingen and co-workers have proposed that cytochrome P-450 can transform lipid hydroperoxides to TBA reactive products (2). Weiss and co-workers have demon-strated that microsomal cytochrome P-450 can interact with cumene hydroperoxide to initiate lipid peroxidation of microsomal lipids (3) and with arachidonic acid hydroperoxides to produce epoxyalcohols and triols (Weiss, R. H., et al., submitted for publication). These observations suggested to us that cytochrome P-450 may be able to interact with lipid hydroperoxides.

115

In this paper, we present our results on the interaction of microsomal cytochrome P-450 with the methyl ester of 15-hydroperoxyeicosa-5,8,11,13-tetraenoic acid (15-HPETE), a model lipid hydroperoxide. We have observed that the addition of the hydroperoxide to rat liver microsomes results in oxygen uptake and the formation of TBA reactive products.

METHODS

Liver microsomes from phenobarbital-treated rats were prepared as described previously (4). The protein concentration of the microsomes was measured by the Biuret method (5). The cytochrome P-450 content of the microsomes was measured by the method of Omura and Sato (6) and was in the range of 2.2-2.8 nmol of cytochrome P-450 per mg protein.

The methyl ester of [1-^{14}C]15-HPETE (about 50 cpm/nmol) was prepared by treatment of [1-^{14}C] arachidonic acid with soybean lipoxygenase (7) followed by treatment of the hydroperoxide with diazomethane. The methyl ester of 15-HPETE was purified by reverse-phase (C$_{18}$) high-performance liquid chromatography using a linear gradient of 50% CH$_3$CN/50% H$_2$0/0.1% acetic acid to 100% CH$_3$CN/0.1% acetic acid with a rate of change of 1.25% per min at a flow rate of 1.0 ml per min. The retention time of the methyl ester of 15-HPETE was 24.9 min, compared to 19.5 min for unmethylated 15-HPETE. The hydroperoxide was stored dry at -80°C under nitrogen and was purified each day before use. The yield of hydroperoxide was about 60%.

Oxygen electrode measurements were carried out using a Clark-type electrode as described previously (4). TBA reactive product formation was assayed by the method of Buege and Aust (8).

RESULTS

Oxygen Uptake and TBA Reactive Product Formation

Addition of the methyl ester of 15-HPETE to liver microsomes, prepared from phenobarbital-treated rats, results in the uptake of oxygen and the formation of TBA reactive products. As shown in Figure 1, the extents of oxygen uptake and TBA reactive product formation are linearly dependent on the concentration of the hydroperoxide. The stoichiometry has been determined to be 1.5 nmol of oxygen taken up and 0.12 nmol of TBA reactive products formed per nmol of hydroperoxide utilized.

Addition of the hydroperoxide to boiled microsomes does not result in lipid peroxidation, indicating the requirement for an active enzyme. The K$_m$'s for oxygen uptake and TBA reactive product formation are 40 μM and 54 μM, respectively,

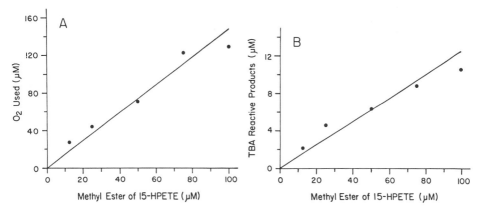

Fig. 1. Extent of (A) oxygen uptake and (B) TBA reactive product formation as a function of concentration of methyl ester of 15-HPETE. Liver microsomes (2 mg/ml), prepared from phenobarbital-treated rats, were suspended in buffer medium containing 50 mM tris-HCl, 150 mM KCl, and 10 mM MgCl₂, pH 7.5, and incubated with variable concentrations of methyl ester of 15-HPETE at 25°C. Oxygen uptake and TBA reactive product formation were measured as described in Methods.

and the V_{max}'s are 265 nmol of oxygen taken up/min/mg protein and 29 nmol of TBA reactive products formed/min/mg protein.

Effect of Cytochrome P-450 Inhibitors

As shown in Table 1, imidazole-containing antimycotic agents (such as clotrimazole, miconazole, and ketoconazole), potent inhibitors of microsomal cytochrome P-450-dependent reactions (9), inhibit the initial rates of oxygen uptake and TBA reactive product formation caused by addition of the methyl ester of

Table 1. Effect of Cytochrome P-450 Inhibitors on Lipid Peroxidation[1]

Inhibitor (μM)	% Inhibition of Initial Rates	
	Oxygen taken up	TBA reactive products formed
Clotrimazole (10)	69	61
Miconazole (10)	57	55
Ketoconazole (100)	64	45
Metyrapone (100)	72	63

[1]Liver microsomes (2 mg/ml) containing inhibitor at indicated concentration were incubated with 100 μM of methyl ester of 15-HPETE at 25°C. Results are expressed as % inhibition of initial rates observed when inhibitor was not present.

15-HPETE to rat liver microsomes. In addition, metyrapone, a general inhibitor of cytochrome P-450 reactions, decreases the initial rates. These results indicate that microsomal cytochrome P-450 interacts with the methyl ester of 15-HPETE to initiate lipid peroxidation.

Inhibition of Lipid Peroxidation by an Antioxidant

Increasing amounts of butylated hydroxyanisole (BHA) causes decreasing amounts of oxygen uptake and TBA reactive product formation to occur when the methyl ester of 15-HPETE (100 μM) is added to liver microsomes (2 mg/ml). At a concentration of 10 μM BHA (a sub-stoichiometric amount relative to the hydroperoxide), about 50% of the lipid peroxidation is inhibited, suggesting that the reaction is an autocatalytic free-radical process. The initial rates are also inhibited in a dose-dependent fashion by BHA.

DISCUSSION

Raleigh and co-workers have demonstrated that X irradiation of fatty acid micelles produces fatty acid hydroperoxides (1). Lipid hydroperoxides formed by irradiation of membrane lipids may be further transformed to products of lipid peroxidation. Yukawa and Nakazawa (10) have shown that radiation-induced lipid peroxidation of liver microsomes decreases the content and activity of cytochrome P-450, suggesting a possible relationship between lipid peroxidation and cytochrome P-450.

We have observed that addition of the methyl ester of 15-HPETE, a model lipid hydroperoxide, to rat liver microsomes results in oxygen uptake and the formation of TBA reactive products, indicative of lipid peroxidation. The inability of boiled microsomes to support lipid peroxidation when treated with the hydroperoxide indicates that an active enzyme is required. The ability of cytochrome P-450 inhibitors to decrease the initial rates of oxygen uptake and TBA reactive product formation suggests that microsomal cytochrome P-450 can interact with the methyl ester of 15-HPETE to initiate lipid peroxidation.

Inhibition of the lipid peroxidation by a sub-stoichiometric concentration of BHA relative to the hydroperoxide suggests that the lipid peroxidation is an autocatalytic free-radical process. It is of interest that BHA does not inhibit the extent of transformation of unmethylated 15-HPETE to an epoxyalcohol and triol (unpublished results), indicating that the epoxyalcohol and triol are not TBA reactive products.

In conclusion, microsomal cytochrome P-450 interacts with the methyl ester of 15-HPETE to initiate microsomal lipid peroxidation, suggesting that cytochrome P-450 may play a role in the transformation of lipid hydroperoxides (produced by the irradiation of membranes) to lipid peroxidation products.

ACKNOWLEDGMENTS

This study was supported in part by grants from the U.S. Public Health Service National Institutes of Health (NIGMS 16488) and The Robert A. Welch Foundation (I-959).

REFERENCES

1. Raleigh, J. A., Kremers, W., and Gaboury, B. Dose-rate and oxygen effects in models of lipid membranes: Linoleic acid. Int. J. Radiat. Biol. 31: 203-213, 1977.

2. Svingen, B. A., Buege, J. A., O'Neal, F. O., and Aust, S. D. The mechanism of NADPH-dependent lipid peroxidation. J. Biol.Chem. 254: 5892-5899, 1979.

3. Weiss, R. H., and Estabrook, R. W. The mechanism of cumene hydroperoxide dependent lipid peroxidation: The function of cytochrome P-450. Arch. Biochem. Biophys., in press.

4. Weiss, R. H., and Estabrook, R. W. The mechanism of cumene hydroperoxide dependent lipid peroxidation: The significance of oxygen uptake. Arch. Biochem. Biophys., in press.

5. Gornall, A. G., Bardawill, C. J., and David, M. M. Determination of serum proteins by means of the biuret reaction. J. Biol. Chem. 177: 751-766, 1949.

6. Omura, T., and Sato, R. The carbon monoxide-binding pigment of liver microsomes. I. Evidence for its hemoprotein nature. J. Biol. Chem. 239: 2370-2378, 1964.

7. Funk, M. O., Isaac, R., and Porter, N. A. Preparation and purification of lipid hydroperoxides from arachidonic and γ-linoleic acids. Lipids 11: 113-117, 1976.

8. Buege, J. A., and Aust, S. D. Microsomal lipid peroxidation. Methods Enzymol. 52: 302-310, 1978.

9. Sheets, J. J., Mason, J. I., Wise, C. A., and Estabrook, R. W. Inhibition of rat liver microsomal cytochrome P-450 steroid hydroxylase reactions by imidazole antimycotic agents. Biochem. Pharmacol. 35: 487-491, 1986.

10. Yukawa, O., and Nakazawa, T. Radiation-induced lipid peroxidation and membrane-bound enzymes in liver microsomes. Int. J. Radiat. Biol. 37: 621-631, 1980.

LEAKAGE OF FATTY ACID RADICALS DURING LIPOXYGENASE CATALYSIS

C. Kemal[1], L.-C. Yuan[1], and K. Laurenzo[2]

[1]Department of Pharmaceutical Research and Technologies
Smith Kline and French Laboratories
Philadelphia, Pennsylvania 19101

[2]Department of Chemistry
University of Florida
Gainesville, Florida 32611

ABSTRACT

The kinetics of soybean lipoxygenase-1 catalyzed oxygenation of linoleate (LH) was studied as a function of the concentrations of LH and antioxidants 2,6-di-tert-butyl-4-methyl-phenol (BHT) and N,N,N',N'-tetramethyl-p-phenylene-diamine (TMPD). The results revealed two distinct oxidative pathways that contribute to O_2 uptake. In one of these, oxygenation of LH occurred at the active site of the enzyme and in the other it occurred in solution. The latter pathway was an autoxidation process initiated by linoleate-derived radicals released by the enzyme during turnover. Product analysis revealed that 13-hydroperoxy-(9Z,11E)-octadeca-9,11-dienoic acid (ZE13-LOOH) was the major (98%) hydroperoxide formed when LH was used at a concentration of 100 μM. At 1.8 mM LH, after consumption of 100-200 μM O_2, significant amounts of three isomers of ZE13-LOOH (EE13-, EZ9-, and EE9-LOOH) also formed (total yield 35%). That these isomers result from solution autoxidation initiated by LH-derived radicals released by the enzyme during turnover was demonstrated by selectively inhibiting solution autoxidation with antioxidants. By quantifying the amount of BHT and TMPD consumed, it was established that one LH-derived radical is released by the enzyme about once every 13 turnovers.

INTRODUCTION

Lipoxygenases (EC 1.13.11.12) catalyze dioxygenation of polyunsaturated fatty acids possessing a cis,cis-1,4-pentadiene unit to give cis,trans conjugated diene hydroperoxides. There is much current interest in these enzymes because the biosynthesis of a number of biologically active eicosanoids (e.g., leukotrienes) involve lipoxygenase catalyzed oxygenation of arachidonic acid (1). Soybean lipoxygenase-1, a non-heme iron dioxygenase, is the best studied example of this class of enzymes. Pentadienyl radical derivative of the fatty acid substrate has long been believed to be an intermediate in the lipoxygenase reaction (2). The purpose of this paper is to show that during soybean lipoxygenase-1 catalyzed oxygenation of linoleate (LH), radicals derived from LH are released into solution.

METHODS

Soybean lipoxygenase-1 (Sigma Chemical Co., St. Louis, MO) was purified using a literature procedure (3). LH($>$99%) was obtained from Nu Chek Prep, Inc. (Elysian, MN). 2,6-Di-tert-butyl-4-methyl-phenol (BHT) and N,N,N',N'-tetramethyl-p-phenylenediamine (TMPD) were purified by sublimation.

In a typical O_2 uptake experiment, 3 μl of 2.4 μM lipoxygenase was added to a 3.0-ml freshly prepared LH solution in an air-saturated 0.1-M borate buffer of pH 9.0 ($\mu = 0.2$ with KCl) containing 5% methanol (v/v). The O_2 disappearance was followed using a Yellow Springs Instruments Model 53 oxygen monitor. For product analysis by high pressure liquid chromatography (HPLC), the reaction was quenched by the addition of 150 μl 1.0 N HCl after the desired amount of O_2 had been consumed. The products and any remaining LH were extracted into dichloromethane and immediately analyzed by HPLC (detection at 229 nm) by the procedure of Teng and Smith (4) using an 8 cm x 6.2 mm Zorbax SIL 3 μm column (DuPont Instruments, Wilmington, DE) and a mobile phase of hexane:ether:acetic acid (1000:150:1).

RESULTS AND DISCUSSION

The experiments described herein were specifically designed to test the possibility that, in the presence of soybean lipoxygenase-1, the conversion of LH to hydroperoxides occurs by two distinct oxidative processes: (a) a "normal" enzymic process, defined minimally as one in which all chemical transformations leading to product occur at the active site of the enzyme, and (b) a solution autoxidation process initiated by LH-derived radicals occasionally released by the enzyme during turnover. At steady state, the latter process is described by equation 1, where R_i is the rate of leakage of radicals by enzyme, k_p is the rate constant for abstraction of a bisallylic hydrogen atom from LH by LOO·, and k_t is the termination rate constant for the reaction of two LOO· radicals (5).

$$d[LOOH]/dt = k_p[LH]R_i^{0.5}/(2k_t)^{0.5} \qquad (1)$$

Since the rate of formation of autoxidation products is proportional to LH concentration (equation 1), it may be predicted that the rate of product formation from autoxidation will increase as LH concentration is increased, even after the rate of product formation from the "normal" enzymic process reaches a saturating value ($K_m = 20\ \mu M$) (6). Due to the increased rate of propagation in micelles (7), it is expected that the contribution to oxygen consumption from autoxidation would be greatest above the cmc of LH (under our conditions, about 200 μM). We have demonstrated the validity of these predictions by showing that as LH concentration is increased, (a) the contribution to O_2 consumption from autoxidation, reflected in the difference between reaction progress curves obtained with and without BHT, increases, and (b) the yield of products expected from autoxidation increases.

At 110 μM LH ([lipoxygenase] $= 2.4$ nM), the reaction progress curves obtained in the presence and absence of 40 μM BHT were, within experimental error, the same (Figure 1A). HPLC analysis of the reaction products indicated that 13-hydroperoxy-(9Z,11E)-octadeca-9,11-dienoate (ZE13-LOOH) was the major hydroperoxide product (98%). Three isomeric hydroperoxides, 13-hydroperoxy-(9E,11E)-octadeca-9,11-dienoate(EE13-LOOH),9-hydroperoxy-(10E,12Z)-octadeca-10,12-dienoate(EZ9-LOOH), and 9-hydroperoxy-(10E,12E)-octadeca-10,12-dienoate(EE9-LOOH), were formed in a total yield of about 2%. At 300 μM and 1.8 mM LH, the rate of O_2 uptake was retarded in the presence of 40 μM BHT by about 10% (Figure 1A) and 30% (Figure 1B), respectively. Product analysis also indicates a substantial contribution to O_2 uptake from solution autoxidation at 1.8 mM LH. After consumption of 100-200 μM LH, the yield EE13 LOOH, EZ9-LOOH, and EE9-LOOH totalled 35% (13/9 ratio $= 74/26$, EZ/EE ratio $= 82/18$) (Figure 1C). In the absence of enzyme, autoxidation of 1.8 mM LH (4 days at 25°C, 5% conversion) produced the four isomers with 13/9 and EZ/EE ratios of 54/46 and 73/27, respectively.

The frequency of radical leakage during lipoxygenase catalysis was estimated by quantifying the amount of BHT and TMPD consumed during turnover. For example, at 1.8 mM LH, although the rate of oxygen consumption was initially slower in the presence of 8 μM BHT than in its absence, after consumption of about 200 μM oxygen, the rate suddenly increased and became identical to that observed with the control run, signaling depletion of BHT (Figure 1B). Using 5-8 μM BHT, the number of oxygen molecules consumed per BHT molecule oxidized was calculated to be 27.7 ± 2.3. The cooxidation of TMPD during lipoxygenase-1 (2.5 nM) catalyzed oxygenation of LH (160 μM) was followed by monitoring the formation of TMPD$^{+\cdot}$ at 565 nm ($\varepsilon = 14,700$ M^{-1} cm^{-1}). From the plot shown in Figure 1D, the rate of TMPD oxidation at infinite TMPD concentration was calculated to be 43 nM/s. Under the same conditions, the rate of LOOH formation was 560 nM/s. From the ratio of these two rates, it was calculated that about 13 LH molecules are oxygenated per TMPD molecule oxidized.

It is well established that BHT consumes two peroxyl radicals (8) while TMPD consumes one. Thus, our finding that one BHT molecule is consumed per about 28 turnovers can be interpreted to mean that 2 radicals are generated per ≤ 28 turnovers. Consistent with this interpretation is our finding that one TMPD is converted to TMPD$^{+\cdot}$ per about 13 turnovers.

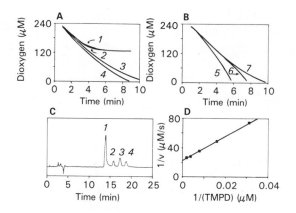

Fig. 1. *A* and *B:* Effects of varying concentrations of BHT and LH on oxygen uptake ([lipoxygenase-1] = 2.4 nM, [O$_2$] = 240 μM). Concentrations (μM) of LH and BHT were (1) 110, 0; (2) 110, 40; (3) 300, 40; (4) 300, 0; (5) 1800, 0; (6) 1800, 8; (7) 1800, 40, respectively. Incubation was at 25°C and pH 9.0 [0.1 M borate, μ = 0.2 with KCl, 5% methanol (v/v)]. *C:* HPLC chromatogram of hydroperoxides produced during incubation of 1.8 mM LH with 2.4 nM lipoxygenase-1. Reaction was quenched after consumption of 100 μM O$_2$. 1 = *ZE*13-LOOH; 2 = *EE*13-LOOH; 3 = *EZ*9-LOOH; 4 = *EE*9-LOOH. *D:* Double reciprocal plot of rate of TMPD$^{+\cdot}$ production against TMPD concentration during lipoxygenase-1 (2.5 nM) catalyzed oxygenation of LH (160 μM). Rate of TMPD oxidation at infinite TMPD concentration was calculated from reciprocal of Y-intercept.

Illustrated in equation 2 is an overall enzymic mechanism (9) that is consistent with the results of this paper. In this mechanism, an intermediate complex of ferrous enzyme and linoleate radical (formed by the reduction of the catalytically active

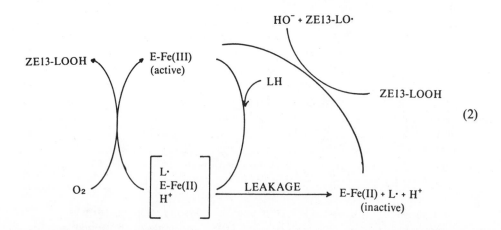

(2)

ferric enzyme with LH) occasionally (once every 28 turnovers) dissociates into its components, leaving the enzyme in the ferrous state which is catalytically inactive (C. Kemal and T. Karali, unpublished results). The oxidation of the latter by *ZE*13-

LOOH results in the formation of an alkoxyl radical, ZE13-LO·, accounting for the observation of two LH-derived radicals released into solution per 28 turnovers. ZE13-LO· has been shown to cyclize to an epoxide containing allylic radical, which couples with O_2 to give a peroxyl radical (10).

The mechanism of equation 2 serves to emphasize that leakage of the fatty acid substrate radical leaves the enzyme in a catalytically inactive state; i.e., the enzyme is turned off as a result of this event and cannot produce any more lipid hydroperoxide until it is turned on again by oxidation of the iron cofactor from ferrous to ferric. Potato lipoxygenase also leaks radicals, about twice as frequently as the soybean enzyme (C. Kemal, unpublished results). It may be that regulation of lipoxygenase activity in this fashion is beneficial despite the potential damage that may result from the radicals that are released into solution.

Note Added In Proof: Recent experiments from our laboratory have shown that at concentrations of oxygen that exist in tissues (2%), radical leakage is a major pathway with both the soybean and potato enzymes. These results suggest that the physiological function of lipoxygenase may be the initiation of nonenzymic lipid peroxidation (e.g., of biological membranes).

ACKNOWLEDGMENT

This work was supported by NIH grant GM32426.

REFERENCES

1. Samuelsson, B. Leukotrienes: Mediators of immediate hypersensitivity reactions and inflammation. Science 220: 568-575, 1983.
2. Gibian, M. J., and Galaway, R. A. Chemical aspects of lipoxygenase reactions. In: "Bioorganic Chemistry." E. E. vanTamelen, ed. Academic Press, New York, 1977, pp. 117-136.
3. Galpin, J. R., Tielens, L. G. M., Veldink, G. A., Vliegenthart, J. F. G., and Kay, C. M. On the interaction of some catechol derivatives with the iron atom of soybean lipoxygenase. FEBS Lett. 69: 179-182, 1976.
4. Teng, J. I., and Smith, L. L. High-performance liquid chromatography of linoleic acid hydroperoxides and their corresponding alcohol derivatives. J. Chromatog. 350: 445-451, 1985.
5. Pryor, W. A. Free radicals in autoxidation and in aging. In: "Free Radicals in Molecular Biology, Aging, and Disease." D. Armstrong, R. S. Sohal, R. G. Cutler, and T. F. Slater, eds. Raven Press, New York, 1984, pp. 13-41.
6. Egmond, M. R., Brunori, M., and Fasella, P. M. The steady-state kinetics of the oxygenation of linoleic acid catalyzed by soybean lipoxygenase. Eur. J. Biochem. 61: 93-100, 1976.

7. Gebicki, J. M., and Allen, A. O. Relationship between critical micelle concentration and rate of radiolysis of aqueous sodium linoleate. J. Phys. Chem. 73: 2443-2445, 1969.

8. Howard, J. A., and Furimsky, E. Arrhenius parameters for reaction of tert-butylperoxy radicals with some hindered phenols and aromatic amines. Can. J. Chem. 51: 3738-3745, 1973.

9. de Groot, J. J. M. C., Veldink, G. A., Vliegenthart, J. F. G., Bolding, J., Wever, R., and van Gelder, B. F. Demonstration by epr spectroscopy of the functional role of iron in soybean lipoxygenase-1. Biochem. Biophys. Acta 377: 71-79, 1975.

10. Dix, T. A., and Marnett, L. J. Hematin-catalyzed rearrangement of hydroperoxylinoleic acid to epoxy alcohols via an oxygen rebound. J. Am. Chem. Soc. 105: 7001-7002, 1983.

THE MAST CELL GRANULE: A PHOSPHOLIPID SOURCE FOR PROSTAGLANDIN SYNTHESIS

S. P. Chock and E. A. Schmauder-Chock

Experimental Hematology Department
Armed Forces Radiobiology Research Institute
Bethesda, Maryland 20814-5145

ABSTRACT

Phospholipid, the source of fatty-acid-derived mediators of the inflammatory response, such as prostaglandins, leukotrienes, thromboxanes, and other eicosanoids, has been localized in the mast cell secretory granule. This matrix-bound granule phospholipid accounts for about 50% of the total content of the mast cell phospholipid. Evidence for the presence of this phospholipid can be observed ultrastructurally in the form of micelles and vesicles when the granule is subjected to water infiltration. It is suggested that the spontaneous sequestration of phospholipid and other hydrophobic elements in response to water influx results in the formation of micelles of various sizes. The coalescence of micelles results in the formation of large spherical micellar aggregates and bilayer vesicles. Since granule matrix is extruded in conjunction with histamine release in radiation injury, this large quantity of matrix-bound phospholipid becomes an available source for eicosanoid synthesis.

INTRODUCTION

Many symptoms of radiation sickness are similar to the symptoms expressed in mast cell anaphylaxis. Histamine release has been implicated in radiation injury (1). The production of eicosanoids such as prostaglandins E_2 and $F_{2\alpha}$ and thromboxane B_2 have been detected in animals exposed to ionizing radiation (2). The formation of eicosanoids has also been observed following mast cell anaphylaxis (3). In spite of the coincidence, there has been no evidence directly linking the secretion of mast cells to the production of eicosanoids. Since arachidonic acid constitutes about 18.2% of the total mast cell phospholipid fatty acid (4), the release of phospholipid in conjunction with histamine release during radiation injury could serve as the source for the production of eicosanoids.

We recently found the presence of a large quantity of phospholipid in the mast cell secretory granule. This granule-bound phospholipid can be seen ultrastructurally in the forms of micellar structures and vesicles of various sizes when the granule is subjected to water infiltration. The extrusion of this huge phospholipid store into the interstitial space in conjunction with histamine release may provoke the synthesis of many fatty acid-derived eicosanoids and chemotactic factors by cells at the site of anaphylaxis.

MATERIALS AND METHODS

Rat peritoneal mast cells were purified according to Sullivan et al. (5). Granules were purified from mast cells by brief sonication pulses similar to those described by Amende and Donlon (6). The granule fraction was obtained by centrifugation of the 198 x g supernatant for 15 minutes at 1000 x g. After washing the granule twice with Hank's balanced salt solution without calcium and magnesium (HBSS), the granule pellet was resuspended in about 1 ml of HBSS. A 10-microliter aliquot was diluted for counting using a particle counter. The phospholipid was extracted exhaustively in a solvent mixture of chloroform:methanol:water (2:1:1). The phospholipid was measured using the micromethod of Vaskovsky et al. (7). Electron microscopy was carried out as described earlier (8). All reagents used were of analytical grade.

RESULTS AND DISCUSSION

Due to difficulty in obtaining purified quiescent mast cell granules, the phospholipid content of the granule has never been carefully evaluated. Purified mast cell granules usually exist in one of two forms: either the membrane-bound electron-dense quiescent form or the swollen dispersed "activated" form which usually contains no perigranular membrane (9). Activated granules have lost much of their matrix phospholipid due to the process known as *de novo* membrane generation (8). Based on ultrastructural observation, our granule preparation can contain up to 50% of activated granules. This preparation contained $1.97 \times 10^{-17} \pm 1 \times 10^{-17}$ (S.D.) mole phospholipid/granule as determined by method of Vaskovsky et al. (7). If we assume that the average phospholipid molecular weight is 800 daltons and that an average mast cell contains 1200 granules (10), the amount of phospholipid found in the granules would total 1.89×10^{-11} g/cell. This value compared to the total phospholipid value of 4.14×10^{-11} g/cell reported for the mast cell (4) means that about 50% of the total mast cell phospholipid was stored in the secretory granules. The presence of such a large amount of phospholipid in the granule was reflected in the large number of hydrophobic micellar structures seen with the electron microscope when the granule was subjected to water infiltration.

Figure 1 is an electron micrograph of a granule after its perigranular membrane has been removed by hypo-osmotic shock. Large bead-like micellar structures are

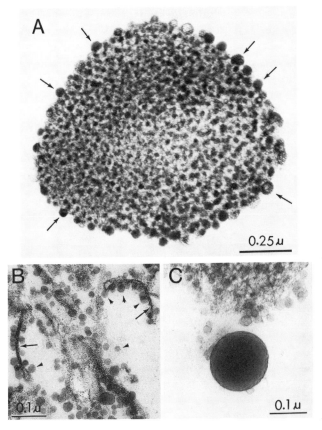

Fig. 1. Formation of micellar structures due to hydration. (A) Granule after membrane removal was exposed to water. Appearance of bead-like micellar structures (arrows) is most apparent at periphery. (B) Unraveling of demembraned granule by vortexing caused formation of small micellar structures (arrowheads) adhering to strands of proteoglycan backbone material (arrows). (C) Large bead-like structure, probably an aggregate of lipid material, can also be seen in close association with smaller micellar structures.

clearly seen in the periphery of the granule where water infiltration predominated (Fig. 1A). When a demembraned granule was further unraveled by vortexing in water, the appearance of small micellar structures was seen throughout and in association with the strand-like granule proteoglycan (or proteo-heparin) backbone material (Fig. 1B). The formation of these hydrophobic micellar structures was the result of the spontaneous sequestration of phospholipid into micelles in the presence of water. The coalescence of many small micelles into large bead-like spherical structures can be seen in Fig. 1C. When the coalescence of micelles results in the formation of bilayer vesicles, their insertion into the perigranular membrane enables the granule to enlarge during activation (8,11). When the granule was extracted with neutral detergent (Brij-56), all the micellar structures disappeared, leaving behind a spherical structure like a ball made of loose narrow ribbons (Fig.

2A). Further extraction of the granule with detergent caused the granule to unravel into strands of ribbon-like structure (Fig. 2B), which might be the proteo-heparin backbone material. This suggests that the granule was constructed by compressing strands of proteoglycan into a spherical structure to which phospholipid and other granule components were bound.

Figure 3A shows an activated granule unraveled after being extruded from a mast cell nearby during the process of rapid freezing and freeze-substitution. The presence of vesicles is also seen in association with the proteo-heparin strand. Under high magnification, small micellar structure can be seen decorating the proteo-heparin strand like a string of pearls (Fig. 3B). This structure might represent the same micellar structure found in the *in vitro* experiment described in Fig. 1B. This further supports our finding that phospholipid is a constituent of the mast cell granule.

Since the bulk of this granule-bound phospholipid was extruded with histamine during anaphylaxis and radiation injury, this large amount of phospholipid might provoke the synthesis of eicosanoids by the various cell types found near the site of anaphylaxis. The phagocytosis of the extruded mast cell granule by neutrophils, macrophages, or eosinophils could also lead to the production of various eicosanoids. The production of the different eicosanoids might serve as chemical mediators of the immune response process, or they might represent a system of cell communication.

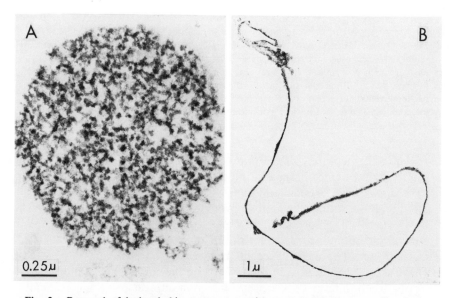

Fig. 2. Removal of hydrophobic components with neutral detergent revealing basic structure of mast cell granule. (A) Granules extracted with 1% Brij-56 followed by brief heating resulted in removal of hydrophobic elements from dispersed granule structure. (B) Further heating and vortexing of detergent-extracted sample resulted in reducing granule matrix into strands of ribbon-like proteo-heparin.

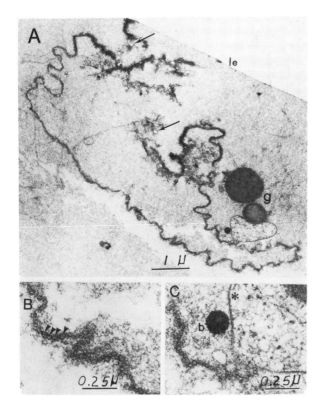

Fig. 3. Unraveling of a cold-activated granule following its extrusion from mast cell during rapid freezing and freeze-substitution. (A) Unraveling of an extruded granule also shows presence of vesicles (arrows) and smaller micellar structures in close association with strand of proteoglycan granule backbone. (B) Under higher magnification, small micellar structures (arrowheads) appear to adhere to strand of proteoglycan-like pearls on string. (C) Large electron-dense bead-like structure (b) of 170 nm can also be seen in association with backbone and a strand of electron-dense material (asterisk).

ACKNOWLEDGMENTS

Part of the data presented here have been submitted to the Department of Biology, The George Washington University, Washington, DC, as partial fulfillment for the degree of Doctor of Philosophy (E. A. S.-C.). Supported by the Armed Forces Radiobiology Research Institute, Defense Nuclear Agency, under work unit MJ00132. Views presented in this paper are those of the authors; no endorsement by the Defense Nuclear Agency has been given or should be inferred. Research was conducted according to the principles enunciated in the "Guide for the Care and Use of Laboratory Animals" prepared by the Institute of Laboratory Animal Resources, National Research Council.

REFERENCES

1. Doyle, T. F., and Strike, T. A. Radiation-released histamine in the rhesus monkey as modified by mast cell depletion and antihistamine. Scientific Report SR75-18. Armed Forces Radiobiology Research Institute, Bethesda, MD, 1975.
2. Donlon, M., Steel, L. K., Helgeson, E. A., Shipp, A., and Catravas, G. N. Radiation-induced alterations in prostaglandin excretion in the rat. Life Sci. 32: 2631-2639, 1983.

3. Metcalfe, D. D., Kaliner, M., and Donlon, M. A. The mast cell. CRC Crit. Rev. Immunol. 3: 23-74, 1981.

4. Strandberg, K., and Westerberg, S. Composition of phospholipids and phospholipid fatty acids in rat mast cells. Mol. Cell. Biochem. 11: 103-107, 1976.

5. Sullivan, T. J., Parker, K. L., Stenson, W., and Parker, C. W. Modulation of cyclic AMP in purified rat mast cells. J. Immunol. 114: 1473-1478, 1975.

6. Amende, L. M., and Donlon, M. A. Isolation of cellular membranes from rat mast cells. Biochim. Biophys. Acta 812: 713-720, 1985.

7. Vaskovsky, V. E., Kostetsky, E. Y., and Vasendin, I. M. A universal reagent for phospholipid analysis. J. Chromatogr. 114: 129-141, 1975.

8. Chock, S. P., and Schmauder-Chock, E. A. Evidence of *de novo* membrane generation in the mechanism of mast cell secretory granule activation. Biochem. Biophys. Res. Commun. 132: 134-139, 1985.

9. Uvnas, B. Mast cell granules. In: "The Secretory Granules." A. M. Poisner and J. M. Trifaro, eds. Elsevier Biomedical Press, Amsterdam, 1982, pp. 357-384.

10. Helander, F. H., and Bloom, G. D. Quantitative analysis of mast cell structure. J. Microsc. 100: 315-321, 1974.

11. Chock, S. P., and Chock, E. S. A two-stage fusion model for secretion. Fed. Proc. 44 (4): 1324, 1985.

RADIATION-INDUCED CHANGES
IN EICOSANOID SYNTHESIS

STUDIES ON EFFECT OF GAMMA RADIATION ON PROSTACYCLIN AND PLATELET-ACTIVATING FACTOR SYNTHESIS AND ON CELL GROWTH USING SMOOTH MUSCLE, FIBROBLAST, AND ENDOTHELIAL CELLS IN CULTURES

M. Menconi, L. Taylor, and P. Polgar

Department of Biochemistry
Boston University School of Medicine
Boston, Massachusetts 02118

ABSTRACT

The effect of gamma radiation on cell division and the synthesis of prostaglandin and platelet-activating factor was studied using homogeneous cultures of smooth muscle, fibroblast, and endothelial cells isolated from the calf pulmonary artery. Exposure of these cultures to doses of up to 30 Gy resulted in some differences in the ability of the cultures to continue their growth. The endothelial cultures, which were the most rapidly proliferating at the time of irradiation, proved the most easily growth inhibited compared to the smooth muscle cells and fibroblasts. On the other hand, the production of prostacyclin, in response to the addition of bradykinin and arachidonate, increased considerably within 30 hours of irradiation in all three cell types. Maximum activation was observed at 10 Gy. Exposure of smooth muscle cells to a trapper of oxygen radicals after irradiation reduced the radiation-induced increase in prostacyclin (PGI_2) synthesis by about 60% in both the bradykinin- and arachidonate-treated cultures. Since the added arachidonate was illustrated to be directly converted to PGI_2 without preliminary esterification, these experiments suggest that the cyclooxygenase complex activity was being affected. However, the synthesis of platelet-activating factor in endothelial cells in response to bradykinin and ionophore A23187 also increased almost twofold in response to irradiation, further suggesting that the action of phospholipase A_2 is also affected by gamma irradiation.

INTRODUCTION

Transient alterations in vascular tone, vascular permeability (1-3), and thrombogenesis (4) have been reported to accompany exposure to ionizing radiation. Vascular endothelial cell function has been postulated to play an important role in these events (1,2,5). Prostacyclin (PGI₂), the principal product of arachidonate synthesized by endothelial cells, is a potent vasodilator (6) and an inhibitor of platelet aggregation (7). Radiation-induced alterations in PGI₂ production by endothelial cells have been reported (8-11) and may be important participants in these functional changes accompanying radiation.

Although the effect of ionizing radiation on endothelial cell function has been investigated extensively, studies on irradiated medial smooth muscle cells and adventitial fibroblasts (the two other major cell types that comprise vascular tissue) are scarce. In the present report we irradiated homogeneous cultures of smooth muscle cells and fibroblasts, as well as endothelial cells, from the bovine pulmonary artery, and examined the effect of radiation exposure on cell growth and the production of prostacyclin and platelet-activating factor.

MATERIALS

Cultured cells were grown on Costar plasticware (Hazleton Dutchland, Inc., Denver, PA). McCoy's 5A modified medium, trypsin-EDTA (10X), Hank's balanced salt solution, sodium bicarbonate solution (7.5%), and penicillin-streptomycin were purchased from Grand Island Biological Company (Grand Island, NY). Hyclone fetal bovine serum (virus screened) was purchased from Sterile Systems, Inc. (Logan, UT). Bradykinin, arachidonate, ionophore A23187, dimethyl sulfoxide, Trizma base, gelatin, and dextran were purchased from Sigma Chemical Co. (St. Louis, MO). Charcoal (decolorizing) was purchased from Fisher Scientific Co. (Medford, MA). Four-hydroxy-2,2,6,6-tetramethylpiperidinoxy radical (HTP) was purchased from Aldrich Chemical Co., Inc. (Milwaukee, WI). Thin layer chromatography (TLC) aluminum plates (silica gel 60 F_{254} pre-coated) were purchased from EM Science, Cherry Hill, NJ.

³H-6-keto-PGF₁$_\alpha$ (150 Ci/mmol) was purchased from Amersham Corp. (Arlington Heights, IL). ³H-hexadecyl-2-acetyl-sn-glyceryl-3-phosphorylcholine-1-0-[hexadecyl-1′,2′-³H(N)], (44 Ci/mmol), ³H-acetic acid (3.4 Ci/mmol), and scintillation fluid (Formula 963) were purchased from New England Nuclear Corp. (Boston, MA). Unlabeled 6-keto prostaglandin F₁$_\alpha$ (6-keto-PGF₁$_\alpha$) was a gift from Dr. John Pike of the Upjohn Company (Kalamazoo, MI).

For the Factor VIII assay, anti-human Factor VIII IgG, goat anti-human IgG labeled with fluorescein isothiocyanate, and goat serum were purchased from Behring Diagnostics (Somerville, NJ), Cappel Labs (Cochranville, PA), and Pel Freez (Rogers, AR), respectively.

METHODS

Cell Cultures

Endothelial cells, medial smooth muscle cells, and adventitial fibroblasts from the bovine pulmonary artery were isolated and placed into culture as previously described (12). The three cell types were identified by morphological, growth, and immunological characteristics. The endothelial cells grew as monolayers, exhibited a cobblestone pattern at confluence, and stained positively for Factor VIII antigen, as detected by immunofluorescence according to the method of Weinberg et al. (13). Factor VIII staining for smooth muscle cells and fibroblasts was negative. Smooth muscle cells were spindle-shaped and grew in multiple layers. Fibroblast cultures grew as monolayers and exhibited concentric whorls at confluence. Contamination of smooth muscle and fibroblast cultures with endothelial cells resulted in the appearance of small but distinct patches of cobblestoned areas within these confluent cultures, which were easily detected under the light microscope. Only cultures found to be homogeneous for each cell type were used for experiments.

Cultures of all three cell types were grown in McCoy's 5A medium supplemented with 20% (vol/vol) fetal bovine serum (HyClone, Logan, UT), streptomycin (50 μg/ml), and penicillin (50 U/ml). Cells were grown at 37°C in a 5% CO_2-95% air humidified incubator. Stock cultures were fed twice a week and were passed weekly into T25 flasks (250,000 cells/T25). For experiments, cells from stock flasks were dissociated with a 0.05% trypsin-0.02% EDTA solution and seeded into 8-cm^2 petri dishes (approximately 60,000 cells/dish). Cultures were grown to confluence and irradiated 2 days later (approximately 10 days after plating). Only early passage cultures ($<$ 20 population doublings) were used for experiments.

Irradiation of Cells

Two days after reaching confluence, cell cultures were exposed to gamma radiation in a Gammacell dual cesium-137 small-animal irradiator (Atomic Energy of Canada). The dose of radiation delivered was controlled by varying the time of exposure at a fixed distance. The radiation dose rate was 1.29 Gy/min. Nonirradiated control cultures were allowed to stand at room temperature during the irradiation procedure.

Determination of Prostacyclin Production

Homogeneous cultures of cells were grown to confluence and then irradiated. The synthesis of PGI_2 was examined at 5 or 29 hours after irradiation under basal conditions and in response to bradykinin and exposure to arachidonic acid (10 μm). Confluent cultures were washed twice with warm medium (serum-free). By thoroughly washing the cultures and using fresh medium for the incubation itself, the contribution to PGI_2 production by any dead, floating cells was eliminated. Thus, only PGI_2 production by viable cells adhering to the culture dish was measured. Serum-free medium containing the various compounds to be tested were then added to start the incubation. Basal production of PGI_2 was determined by incubation of the cultures in serum-free medium containing no additions. Incubations were performed for 30 min at 37°C in a humidified 95% air-5% CO_2 incubator. Incubations were stopped by removing the incubation medium. PGI_2 was measured by radioimmunoassay of 6-keto-$PGF_{1\alpha}$ in aliquots of incubation medium that had been stored frozen. Extraction of the medium was not necessary (14). Values for PGI_2 production were expressed as nanograms 6-keto-$PGF_{1\alpha}$/ml of incubation medium.

Radioimmunoassay for 6-Keto-$PGF_{1\alpha}$

Prostacyclin production was determined by radioimmunoassay for 6-keto-$PGF_{1\alpha}$. The specificity and cross-reactivity of the 6-keto-$PGF_{1\alpha}$ antibody was previously described (15). The reaction was carried out in a total volume of 0.5 ml: 0.2 ml gel-tris buffer (1.2 g Trizma base, 8.3 g NaCl, 1 g gelatin/liter, pH 7.4), 0.1 ml sample or standard, 0.1 ml tracer, and 0.1 ml antibody. Standards were diluted in the same medium as the samples. A standard curve was prepared in the range of 0.5 pmol/0.1 ml to 20 pmol/0.1 ml. Tracer (^{3}H-6-keto-$PGF_{1\alpha}$) was diluted in gel-tris buffer to 10,000 cpm/0.1 ml. Antibody was diluted in gel-tris buffer so that it bound approximately 45% of the tracer added in the absence of unlabeled 6-keto-$PGF_{1\alpha}$. The assay was incubated for 2 hours at room temperature or overnight at 4°C. Following the incubation, the assay tubes were placed in an ice bath, and a slurry of 1.0% charcoal and dextran T70 in water was added. After centrifuging the tubes at 1800 x g for 10 min (at 4°C), the supernatants were aspirated and transferred to scintillation vials containing 3 ml of scintillation fluid. The radioactivity was determined in a β-spectrometer.

Conversion of ^{3}H-Arachidonic Acid to Prostaglandins

For determination of the arachidonate metabolites by high-performance liquid chromatography (HPLC), cells were prelabeled with ^{3}H-arachidonic acid for 24 hours (2 μCi/20 cm^2 dish). Eighty-nine percent of the radioactivity was taken up during this time. These cultures were then washed to eliminate unincorporated label and incubated as described above.

Extraction and Separation of ^{3}H-Arachidonic Acid and ^{3}H-Prostaglandins

Extraction was carried out in 20-ml screw-capped test tubes with Teflon-lined caps. Samples were acidified (pH 3.5-4.5) by the addition of 2 N hydrochloric acid. Equal volumes of ethyl acetate were then added and the tubes agitated vigorously on a vortex mixer. For some of the samples, 1 ml of methanol was added to dissolve protein contained in the medium. The tubes were mixed again and centrifuged at 500 x g for 10 min. The organic layer was removed. The aqueous layer was extracted a second time with an equal volume of ethyl acetate. The ethyl acetate fractions were pooled. Immediately before HPLC, the samples were dried under nitrogen and resuspended in an appropriate volume of solvent.

High-Performance Liquid Chromatography

Aliquots of the extracted samples were analyzed on a Waters (Millipore Corp., Milford, MA) HPLC system equipped with a μBondapak C18 column. ^{3}H-prostaglandins in the eluate were detected with a Berthold HPLC radioactivity monitor (Model LB 505, Berthold Analytical, Nashua, NH) attached to the Waters HPLC unit. N_2-dried samples were dissolved in 35% (v/v) acetonitrile in distilled water adjusted to pH 3.5 with glacial acetic acid. The separation program was a linear gradient of acetonitrile and water (pH 3.5). The gradient began at 35% acetonitrile/65% water and ended with 100% acetonitrile over a 15-min period. Then 100% acetonitrile was maintained over the next 10 min. The flow rate was 1.5 ml/min. ^{3}H-prostaglandin E_2, ^{3}H-prostaglandin $F_{2\alpha}$, ^{3}H-6-keto-PGF$_{1\alpha}$, and ^{3}H-thromboxane B_2 were used to establish retention times. A calibration curve was created by injecting known amounts of radioactive prostaglandins. After the run of an unknown sample, the areas of the prostaglandin peaks were integrated and converted into ^{3}H-counts/min from the calibration curve.

Incorporation of ^{3}H-Acetate Into Platelet-Activating Factor

The incorporation of ^{3}H-acetate into platelet-activating factor by bovine pulmonary artery endothelial cells was determined according to the procedure of McIntyre et al. (16). Irradiated and nonirradiated confluent cells grown on 20-cm^2 petri dishes were incubated with ^{3}H-acetate (50 μCi/dish) in HEPES buffered Hank's balanced salt solution, pH 7.4, containing bradykinin or ionophore A23187. After 10 min at room temperature, the incubation was stopped by removing the medium and adding methanol/water (1:2) containing 50 mM acetic acid. The cells were scraped into this solution and transferred to screw-capped tubes containing chloroform. After 2 hours at room temperature, chloroform and water were added to split the monophase. The chloroform layer was dried under nitrogen and the lipids dissolved in small amounts of chloroform:methanol (2:1). The lipids were separated by TLC on alumina-backed silica-coated TLC plates in chloroform/methanol/acetic acid/water (25:14:4:2). Radioactivity was located by autoradiography, and the radioactive spots were cut from the chromatogram and quantitated.

RESULTS

The effect of gamma radiation on the growth of homogeneous cultures of endothelial cells, smooth muscle cells, and adventitial fibroblasts is illustrated in Table 1. After approaching confluence, the cells were irradiated with doses of 5-30 Gy. The cells were then induced to further growth by exposure to culture medium containing 20% fresh fetal bovine serum. The growth of the irradiated cells was compared to duplicate, nonirradiated cultures. Over a 4-day span, the nonirradiated endothelial cells multiplied 4.5-fold. Exposure to 5 Gy reduced growth to 1.4-fold. At 20 Gy, all growth ceased. In contrast, the non-irradiated smooth muscle cells divided more slowly over a 4-day span, increasing 2.4-fold. Exposure of these cells

Table 1. Effect of Gamma Radiation on Growth of Vascular Endothelial Cells, Smooth Muscle Cells, and Fibroblasts

		Cell Number (x 10^5)		
Day	Gy	Endothelial	Smooth Muscle	Fibroblast
4	0	6.60 ± 0.04	7.69 ± 0.16	6.28 ± 0.18
8	0	27.1 ± 0.04	18.4 ± 0.35	10.5 ± 0.21
	5	9.45 ± 0.05	11.0 ± 0.07	7.90 ± 0.03
	10	7.36 ± 0.11	10.6 ± 0.10	7.81 ± 0.09
	20	6.44 ± 0.02	10.2 ± 0.20	7.62 ± 0.13
	30	6.39 ± 0.07	10.1 ± 0.15	7.36 ± 0.08

Homogeneous cultures of endothelial cells, smooth muscle cells, and fibroblasts from bovine pulmonary artery were grown in 8-cm² petri dishes. When cells had approached confluence (day 4), cultures were irradiated with indicated doses of gamma radiation. Four days later, dishes were washed twice with Hank's balanced salt solution, treated with 0.05% trypsin, and then counted using a Coulter Counter. Under phase contrast microscope, no differences could be observed between irradiated and nonirradiated cultures. Nonadherent cells were not present in irradiated cultures. Values are expressed as mean ± SEM of three separate counts.

to 5 Gy resulted in a reduction in growth to 1.4-fold. However, little further drop in growth was observed up to 30 Gy. The growth response to irradiation of the adventitial fibroblasts was very similar to that of the smooth muscle cells.

We next examined the effect of gamma radiation on prostaglandin production by the three cell types. All cells isolated from the calf pulmonary artery produce predominantly PGI_2 (12). Table 2 illustrates the effect of gamma radiation on PGI_2 production (assayed as 6-keto-$PGF_{1\alpha}$) by confluent endothelial monolayers at 29 hours postirradiation. Nonirradiated cultures synthesized 3.2 ng/ml PGI_2 under unstimulated (basal) conditions and 73.1 ng/ml in response to 5 x 10^{-6} M bradykinin. However, after exposure to 4 Gy, bradykinin-stimulated production

Table 2. Effect of Gamma Radiation on PGI_2 Production by Calf Pulmonary Artery Endothelial Cells

	6-Keto-$PGF_{1\alpha}$ (ng/ml)	
Dose (Gy)	Control	Bradykinin
0	3.2 ± 1.8	73.1 ± 6.8
4	4.4 ± 0.6	102.1 ± 5.3
10	7.7 ± 1.1	116.3 ± 1.2
20	7.8 ± 1.8	105.3 ± 2.4

Homogeneous cultures of endothelial cells were grown in 8-cm² petri dishes to confluence and then exposed to indicated doses of radiation. Twenty-four hours after irradiation, medium was removed and replaced with McCoy's medium containing 1% fetal bovine serum. Five hours later, cultures were incubated for 30 min in serum-free McCoy's medium containing no additions or bradykinin (5 μM). At end of incubation period, media were removed and assayed for 6-keto-$PGF_{1\alpha}$ by RIA. Values are expressed as mean ± SEM of three separate determinations.

of PGI$_2$ by endothelial cultures increased to 102.1 ng/ml. At doses of 10 and 20 Gy, basal levels of PGI$_2$ more than doubled to over 7 ng/ml, while the bradykinin-stimulated production of PGI$_2$ increased to approximately the same levels as that of cultures irradiated with 4 Gy (over 100 ng/ml).

The effect of gamma radiation on PGI$_2$ production by medial smooth muscle cells is shown in Table 3. Treatment of the cultures with radiation increased the synthesis of PGI$_2$ above the corresponding non-irradiated cultures. For example, upon irradiation with 5 Gy, PGI$_2$ production by control (unstimulated) cultures increased twofold from 0.66 to 1.11 ng/ml, bradykinin-stimulated PGI$_2$ production increased threefold from 1.74 to 5.28 ng/ml, and arachidonate-stimulated PGI$_2$ production increased twofold from 4.35 to 8.23 ng/ml. Doses of 10 and 30 Gy radiation resulted in increases in PGI$_2$ production over nonirradiated cultures that were similar to the increases observed at 5 Gy.

Similar effects of radiation on PGI$_2$ production were observed in cultures of adventitial fibroblasts (Table 4). Irradiation of fibroblast cultures with 5 Gy and above increased PGI$_2$ production over that of cultures not exposed to radiation. The treatment of cultures with 5 Gy of radiation elevated levels of PGI$_2$ in unstimulated and arachidonate-stimulated cultures twofold, while elevating PGI$_2$ levels in bradykinin-stimulated cultures threefold. As the dose of radiation was increased from 5 to 30 Gy, PGI$_2$ production increased from 0.16 to 0.56 ng/ml in unstimulated cultures, from 0.72 to 1.78 ng/ml in bradykinin-stimulated cultures, and from 4.20 to 5.32 ng/ml in arachidonate-treated cultures.

One of the mechanisms by which radiation influences PGI$_2$ synthesis may involve an increased lipid peroxidation of cell membrane lipids or formation of oxygen-derived free radicals induced by ionizing radiation. We proceeded to examine the effect of HTP [an oxygen radical scavenger (17)] on the radiation-stimulated

Table 3. Effect of Gamma Radiation on PGI$_2$ Production by Calf Pulmonary Artery Smooth Muscle Cells

Dose (Gy)	6-Keto-PGF$_{1\alpha}$ (ng/ml)		
	Control	Bradykinin	Arachidonate
0	0.66	1.74	4.35
5	1.11	5.28	8.23
10	1.00	4.83	12.4
30	1.84	6.50	9.10

Homogeneous cultures of smooth muscle cells were grown in 8-cm^2 petri dishes to confluence and then exposed to indicated doses of radiation. Twenty-four hours after irradiation, medium was removed and replaced with McCoy's medium containing 1% fetal bovine serum. Five hours later, cultures were incubated for 30 min in serum-free McCoy's medium containing no additions, bradykinin (5 μM), or arachidonate (10 μM). At end of incubation period, media were removed and assayed for 6-keto-PGF$_{1\alpha}$ by RIA. Values are expressed as mean of two separate determinations. Variability between duplicates did not exceed 5%.

Table 4. Effect of Gamma Radiation on PGI$_2$ Production by Calf Pulmonary Artery Fibroblasts

	6-Keto-PGF$_{1\alpha}$ (ng/ml)		
Dose (Gy)	Control	Bradykinin	Arachidonate
0	0.08	0.20	2.38
5	0.16	0.72	4.20
10	0.20	1.04	5.36
30	0.56	1.78	5.32

Homogeneous cultures of fibroblasts were grown in 8-cm^2 petri dishes to confluence and then exposed to indicated doses of radiation. Twenty-four hours after irradiation, medium was removed and replaced with McCoy's medium containing 1% fetal bovine serum. Five hours later, cultures were incubated for 30 min in serum-free McCoy's medium containing no additions, bradykinin (5 μM), or arachidonate (10 μM). At end of incubation period, media were removed and assayed for 6-keto-PGF$_{1\alpha}$ by RIA. Values are expressed as mean of two separate determinations. Variability between duplicates did not exceed 5%.

production of PGI$_2$ in smooth muscle cells. As seen in Table 5, a dose of 10 Gy substantially increased the smooth muscle cell production of PGI$_2$ under nonstimulatory conditions and in response to arachidonate and bradykinin. When added to the cultures alone, HTP had very little effect on PGI$_2$ synthesis. However, when HTP was administered to the cultures that had been irradiated with 10 Gy, the radiation-induced increase in PGI$_2$ production was significantly reduced. For example, addition of HTP to the irradiated cultures reduced the production of PGI$_2$ from 1.47 to 1.00 ng/ml in response to bradykinin. Similarly, PGI$_2$ synthesis by irradiated cultures in response to arachidonate decreased from 4.78 to 3.08 ng/ml when HTP was included in the incubation.

Table 5. Effect of a Trapper of Oxygen Radicals on Radiation-Induced PGI$_2$ Synthesis by Calf Pulmonary Artery Smooth Muscle Cells

	6-Keto-PGF$_{1\alpha}$ (ng/ml)		
Treatment	Control	Bradykinin	Arachidonate
0 Gy	0.39 ± 0.05	0.58 ± 0.07	2.85 ± 0.06
+ HTP	0.53 ± 0.05	0.65 ± 0.03	2.49 ± 0.32
10 Gy	1.18 ± 0.12	1.47 ± 0.14	4.78 ± 0.69
+ HTP	0.84 ± 0.06	1.00 ± 0.04	3.08 ± 0.32

Homogeneous cultures of smooth muscle cells were grown in 8-cm^2 petri dishes. After reaching confluence, half of plates were exposed to 10 Gy of gamma radiation and other half were left untreated. Cells were treated and incubated as described in Table 3. HTP (2.5 mg/ml) was dissolved in Me$_2$SO so that final concentration of Me$_2$SO in incubation medium was 1%. HTP at concentration used did not affect cell viability. Viability was assessed by comparing replating efficiency of control and HTP-treated cultures. Results are expressed as mean of triplicate cultures ± SEM.

Table 6. Effect of Gamma Radiation on Synthesis of Platelet-Activating Factor
in Pulmonary Artery Endothelial Cells

| Treatment | Counts Per Minute | |
	Bradykinin	Ionophore
Nonirradiated	1238 ± 92	1261 ± 142
Irradiated (10 Gy)12	1809 ± 34	2183 ± 190

Endothelial cells were grown to confluence on 20-cm² petri dishes. Half of dishes were irradiated
with 10 Gy and other half were left untreated. Five hours after irradiation, cultures were incubated
with ³H-acetate (50 μCi/dish) in HEPES buffered Hank's balanced salt solution (pH 7.4) containing
bradykinin (50 nM) or ionophore (10 μM) for 10 min at room temperature. Platelet-activating factor
was extracted from cell monolayer, separated by thin layer chromatography, and quantitated as described
in Methods section.

In the above experiments, PGI_2 synthesis was determined in response to high
(10 μM) concentrations of free arachidonate as well as to bradykinin in order to
separate the action of phospholipase A_2 from that of the endoperoxide-prostacyclin
synthetase complex. To ensure that indeed the exposure of the cells to arachidonate
resulted in PGI_2 synthesis free from the action of phospholipase A_2, endothelial
cells were prelabeled with ³H-arachidonate. Labeled and total synthesis of PGI_2
was then determined in response to either bradykinin or 10 μM arachidonate.
Determination of labeled (cpm)/total (ng) PGI_2 production in duplicate cultures
illustrated that whereas the values for bradykinin were 270 and 280, the values
for treatment with arachidonate were approximately 20-fold less: 14.6 and 12.7.
This illustrated that little if any arachidonate-stimulated PGI_2 production involves
the release of arachidonate from lipid stores.

A possible effect of irradiation on the action of phospholipase A_2 was also explored.
The production of platelet-activating factor requires the action of phospholipase
A_2 but not cyclooxygenase. Endothelial cells are known to synthesize platelet-
activating factor in response to bradykinin and ionophore A23187 (16,18,19). As
shown in Table 6, endothelial cells were subjected to 10 Gy. The production of
platelet-activating factor was then determined in the irradiated and corresponding
nonirradiated cultures. Platelet-activating factor production was found to increase
by about 50% in the bradykinin-stimulated cultures. Treatment with ionophore
A23187 increased the production of platelet-activating factor in the irradiated cells
by almost twofold.

DISCUSSION

The literature contains conflicting reports of both increased and reduced synthesis
of PGI_2 by irradiated endothelial cells and vascular tissue. For example, Eldor
et al. (8) found an immediate increase in PGI_2 production in irradiated endothelial

cultures and attributed this increase to cell destruction. They also observed a decrease in PGI_2 production in the remaining adherent cells 2-21 days postirradiation. Other investigators (9-11) have reported an increase in PGI_2 production by cultured bovine endothelial cells within 24 hours of irradiation. Sinzinger et al. (20) reported an increase in PGI_2 synthesis in aortas excised from rabbits immediately after irradiation. They then observed a decreased production of PGI_2 by vessels excised weeks after radiation treatment. In contrast, Allen et al. (21) found a reduced capacity for PGI_2 production by irradiated vascular rings from human umbilical arteries. The inconsistency of these results may simply be related to the temporal relationship between radiation-induced endothelial cell damage and the subsequent alteration in PGI_2 production. In any case, these results illustrate that radiation does affect PGI_2 production.

In this report, we extended our former observation that gamma radiation has a short-term stimulatory effect on PGI_2 production by endothelial cells (9) to studies on the irradiation of pulmonary artery smooth muscle cells and adventitial fibroblasts. Reports from several laboratories (22-24) have indicated that smooth muscle cells may not be as sensitive to gamma irradiation in terms of growth as are endothelial cells. If this were the case, would prostacyclin synthesis also be more resistant to alteration upon irradiation? Our results support somewhat less sensitivity of smooth muscle cells (and fibroblasts) to growth inhibition by gamma radiation than of endothelial cells. This decrease in sensitivity may be related to the slower growth rate of the smooth muscle cells and fibroblasts, suggesting that the destruction of the ability of these cells to proliferate is most sensitive during active division. On the other hand, all three vascular cell types responded to gamma irradiation with increased synthesis of prostacyclin. The relative increase in PGI_2 synthesis appears similar in the three cell types. According to our results reported formerly (9), the cyclooxygenase complex (endoperoxide synthetase and prostacyclin synthetase) appears to become activated upon irradiation. Experiments with smooth muscle cells and fibroblasts further confirm this observation. As illustrated in the Results section, the addition of arachidonate to these cells essentially bypasses the requirement for the release of arachidonate from lipid stores of the cells, thereby bypassing the action of phospholipase A_2. In this case, the addition of arachidonate to the irradiated smooth muscle cells resulted in a threefold increase in PGI_2 synthesis after exposure to 20 Gy. Irradiated fibroblasts displayed a slightly smaller but similar activation. The use of a trapper of oxygen radicals (HPT) did not entirely block this activation but reduced it considerably. However, irradiation may also be activating the action of phospholipase A_2, as suggested by the increase in PAF synthesis when endothelial cells were irradiated at 10 Gy and then stimulated with either bradykinin or ionophore A23187.

ACKNOWLEDGMENT

This work was supported by National Institutes of Health Grants AG05007 and HL25776.

REFERENCES

1. Rubin, P., and Casarett, G. W. Radiation effects on fine vasculature and interstitial connective tissue. In: "Clinical Radiation Pathology," Volume I. Saunders, Philadelphia, 1968, pp. 43-51.

2. Law, M. P. Radiation-induced vascular injury and its relation to late effects in normal tissues. Adv. Radiat. Biol. 9: 37-73, 1981.

3. Gross, N. J. Experimental radiation pneumonitis. IV. Leakage of circulatory proteins onto the alveolar surface. J. Lab. Clin. Med. 95: 19-31, 1980.

4. Maisin, J. R. The ultrastructure of the lung of mice exposed to a supra-lethal dose of ionizing radiation on the thorax. Radiat. Res. 44: 545-564, 1970.

5. Hopewell, J. W. The importance of vascular damage in the development of late radiation effects in normal tissues. In: "Radiation Biology in Cancer Research." R. E. Meyn and H. R. Withers, eds. Raven, New York, 1980, pp. 449-459.

6. Bunting, S., Gryglewski, R., Moncada, S., and Vane, J. R. Arterial walls generate from prostaglandin endoperoxides a substance (prostacyclin X) which relaxes strips of mesenteric coeliac arteries and inhibits platelet aggregation. Prostaglandins 12: 897-913, 1976.

7. Moncada, S., and Vane, J. R. Arachidonic acid metabolites and the interaction between platelets and blood-vessel walls. New Engl. J. Med. 300: 1142-1147, 1979.

8. Eldor, A., Vlodavsky, I., HyAm, E., Atzmon, R., and Fuks, Z. The effect of radiation on prostacyclin (PGI_2) production by cultured endothelial cells. Prostaglandins 25: 263-279, 1983.

9. Hahn, G. L., Menconi, M. J., Cahill, M., and Polgar, P. The influence of gamma radiation on arachidonic acid release and prostacyclin synthesis. Prostaglandins 25: 783-791, 1983.

10. Rubin, D. B., Drab, E. A., Ts'ao, C.-H., Gardner, D., and Ward, W. F. Prostacyclin synthesis in irradiated endothelial cells cultured from bovine aorta. J. Appl. Physiol. 58: 592-597, 1985.

11. Friedman, M., Saunders, D. S., Madden, M. C., Chaney, E. L., and Kwock, L. The effects of ionizing radiation on the pulmonary endothelial cell uptake of α-aminoisobutyric acid and synthesis of prostacyclin. Radiat. Res. 106: 171-181, 1986.

12. Menconi, M., Hahn, G., and Polgar, P. Prostaglandin synthesis by cells comprising the calf pulmonary artery. J. Cell. Physiol. 120: 163-168, 1984.

13. Weinberg, K., Douglas, W. H. J., MacNamee, D. R., Lanzillo, J. J., and Fanburg, B. L. Angiotensin I-converting enzyme localization on cultured fibroblasts by immunofluorescence. In Vitro 18: 400-406, 1982.

14. Polgar, P., Taylor, L., and Downing, D. Unsaturated fatty acid effect on cyclic AMP levels in human embryo lung fibroblasts. Prostaglandins 18: 43-52, 1979.

15. Polgar, P., and Taylor, L. Stimulation of prostaglandin synthesis by ascorbic acid via hydrogen peroxide formation. Prostaglandins 19: 696-700, 1980.

16. McIntyre, T. M., Zimmerman, G. A., Satoh, K., and Prescott, S. M. J. Cultured endothelial cells synthesize both platelet-activating factor and prostacyclin in

response to histamine, bradykinin and adenosine triphosphate. Clin. Invest. 76: 271-280, 1985.

17. Corey, E. J., Albright, J. O., Barton, A. E., and Haschimoto, F. Chemical and enzymatic syntheses of 5-HPETE, a key biological precursor of slow-reacting substance of anaphylaxis (SRS), and 5-HETE. J. Am. Chem. Soc. 102: 1435-1436, 1980.

18. Camussi, G., Aglietta, M., Malavasi, F., Tetta, C., Piacibello, W., Sanavio, F., and Bussolino, F. The release of platelet-activating factor from human endothelial cells in culture. Immunology 131: 2397-2403, 1983.

19. Prescott, S. M., Zimmerman, G. A., and McIntyre, T. M. Human endothelial cells in culture produce platelet-activating factor (1-alkyl-2-acetyl-sn-glycero-3-phosphocholine) when stimulated with thrombin. Proc. Natl. Acad. Sci. (USA) 81: 3534-3538, 1984.

20. Sinzinger, R., Firbas, W., and Cromwell, M. Radiation induced alterations in rabbit aortic prostacyclin formation. Prostaglandins 24: 323-329, 1982.

21. Allen, J. B., Sagerman, R. H., and Stuart, M. J. Irradiation decreases vascular prostacyclin formation with no concomitant effect on platelet thromboxane production. Lancet 2: 1193-1196, 1981.

22. Fischer-Dzoga, K., Dimitrievich, G. S., and Griem, M. L. Radiosensitivity of vascular tissue. II. Differential radiosensitivity of aortic cells in vitro. Radiat. Res. 99: 536-546, 1984.

23. Johnson, L. K., Longenecker, J. P., and Fajardo, L. F. Differential radiation response of cultured endothelial cells and smooth muscle myocytes. Anal. Quant. Cytol. 4: 188-198, 1982.

24. Hirst, D. G., Denekamp, J., and Hobson, B. Proliferation studies of the endothelial and smooth muscle cells of the mouse mesentery after irradiation. Cell Tissue Kinet. 13: 91-104, 1980.

PROSTACYCLIN AND THROMBOXANE PRODUCTION BY IRRADIATED RAT LUNG

W. F. Ward and J. M. Hinz

Department of Radiology
Northwestern University Medical School
Chicago, Illinois 60611

ABSTRACT

The production of prostacyclin (PGI_2) and thromboxane (TXA_2) by irradiated rat lung and the concentration of these prostanoids in the bronchoalveolar lavage (BAL) fluid were measured in rats sacrificed at 2 or 6 months after a single dose of 10-30 Gy of cobalt-60 gamma rays to the right hemithorax. Lung PGI_2 and TXA_2 production increased linearly with increasing radiation dose at rates of 3.1% and 4.3% per Gy, respectively, at 2 months and 21.7% and 6.1% per Gy, respectively, at 6 months. At both times, PGI_2 and TXA_2 production tended to plateau at doses of 25 Gy or higher. This plateau usually disappeared in the presence of exogenous arachidonic acid, suggesting that the Vmax for PGI_2 production at 2 and 6 months and for TXA_2 at 6 months was substrate-limited rather than enzyme-limited. At 2 months after irradiation, the modifying agents D-penicillamine (an inhibitor of collagen cross-linking) and Captopril (an angiotensin-converting enzyme inhibitor) exhibited significant dose-reduction factors of 1.3-1.8 for PGI_2 production and 1.3-1.4 for TXA_2 production. In the shielded left lung, PGI_2 production increased slightly but significantly with increasing contralateral dose at 2 months postirradiation. The concentration of TXA_2 in the BAL fluid increased linearly with increasing radiation dose at 2 months after exposure, whereas PGI_2 concentration was independent of dose. At 6 months, the reverse was true. These data demonstrate that hemithorax irradiation produces a dose-dependent increase in pulmonary PGI_2 and TXA_2 production for at least 6 months, and suggest that BAL data may serve as an index of radiation-induced pulmonary damage.

INTRODUCTION

Chronic, progressive fibrosis is a major dose-limiting complication of radiotherapy in critical organs such as lung, heart, and gut. Attempts to ameliorate radiation-induced tissue fibrosis have been hampered by incomplete understanding of its pathogenesis. The traditional hypothesis that vascular damage causes radiation fibrosis is now under challenge (1,2). Therefore, information concerning the radioresponsiveness of vascular endothelial cells and of the relationship between endothelial injury and tissue fibrosis should contribute to the resolution of this controversy.

Among its many metabolic activities, the pulmonary endothelium generates both prostacyclin (PGI$_2$) and thromboxane (TXA$_2$). PGI$_2$, the major endothelial prostanoid, is a potent inhibitor of platelet aggregation and a vasodilator (3). TXA$_2$, in contrast, promotes platelet aggregation and may mediate events in vascular permeability and inflammation (4). Since some of the activities of these two eicosanoids are mutually antagonistic, the PGI$_2$/TXA$_2$ production ratio may be as important as their individual rates of generation in determining net hemodynamic and thrombogenic status of the lung (5).

Perturbations in PGI$_2$ and TXA$_2$ production are a well-documented and apparently nonspecific response of the pulmonary endothelium to injury, since these reactions can be induced by a variety of insults, including ionizing radiation (6-9), bleomycin (10), and the pneumotoxic plant alkaloid monocrotaline (11). Radiation-induced alterations in pulmonary prostaglandin metabolism develop quickly (within days) and, at least in the case of elevated PGI$_2$ and TXA$_2$ production, persist for more than 6 months after high doses (25 Gy) to the thorax (6-9). The role of these reactions in the development of and recovery from both acute pulmonary responses (such as increased vascular permeability, abnormal vasomotor tone, and thrombogenesis) and chronic responses (such as fibrosis) remains to be clarified.

In the present study, we examined the effect of time, dose, and dose-modifying agents on the production of PGI$_2$ and TXA$_2$ by irradiated rat lung. Both basal prostaglandin production and production stimulated by exogenous arachidonate were analyzed in rats sacrificed at 2 or 6 months after single doses of 10-30 Gy of gamma rays to the right hemithorax. Two months represents the time of early lung fibrosis and 6 months, the time of peak fibrosis in this model (12). The modifying agents studied were D-penicillamine, an inhibitor of collagen cross-linking, and Captopril, an inhibitor of angiotensin-converting enzyme activity. Penicillamine was selected because we have demonstrated previously that it ameliorates radiation injury in rat lung (8,12,13). Captopril was evaluated because it, like penicillamine, reduces cardiopulmonary damage in monocrotaline-treated rats (14). We also measured PGI$_2$ and TXA$_2$ concentrations in the bronchoalveolar lavage fluid, in order to determine whether data from a minimally invasive procedure might reflect or predict lung status after irradiation. While this study is directed primarily at a problem in radiotherapy (i.e., strategies to increase lung tolerance), information obtained in this model should be applicable to other radiobiological settings.

METHODS

Male Sprague-Dawley rats (Harlan Industries, Madison, WI), weighing 350-400 g, were housed at $23° \pm 1°C$ and received standard lab chow (Ralston-Purina, St. Louis, MO) ad libitum. All drinking water contained 0.5 g oxytetracycline (Sigma Chemical Co., St. Louis, MO) per liter to prevent outbreaks of respiratory infections.

Animals receiving modifying agents consumed powdered feed containing 0.05% (w/w) D-penicillamine (Sigma Chemical Co., St. Louis, MO) or 0.12% (w/w) Captopril (Squibb, Inc., Princeton, NJ) continuously after irradiation. Food consumption was measured periodically and was fairly constant at 20 g per rat per day, resulting in daily drug doses of 25 mg/kg penicillamine and 60 mg/kg Captopril.

Animals were anesthetized with sodium pentobarbital (35 mg/kg, i.p.) and then received a single dose of 0-30 Gy of cobalt-60 gamma rays to a 3.5 cm² right hemithorax port at a rate of approximately 3.0 Gy/min as described previously (13). Two or 6 months later, the animals were anesthetized with pentobarbital and exsanguinated from the abdominal aorta. The thoracic organs were dissected en bloc, and the left lung was ligated with thread at the main bronchus and removed. The right lung was lavaged three times with 4.0-ml aliquots of Dulbecco's phosphate-buffered saline (PBS, Grand Island Biological, Grand Island, NY) with glucose (1 mg/ml), resulting in a pooled bronchoalveolar lavage (BAL) fluid sample of approximately 10 ml. Aliquots (0.5 ml) of the BAL sample were mixed with acetylsalicylic acid to a final concentration of 2.0 mM, and were left at room temperature for 1 hr. Then the samples were frozen in liquid N₂ and stored at -70°C. The inferior lobe of the right lung and a comparable area of the left lung were minced with a scalpel, blotted gently with filter paper, and weighed. The tissue (50-100 mg) then was incubated in 3.0 ml of PBS or PBS containing 163 nM sodium arachidonate (Nu-Chek Prep, Inc., Elysian, MN) at 37°C for 10 min. Aliquots (0.1 ml) of the incubation medium then were added to acetylsalicylic acid (final concentration 2 mM), placed at room temperature for 1 hour, frozen in liquid N₂, and stored at -70°C.

PGI₂ and TXA₂ concentration in the culture medium and the BAL fluid were determined by radioimmunoassay of their stable metabolites 6-keto-PGF$_{1\alpha}$ and TXB₂, respectively (New England Nuclear, Boston, MA). Data were expressed as ng of prostanoid produced per mg wet weight of lung mince during the 10-min incubation, or as pg per ml of BAL fluid.

Dose-response curves were obtained by linear regression analysis. The effect of time, dose, and modifying agents was determined by multiple analysis of variance, and the significance of differences between group means was determined by the Newman-Keuls test (15).

RESULTS

Irradiated Right Lung (Table 1)

Right lung PGI₂ and TXA₂ production increased linearly with increasing radiation dose at rates of 3.1% and 4.3% per Gy, respectively, at 2 months postirradiation and 21.7% and 6.1% per Gy, respectively, at 6 months (Figure 1). The steeper response slope for prostacyclin production at the latter time was attributable to a decline to normal production rates after the lower doses (10-15 Gy) and a progressive

Table 1. Prostacyclin (PGI$_2$) and Thromboxane (TXA$_2$) Production by the Irradiated Right Lung

Time (mo)	Arachidonate (nM)	Response	Right Hemithorax Dose (Gy)					
			0	10	15	20	25	30
Two	0	PGI$_2$	0.98 ± 0.13[1]	1.19 ± 0.15	1.44 ± 0.16	1.85 ± 0.18	1.91 ± 0.25	1.92 ± 0.20
		TXA$_2$	0.35 ± 0.05	0.35 ± 0.02	0.48 ± 0.05	0.52 ± 0.06	0.72 ± 0.10	0.64 ± 0.13
		PG/TX	3.0 ± 0.5	3.4 ± 0.3	3.3 ± 0.4	3.8 ± 0.4	3.1 ± 0.6	4.1 ± 0.7
	163	PGI$_2$	1.90 ± 0.28	1.73 ± 0.20	2.06 ± 0.28	2.52 ± 0.19	2.79 ± 0.30	2.99 ± 0.45
		TXA$_2$	0.49 ± 0.05	0.57 ± 0.07	0.72 ± 0.13	0.93 ± 0.18	1.06 ± 0.13	1.03 ± 0.17
		PG/TX	3.9 ± 0.4	3.1 ± 0.2	3.0 ± 0.4	3.6 ± 0.6	2.9 ± 0.4	3.3 ± 0.5
Six	0	PGI$_2$	0.84 ± 0.13	0.85 ± 0.10	0.87 ± 0.08	1.58 ± 0.20	2.93 ± 0.72	2.97 ± 0.57
		TXA$_2$	0.40 ± 0.05	0.39 ± 0.05	0.48 ± 0.06	0.74 ± 0.08	0.85 ± 0.14	0.82 ± 0.08
		PG/TX	2.1 ± 0.2	2.4 ± 0.4	1.9 ± 0.2	2.3 ± 0.4	3.8 ± 1.1	3.2 ± 0.6
	163	PGI$_2$	0.94 ± 0.23	1.08 ± 0.16	1.24 ± 0.15	1.93 ± 0.34	3.43 ± 1.06	3.76 ± 1.20
		TXA$_2$	0.46 ± 0.04	0.50 ± 0.05	0.60 ± 0.07	0.72 ± 0.08	0.83 ± 0.11	0.89 ± 0.10
		PG/TX	3.2 ± 0.4	2.1 ± 0.3	2.5 ± 0.4	2.7 ± 0.4	5.1 ± 1.6	4.6 ± 1.0

[1] ng/mg wet weight/10 min; mean ± SEM; n = 7-11

increase in production after the higher doses (25-30 Gy) between the second and sixth months after irradiation. At both autopsy times, basal PGI₂ and TXA₂ production tended to plateau at doses of 25-30 Gy. Exogenous arachidonate (163 nM) eliminated this plateau in PGI₂ production at both autopsy times and in TXA₂ production at 6 months. At 2 months, exogenous arachidonate elevated both PGI₂ and TXA₂ production by a factor of 1.4 to 1.8 over the basal rate. At 6 months, however, excess substrate increased basal PGI₂ production by a factor of only 1.2-1.4 and had no significant effect on basal TXA₂ Production (Figure 1). The prostacyclin/thromboxane production ratio (PG/TX) increased slightly but not significantly with increasing radiation dose at both 2 and 6 months in the absence of exogenous arachidonate and at 6 months in the presence of excess substrate.

Right lung PGI₂ and TXA₂ production in penicillamine- and Captopril-treated rats also increased linearly with increasing dose at 2 months postirradiation; however, the slopes of the dose-response curves in animals given either agent were 2-5 times less steep than were the control curves (Figure 2). Penicillamine dose-reduction

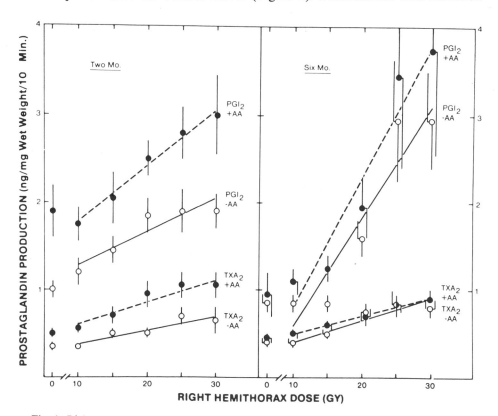

Fig. 1. Right lung prostacyclin (PGI₂) and thromboxane (TXA₂) production as a function of radiation dose in rats sacrificed 2 months (left panel) or 6 months (right panel) after single dose of cobalt-60 gamma rays to right hemithorax. Minced lung was incubated for 10 min in absence (open circles, solid line) or presence (solid circles, broken line) of 163 nM sodium arachidonate (AA). Mean ± SEM; n = 7-11. Lines were fitted by linear regression analysis (r = 0.90-0.99).

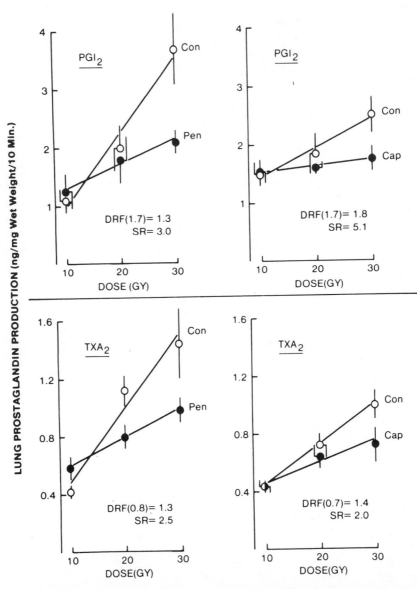

Fig. 2. Right lung PGI₂ (upper panels) and TXA₂ (lower panels) production as a function of radiation dose in rats sacrificed 2 months after single dose of cobalt-60 gamma rays to right hemithorax. Animals received control feed (Con), or feed containing D-penicillamine (Pen, 25 mg/kg/day, left panels, solid circles), or Captopril (Cap, 60 mg/kg/day, right panels, solid circles) continuously after irradiation. Mean ± SEM; n = 6. Lines were fitted by linear regression analysis (r = 0.97-0.99). Modifier effect was quantitated as dose-reduction factor (DRF), defined as ratio of isoeffective doses in presence and absence of modifier at level of response indicated in parentheses. Note also ratio of response curve slopes (SR).

factors (DRF), calculated as the ratio of isoeffective doses in drug-treated and control animals, were 1.26 and 1.27 for PGI_2 and TXA_2 production, respectively ($p < 0.05$), while the Captopril DRF values were 1.77 and 1.37, respectively ($p < 0.05$). However, neither modifier had a significant effect on the PG/TX production ratio.

Shielded Left Lung (Table 2)

Two months after irradiation, PGI_2 production by the shielded left lung increased linearly with increasing contralateral radiation dose at a rate of 1.8% per Gy without exogenous arachidonate and 1.4% per Gy in the presence of excess substrate. In contrast, both basal and stimulated TXA_2 production by the shielded left lung exhibited a nonsignificant dose-independent increase at this time. By 6 months, left lung PGI_2 production with or without excess arachidonate was independent of radiation dose, but remained significantly ($p < 0.05$) higher than the sham-irradiated control values when all doses were pooled. TXA_2 production by the left lung was not influenced significantly by either contralateral dose or excess arachidonate at 6 months.

Bronchoalveolar Lavage (BAL) Fluid (Table 3)

At 2 months postirradiation, 80%-100% of the BAL samples from the right lung contained a measurable concentration of TXA_2, whereas only 44%-63% of the samples were positive for PGI_2. The concentration of TXA_2 in the BAL fluid increased linearly with increasing dose at a rate of 3.0% per Gy, while PGI_2 concentration was independent of dose. At 6 months, in contrast, the percentage of PGI_2 positive samples increased from 63% at 0 Gy to 100% at 30 Gy. The concentration of PGI_2 in the BAL fluid increased with increasing dose up to 25 Gy and then decreased significantly. While all 6-month BAL samples were positive for TXA_2, the concentration of this prostanoid was independent of radiation dose at the latter autopsy time. At both times, the BAL data varied widely within treatment groups, and the PG/TX concentration ratio in the BAL fluid was the reverse of the lung ratio.

DISCUSSION

These data demonstrate that ionizing radiation produces a dose-dependent increase in PGI_2 and TXA_2 production by rat lung. This response is evident by 2 months after irradiation, coincident with the onset of fibrosis (12,13), and persists for at least 6 months. At both autopsy times, prostaglandin production rates plateau at the higher radiation doses. The fact that exogenous arachidonate often eliminates this plateau suggests that the maximum rate of PGI_2 and TXA_2 production by irradiated rat lung may be substrate-limited rather than enzyme-limited. If this is a fact, then radiation studies that routinely add arachidonate to the system might be expected to yield different prostaglandin metabolic data than those that do not. The present data also indicate that the ability of excess arachidonate to stimulate

Table 2. Prostacyclin (PGI_2) and Thromboxane (TXA_2) Production by Shielded Left Lung

Time (mo)	Arachidonate (nM)	Response	Right Hemithorax Dose (Gy)					
			0	10	15	20	25	30
Two	0	PGI_2	1.78 ± 0.34[1]	2.08 ± 0.20	2.01 ± 0.25	2.71 ± 0.39	2.16 ± 0.20	2.87 ± 0.28[2]
		TXA_2	0.66 ± 0.10	0.57 ± 0.06	0.85 ± 0.11	0.86 ± 0.14	0.85 ± 0.11	0.76 ± 0.08
		PG/TX	2.8 ± 0.4	4.0 ± 0.6	2.6 ± 0.3	3.5 ± 0.4	2.9 ± 0.4	4.0 ± 0.5
	163	PGI_2	2.79 ± 0.44	3.16 ± 0.49	3.28 ± 0.49	3.33 ± 0.31	3.57 ± 0.33	4.09 ± 0.44[2]
		TXA_2	0.86 ± 0.10	0.97 ± 0.12	1.27 ± 0.16	1.30 ± 0.20	1.16 ± 0.12	1.12 ± 0.13
		PG/TX	3.2 ± 0.3	3.3 ± 0.3	2.6 ± 0.2	3.1 ± 0.4	3.3 ± 0.4	3.9 ± 0.3
Six	0	PGI_2	1.47 ± 0.19	2.15 ± 0.45	2.07 ± 0.27	1.64 ± 0.16	1.68 ± 0.22	2.08 ± 0.32
		TXA_2	0.67 ± 0.09	0.65 ± 0.06	0.80 ± 0.07	0.74 ± 0.06	0.60 ± 0.04	0.79 ± 0.08
		PG/TX	2.3 ± 0.2	3.3 ± 0.5	2.6 ± 0.3	2.3 ± 0.3	2.9 ± 0.4	2.9 ± 0.6
	163	PGI_2	2.52 ± 0.47	3.15 ± 0.41	3.25 ± 0.32	2.97 ± 0.20	2.77 ± 0.22	3.01 ± 0.36
		TXA_2	0.80 ± 0.12	0.78 ± 0.10	0.80 ± 0.09	1.03 ± 0.15	0.78 ± 0.08	0.78 ± 0.05
		PG/TX	3.2 ± 0.3	4.3 ± 0.5	4.5 ± 0.6	3.3 ± 0.5	3.9 ± 0.5	4.1 ± 0.7

[1] ng/mg wet weight/10 min; mean ± SEM; n = 7-11
[2] Different from Gy, $p < 0.05$

Table 3. Prostacyclin (PGI₂) and Thromboxane (TXA₂) Concentration in Bronchoalveolar Lavage (BAL) Fluid From Irradiated Lung

Time (mo)	Response	Right Hemithorax Dose (Gy)					
		0	10	15	20	25	30
Two	PGI₂	141 ± 27[1] (71)[2]	126 ± 39 (63)	173 ± 26 (63)	135 ± 26 (44)	86 ± 21 (44)	99 ± 28 (45)
	TXA₂	246 ± 81 (88)	361 ± 97 (100)	448 ± 74[3] (80)	437 ± 130[3] (89)	467 ± 101[3] (100)	497 ± 69[3] (90)
Six	PGI₂	77 ± 13[1] (63)	190 ± 51[3] (75)	129 ± 13 (67)	232 ± 46[3] (88)	452 ± 133[3] (91)	196 ± 24[3] (100)
	TXA₂	244 ± 92 (100)	267 ± 33 (100)	201 ± 31 (100)	243 ± 41 (100)	355 ± 112 (100)	304 ± 56 (100)

[1] pg/ml of BAL fluid for all samples; mean ± SEM; n = 7-11
[2] Percent positive samples
[3] Different from 0 Gy, p < 0.05

basal PGI_2 production decreases between the second and sixth months after irradiation, and its stimulatory effect on TXA_2 production is completely abolished during that time. Minced rat lung produces several times as much PGI_2 (PG) as TXA_2 (TX), and the PG/TX production ratio does not change significantly in response to either radiation or arachidonate in the present study. Considering the biological activities of these two prostanoids, an increase in the PG/TX production ratio seems to be an appropriate response on the part of the injured lung; yet the increase in PG/TX observed in this study was small, variable, and not significant statistically.

Both penicillamine and Captopril ameliorate the radiation-induced increase in lung PGI_2 and TXA_2 production, and reduce the slope of the response curves by a factor of 2-5. With either agent, higher doses are required to produce an effect on prostaglandin production equivalent to that seen in irradiated control animals. Penicillamine previously has been reported to reduce radiation injury in rat lung (8,12,13), but the present data are the first to quantitate this phenomenon with DRF values. To our knowledge, this study also is the first to demonstrate that angiotensin-converting enzyme inhibitors ameliorate radiation damage in normal tissue. Captopril, like penicillamine, now appears to be an effective therapeutic agent in both the radiation and monocrotaline models of pulmonary injury in rats (14), although the mechanism of action of the two modifiers is not clear.

The presence of a dose-dependent increase in PGI_2 production by the shielded left lung at 2 months after contralateral irradiation is a curious finding but one that is consistent with our previous observations on cellular and biochemical changes in the left lung after right hemithorax irradiation in this model (8,16). These reactions are similar in kind but reduced in magnitude compared to the right lung response, and they largely disappear by 6 months after irradiation. TXA_2 production by the shielded left lung also increases at 2 months, but unlike PGI_2 production, this increase is neither dose-dependent nor significant. Left lung reactions in this model appear to be the result of either the scatter dose from the right hemithorax port, an abscopal effect of radiation, or part of a compensatory response. It is not clear which alternative(s) is correct, although it is difficult to imagine that the scatter dose after 30 Gy (i.e., <1.5 Gy) could increase left lung PGI_2 production by 60% (Table 2), when direct exposure to 10 Gy increases that production by only 20% (Figure 1).

The BAL data demonstrate that information obtained from a minimally invasive procedure can reflect or even predict the severity of radiation damage in the lung. However, PGI_2 and TXA_2 concentrations in the BAL fluid vary too widely within treatment groups for these data alone to be a reliable index of lung injury. Thus it may be advantageous to develop a spectrum of minimally invasive markers of lung status after irradiation, including BAL data (17), pulmonary arterial perfusion scans (7,13,18,19), and analysis of urinary excretion products (20,21).

ACKNOWLEDGMENTS

This study was supported in part by PHS grant HL25106 from the National Heart, Lung and Blood Institute, DHHS. Captopril was generously donated by E. R. Squibb, Inc., Princeton, NJ.

REFERENCES

1. Law, M. P. Radiation-induced vascular injury and its relation to late effects in normal tissues. Adv. Radiat. Biol. 9: 37-73, 1981.
2. Withers, H. R., Peters, L. J., and Kogelnick, H. D. The pathobiology of late effects of irradiation. In: "Radiation Biology in Cancer Research." R. E. Meyn and H. R. Withers, eds. Raven Press, New York, 1980, pp. 439-448.
3. Hirose, T. Prostacyclin as a modulator of acute lung injury. In: "The Pulmonary Circulation and Acute Lung Injury." S. I. Said, ed. Futura Publishing, Mt. Kisco, New York, 1985, pp. 455-467.
4. Demling, R. H. Role of prostaglandins in acute pulmonary microvascular injury. Ann. N.Y. Acad. Sci. 384: 517-534, 1982.
5. Saldeen, P., and Saldeen, T. 6-Keto-prostaglandin $F_{1\alpha}$/thromboxane B_2 ratio in vascular and lung tissue. Thromb. Res. 30: 643-650, 1983.
6. Steel, L. K., and Catravas, G. N. Radiation-induced changes in production of prostaglandins $F_{2\alpha}$, E, and thromboxane B_2 in guinea pig parenchymal lung tissues. Int. J. Radiat. Biol. 42: 517-530, 1982.
7. Ts'ao, C., Ward, W. F., and Port, C. D. Radiation injury in rat lung. I. Prostacyclin (PGI₂) production, arterial perfusion, and ultrastructure. Radiat. Res. 96: 284-293, 1983.
8. Ward, W. F., Molteni, A., Ts'ao, C., and Solliday, N. H. Radiation injury in rat lung. IV. Modification by D-penicillamine. Radiat. Res. 98: 397-406, 1984.
9. Heinz, T. R., Schneidkraut, M. J., Kot, P. A., Ramwell, P. W., and Rose, J. C. Radiation-induced alterations in cyclooxygenase product synthesis by isolated perfused rat lungs. Prog. Biochem. Pharmacol. 20: 74-83, 1985.
10. Ward, W. F., Molteni, A., Ts'ao, C., and Solliday, N. H. Bleomycin-induced pulmonary endothelial dysfunction in mice: Failure of indomethacin to ameliorate damage. Fed. Proc. 44: 1263, 1985.
11. Molteni, A., Ward, W. F., Ts'ao, C., Port, C. D., and Solliday, N. H. Monocrotaline-induced pulmonary endothelial dysfunction in rats. Proc. Soc. Exp. Biol. Med. 176: 88-94, 1984.
12. Ward, W. F., Shih-Hoellwarth, A., and Tuttle, R. D. Collagen accumulation in irradiated rat lung: Modification by D-penicillamine. Radiology 146: 533-537, 1983.
13. Ward, W. F., Shih-Hoellwarth, A., Port, C. D., and Kim, Y. T. Modification of radiation-induced pulmonary fibrosis in rats. Radiology 131: 751-758, 1979.
14. Molteni, A., Ward, W. F., Ts'ao, C., Solliday, N. H., and Dunne, M. Monocrotaline-induced pulmonary fibrosis in rats: Amelioration by Captopril and penicillamine. Proc. Soc. Exp. Biol. Med. 180: 112-120, 1985.

15. Zar, J. H. Biostatistical Analysis. Prentice-Hall, Englewood Cliffs, NJ, 1974, pp. 121-181.
16. Port, C. D., and Ward, W. F. The ultrastructure of radiation injury in rat lung: Modification by D-penicillamine. Radiat. Res. 92: 61-82, 1982.
17. Ts'ao, C., and Ward, W. F. Plasminogen activator activity in lung and alveolar macrophages of rats exposed to graded single doses of gamma rays to the right hemithorax. Radiat. Res. 103: 393-402, 1985.
18. Ward, W. F. Radiation-induced pulmonary arterial perfusion defects: Modification by D-penicillamine. Radiology 139: 201-204, 1981.
19. Ward, W. F., Molteni, A., Solliday, N. H., and Jones, G. E. The relationship between endothelial dysfunction and collagen accumulation in irradiated rat lung. Int. J. Radiat. Oncol. Biol. Phys. 11: 1985-1990, 1985.
20. Donlon, M., Steel, L., Helgeson, E. A., Shipp, A., and Catravas, G. N. Radiation-induced alterations in prostaglandin excretion in the rat. Life Sci. 32: 2631-2639, 1983.
21. Schneidkraut, M. J., Kot, P. A., Ramwell, P. W., and Rose, J. C. Urinary prostacyclin and thromboxane levels after whole-body gamma irradiation in the rat. Adv. Prostaglandin Thromboxane Leukotriene Res. 12: 107-111, 1983.

EICOSANOID CHANGES IN SKIN FOLLOWING ULTRAVIOLET LIGHT IRRADIATION

V. A. Ziboh and B. Burrall

Department of Dermatology
University of California
Davis, California 95616

ABSTRACT

The incubation of homogenates prepared from UVB-irradiated guinea pig epidermis (24-72 hr) with [^{14}C]-arachidonic acid ([^{14}C]-AA) resulted in decreased transformation of [^{14}C]-AA into the cyclooxygenase products (PGD$_2$, PGE$_2$, and PGF$_{2\alpha}$ while the incorporation of ^{14}C into lipoxygenase products (15-HETE and 12-HETE) increased. Investigation into the selective inhibition of the cyclooxygenase pathway revealed that the *in vitro* transformation of [^{14}C]-AA into [^{14}C]-cyclooxygenase products by the 105,000 x g particulate fraction prepared from normal unirradiated guinea pig epidermis was inhibited by the 105,000 x g cytoplasmic extract prepared from a 24-hr postirradiation guinea pig epidermis. These latter data imply (a) that an endogenous inhibitor(s) of the cyclooxygenase pathway is generated and released into the cytoplasm during UVB irradiation and (b) that it is likely that this selective inhibition of the cyclooxygenase pathway may contribute at least in part to the increased lipoxygenase products in the 24-hr postirradiation skin specimens and possibly the recognized prolonged UVB-induced inflammatory process.

INTRODUCTION

Human skin and animal skin respond with striking changes when exposed to ultraviolet (UV) radiation of various wavelengths (UVA, 320-400 nm; UVB, 290-320 nm; and UVC, 250-290 nm). These changes include epidermal cell death, increase in mitotic index, and hyperplasia. Notable among these changes are vascular responses of vasodilation, altered permeability, and cellular exudation. Evidence that products of arachidonic acid (prostaglandins) may be involved in cutaneous erythema came from an early observation (1) that a single topical application of indomethacin to skin exposed to UVB or sunlight relieves the redness, the elevated skin temperature, and the increased pain perception. Subsequent studies by Snyder (2) indicated that repeated application of indomethacin to human and guinea pig skin in the 48-hr period following exposure to UVB radiation does not improve the results following a single treatment. Interestingly, these studies also revealed that although indomethacin diminishes the erythema and pain perception, it does not reverse the UVB-induced cell death and altered DNA synthesis in the skin. Similar indomethacin effects on the levels of arachidonic acid (AA) and prostaglandins (PG's) E_2 and $F_{2\alpha}$ in human skin after UVB and UVC irradiation have been reported (3).

These inhibitor studies were followed by measurements of levels of prostanoids released into suction blister fluids after UVB irradiation (4-7). In these studies, correlation between PG's and erythema was demonstrated during the first 24 hr. At 48 hr, the PG levels had returned to normal while erythema was still discernable. Similar findings were reported after UVC irradiation (8).

Since the levels of metabolites detected at any time during the course of an inflammatory reaction may not truly reflect the capacity of the injured tissue to generate all or certain metabolites being measured, it became apparent that determination of the capacity of irradiated tissue to generate eicosanoid was more informative. One such study was by Ruzicka et al. (9), who irradiated hairless albino HRS/J mice with 5 J/cm^2 UVA (in the presence or absence of 8-methoxy psoralen), 78 mJ/cm^2 UVB, or 430 mJ/cm^2 UVC. At 24 hr after incubation, animals irradiated with UVA alone revealed no visible change while those irradiated in the presence of 8-methoxy psoralen demonstrated erythema with marked edema. Those irradiated with UVB demonstrated erythema alone, and those irradiated with UVC demonstrated slight erythema and edema. The major findings after the incubation of enzyme preparations from 24-hr postirradiation animals with exogenous [^{14}C]-arachidonic acid ([^{14}C]-AA) revealed (a) marked increase in dermal hydroxyeicosatetraenoic acid (HETE) production in animals treated with 8-methoxy psoralen and (b) marked decrease in prostaglandin D_2 (PGD$_2$) and increase in prostaglandin E_2 (PGE$_2$) in UVB-irradiated animals. Although these findings revealed the capacity of the injured tissues to generate eicosanoids 24 hr after irradiation, remaining unanswered are (a) why indomethacin fails to suppress the UVB-induced erythema and prostanoids generated beyond 24 hr postirradiation and (b) whether noncyclooxygenase products are the major metabolites of AA generated several days after UVB exposure.

This report describes a progressive series of studies in our laboratory to determine the metabolism of AA by enzyme preparations from guinea pig epidermis at various time intervals after UVB irradiation.

MATERIALS AND METHODS

[^{14}C]-AA (60 mCi/mmole) was purchased from Amersham (Arlington, IL). The radiolabel was purified by column chromatography and checked by thin layer chromatography (TLC). Unlabeled AA (99% pure) and L-tryptophan were purchased from Sigma (St. Louis, MO). Hemoglobin was purchased from Eastman Kodak (Rochester, NY). All other reagents and solvents were of analytical grade.

Animals

Male Hartley albino guinea pigs (weighing 300-400 g) were purchased from E. Z. H. Caviary (Williams, CA). Hair from the animals was removed by depilation of the back with sodium thioglycollate (Nair).

Light Source for Irradiation

Two FS 20 sunlamp bulbs (Westinghouse-North American Phillips Lighting, Bloomfield, NJ) were used to irradiate the backs of animals with three minimal erythema doses of ultraviolet B light at an irradiance of 0.8 mW/cm^2.

Epidermal Enzyme Preparations and Incubations

At various times after irradiation, the animals were sacrificed by cervical dislocation. The skin from the dorsum was rapidly removed, followed by removal of superficial 0.2-mm slices with a skin keratome (Cappel Laboratories, Cochranville, PA). These sheets yielded approximately 85% epidermis and 15% dermis by histologic inspection. The specimens were minced and placed in Tris-HCl buffer (pH 8.0) containing EDTA (0.1 mM). The tissues were homogenized with a Polytron (Brinkman Instruments, Westbury, NY) and the resulting homogenates subjected to centrifugation in a Sorvall RC-5B superspeed centrifuge (DuPont, Wilmington, DE), at 800 x g to remove the cell debris and nuclei. The crude supernatant obtained was then centrifuged at 20,000 x g to obtain a supernatant that contained both the microsomal and the high-speed supernatant enzymes. The supernatant fraction was used for the initial incubations.

To localize the subcellular biosynthetic site of the eicosanoids, the 20,000 x g supernatant was subjected to further centrifugation at 105,000 x g for 60 min in a Beckman L5-50E ultracentrifuge (Palo Alto, CA). This resulted in the separation of the high-speed particulate fraction (microsomal pellet) from the cytoplasmic soluble components. Protein determinations in all the fractions were determined by modification of the method of Lowry et al. (10).

In a typical incubation, the 20,000 x g supernatant fraction (which contained the microsomal and cytoplasmic fractions) and approximately 7-10 mg protein were incubated with 4.0 μCi of [^{14}C]-AA in a medium containing 8.0 μg (6.7 μM) of unlabeled AA (this cold AA was added to nullify any effect of endogenous pool size). L-tryptophan (2.0 mM), hemoglobin (0.43 μM), and reduced glutathione (2.0 mM) were added to the incubation mixture as necessary cofactors. Each mixture was incubated for 20 min at 37°C. Termination of incubation and extraction of radioactive metabolites were with four volumes of acidified ethylacetate/methanol/ 0.4 M citric acid (15:2:1 v/v/v).

In experiments to ascertain why prostaglandin biosynthesis was decreased by enzyme preparations from postirradiation epidermis, we evaluated the effect of the varying amounts of 105,000 x g supernatant fraction prepared from a 24-hr postirradiation epidermis on a standard biosynthesis of [^{14}C]-PGD$_2$ (the major prostanoid by guinea pig epidermis as described above by 105,000 x g pellet enzyme preparations from normal unirradiated guinea pig epidermis). Control experiments contained varying amounts of 105,000 x g supernatant from normal unirradiated epidermis.

Extraction of radiometabolites from incubation mixture. Total [^{14}C]-labeled lipids from the incubation mixtures were extracted into an acidified ethylacetate/methanol/citric acid mixture as described above and dried in a rotary evaporator. The extracted [^{14}C]-lipids were resuspended in chloroform/methanol (1:1) with added authentic standards of phospholipids (phosphatidylcholine, phosphatidylethanolamine, and phosphatidylinositol), prostaglandins (PGE$_2$, PGF$_{2\alpha}$, PGD$_2$, and 6-keto-PGF$_{1\alpha}$), hydroxy fatty acid, and AA. An aliquot of the lipid mixture was subjected to chromatography on activated TLC glass plates coated with silica gel G-60 (Merck, Darmstadt, Germany). Total ^{14}C-phospholipids, ^{14}C-cyclooxygenase products, the nonpolar [^{14}C]-lipoxygenase products, and unreacted [^{14}C]-AA were separated from one another in the solvent system of ethylacetate/2,2,4-trimethylpentane/acetic acid/H$_2$O (165:75:30:150 v/v/v/v). The silica gel fractions containing the cyclooxygenase or lipoxygenase products were scraped off separately into a sintered funnel and then eluted with a mixture of chloroform/methanol (1:1). Each eluted fraction was dried in a rotary evaporator.

Identification of Radiometabolites

Cyclooxygenase products. For identification of the [^{14}C]-cyclooxygenase products, portions of the eluted [^{14}C]-cyclooxygenase products that co-eluted with authentic prostaglandins were first derivatized by treating with excess of p-bromophenacyl bromide in the presence of N,N-diisopropylethylamine as catalyst (11). The resultant prostaglandin p-bromophenacyl esters were effectively separated on a reverse-phase ultrasphere-octadecylsilica column (Altex, Irvine, CA) high-performance liquid chromatography (HPLC) with a mobile phase of acetonitrile/water (50:50 v/v). The radioactive effluents were collected by a Gilson fraction collector and the collected ^{14}C quantitated by scintillation counting.

Lipoxygenase products. For identification of the [^{14}C]-lipoxygenase products, two chromatographic systems were used: TLC and HPLC. A portion of the eluted ^{14}C was co-eluted with the added standard hydroxy fatty acid and applied to activated TLC plates, and the radioactive products were separated in the solvent system of ethyl acetate/petroleum ether/acetic acid (50:50:1 v/v/v). This developing solvent system effectively separated the monohydroxyeicosatetraenoic acids (mono-HETE's) from the dihydroxyeicosatetraenoic acids (di-HETE's). The silica gel on the plate was scraped in 5-mm portions into scintillation vials containing 12 ml Econofluor (New England Nuclear, Boston, MA) and counted in the scintillation spectrometer.

The other portion of the ^{14}C extract, which contained [^{14}C]-lipoxygenase products, was dissolved in solvent A (hexane/ethanol/acetic acid, 993:7:1 v/v/v) and applied to a microporasil normal-phase HPLC column (10 μ, 4.1 mm x 30 cm), which had previously been equilibrated with a mixture of 90% solvent A and 10% solvent B (hexane/ethanol/acetic acid, 899:100:1 v/v/v) as described by Sraer et al. (12). The chromatographic separation was performed isocratically at a flow rate of 0.8 ml/min for 20 min followed by an increase in flow rate to 1.5 ml/min for the next 50 min. The ^{14}C metabolites were monitored by co-chromatography with authentic 12-HETE, 15-HETE, and 5-HETE.

Endogenous Arachidonic Acid Measurement

To determine whether free endogenous AA was distributed in the epidermal subcellular preparations, we extracted the 20,000 x g fraction (which was used for the PG biosynthetic studies) with chloroform/methanol (2:1) to obtain all the lipids. Similarly, we extracted the 105,000 x g supernatant fraction (which was used in the inhibitor studies) with the same solvent mixture. Both fractions were dried separately under N_2 and then chromatographed on TLC plates to separate any free AA from the other lipids. The eluted free AA was methylated with an ether-diazomethane solution. The resulting AA methyl ester was quantitated by gas liquid chromatography (GLC) on the Hewlett-Packard instrument (Model 7530A, Avondale, PA). Quantitation of the AA was by comparison with a known amount of an internal standard, heptadecanoic acid (Sigma, St. Louis, MO), which was added to the respective fractions before extraction. In separate experiments, we also determined the total amount of AA esterified to the epidermal complex lipids, which included phospholipids, and neutral lipids. First we extracted these lipids from the subcellular fractions, dried them under N_2, and then subjected them to transesterification with 6% HCl solution in methanol for 16 hr at 74°-78°C. The resulting AA methyl ester was identified and quantitated by GLC, as reported above.

RESULTS

Histological Evaluations of Epidermal Strips After UVB Irradiation

The irradiated sites on the dorsal side of the guinea pig skin revealed mild erythema as early as 4 hr after irradiation. Histological evaluation of the skin keratome slices at 0, 24, and 48 hr are shown in Fig. 1 (A, B, and C). At 24 hr postirradiation, the time of peak erythema, the epidermal specimen was microscopically characterized by numerous dyskeratotic cells with pyknotic nuclei, with mild infiltration of lymphocytes and an occasional polymorphonuclear cell (PMN) as shown in Fig. 1B. At 48 hr (Fig. 1C), erythema was moderately decreased, accompanied by a diminished number of the dyskeratotic cells with pyknotic nuclei and the lymphocytic-PMN infiltrates. By 72 hr (data not shown), the epidermal specimen was similar to the preirradiated specimen, with no noticeable cellular damage and infiltrates. Mast cells were not detectable in our keratome epidermal specimens. Therefore,

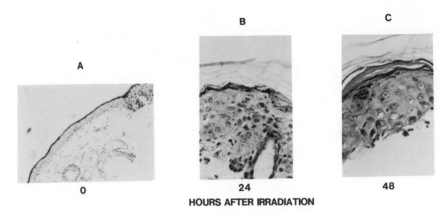

Fig. 1. Histologic evaluations of guinea pig epidermis after exposure to UVB radiation. Histologic evaluations in epidermal slices at 1 (A), 24 (B), and 48 (C) hr after irradiation.

the guinea pig skin seems to be a useful animal model for evaluating the effect of UVB irradiation, as previously noted by other investigators.

Decreased Transformation of AA Into Prostanoids by Irradiated Skin Enzymes

The p-bromophenacyl esters of the [14C]-prostaglandins were separated by HPLC. Fig. 2 summarizes the results obtained from the biosynthesis of the [144C]-PG's from [14C]-AA by the 20,000 x g epidermal preparations at various times after irradiation.

The data demonstrated that the ability of the 20,000 x g epidermal enzyme preparations to transform [14C]-AA into [14C]-cyclooxygenase products diminished in preparations taken from irradiated skin after 24, 48, or 72 hr. Interestingly, the biosynthesis of [14C]-PGD$_2$ from [14C]-AA was the major [14C]-PG metabolite generated by normal unirradiated guinea pig epidermal preparations. At 48 hr and beyond, only negligible transformation of [14C]-AA into all the [14C]-cyclooxygenase products could be demonstrated. Thus our observation of the transformation of [14C]-AA into predominantly [14C]-PGD$_2$ by a normal unirradiated epidermal preparation is consistent with a published report by Ruzicka and Printz (13). However, after UVB irradiation, the transformation of [14C]-cyclooxygenase products was inhibited. These data differ from those of Ruzicka et al. (9); although the latter authors reported inhibition of AA transformation into PGD$_2$, they observed elevation of PGE$_2$ skin preparations from UVB-irradiated hairless mice.

Increased Transformation of AA Into Lipoxygenase Products by Irradiated Skin Enzymes

In contrast to the decreased capacity of the UVB-irradiated epidermal preparations to biosynthesize cyclooxygenase products from [14C]-AA, the biosynthesis of total

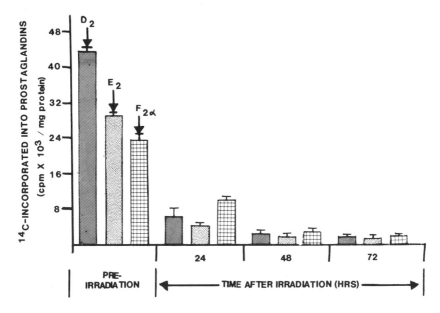

Fig. 2. Metabolism of [14C]-AA into cyclooxygenase products by enzyme preparations from nonirradiated and irradiated guinea pig epidermis. Each 20,000 x g supernatant fraction (7-10 mg protein) prepared from irradiated or nonirradiated epidermis was incubated with [14C]-AA (2 μCi) and cold AA (6.7 μM) in presence of cofactors as described under Materials and Methods.

[14C]-lipoxygenase products from [14C]-AA was elevated in incubations with irradiated skin preparations. A time course of the total [14C]-lipoxygenase products separated by TLC (as described in the Materials and Methods section) revealed that most of the 14C radioactivity in the 24-hr postirradiation skin preparations co-chromatographed with the authentic monohydroxy-eicosatetraenoic acids (Fig. 3).

Although a gradual decrease in the formation of these [14C]-mono-HETE's was evident at 48 and 72 hr postirradiation, the levels were still above those of the control (unirradiated) epidermal preparations. Interestingly, the generation of the lipoxygenase products paralleled the pattern of erythema development in the irradiated animals. To ascertain the nature of the predominant monohydroxy acids, the pooled [14C]-monohydroxy acids (mono-HETE's) fraction was dissolved in solvent with authentic standards of 5-HETE, 12-HETE, and 15-HETE. The mixture was separated into component parts by co-migration of the 14C metabolite with authentic standards by the normal-phase HPLC chromatography as described in the Materials and Methods section. The HPLC tracing revealed that most of the 14C co-eluted with 15-HETE.

To rule out the possibility that the decreased [14C]-PG synthesis (as shown in Fig. 2) by the irradiated epidermal preparations was due to dilution of the [14C]-AA substrate by the nonradioactive AA released during the irradiation of skin,

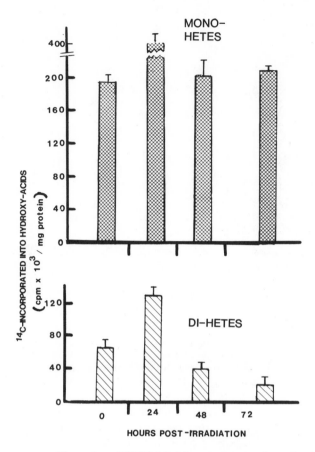

Fig. 3. Time course of formation of [14C]-labeled lipoxygenase products. Incubations of 20,000 x g supernatant fractions (7-10 mg protein) with [14C]-AA are as described in Materials and Methods section.

we measured the total tissue free AA released in the irradiated and the nonirradiated 20,000 x g preparations by GLC as described in Materials and Methods. Data from these analyses revealed no significant difference between the levels of free AA in the control unirradiated and the irradiated skin preparations used in the incubations (0.72 and 0.76 μg AA/mg protein, respectively). To further minimize this possibility, nonradioactive AA (8.0 g) was routinely added to the control and the irradiated enzyme incubations to offset any possible dilution effect. Therefore, the total AA (radioactive and nonradioactive) in each incubation mixture did not differ significantly between the unirradiated and irradiated epidermal preparations and should not influence data in these biosynthetic studies. Therefore, it is reasonable to conclude that the decreased capacity of postirradiation epidermal preparations to transform [14C]-AA into [14C]-PG's could not be due to dilution by nonradioactive AA but could possibly be due to other factors or modulators.

Inhibition of AA Transformation Into Prostanoids by a Soluble Factor Released After Irradiation

To investigate whether an endogenous inhibitor(s) of prostanoid biosynthesis was generated after irradiation of the skin, we investigated whether this factor(s) was released into the soluble 105,000 x g cytoplasmic fraction. Although we do not rule out the possibility that such inhibitors could be contributed by invading inflammatory cells (PMN), we specifically tested whether or not the cytoplasmic preparations from the postirradiation epidermis had any modulating effect on the biosynthesis of PG's from [14C]-AA by enzyme preparations from normal unirradiated guinea pig epidermis.

To achieve this objective, the 105,000 x g pellet (microsomal) fraction and the 105,000 x g supernatant (cytoplasmic) fraction, which were obtained from the 20,000 x g supernatant, were used in subsequent experiments. In control studies, similar fractionations of the normal unirradiated epidermis were prepared. A microsomal fraction prepared from the unirradiated epidermis was used as the basal enzyme source to generate [14C]-prostanoids from [14C]-AA. In typical incubations, varying portions of the 105,000 x g cytoplasmic fraction from 24-hr postirradiation skin and normal unirradiated skin were added to the 105,000 x g pellet (which contained the particulate cyclooxygenase enzymes) from normal unirradiated epidermis in a medium containing [14C]-AA (2 μCi) and appropriate cofactors (as described under Materials and Methods). Transformation of [14C]-AA into [14C]-PGD$_2$ (the major [14C]-AA cyclooxygenase metabolite in the unirradiated guinea pig epidermis) was ascertained after derivatization of [14C]-PGD$_2$ metabolite into the p-bromophenacyl ester and separation of HPLC as reported in the Materials and Methods section. Data from these studies (shown in Fig. 4) revealed that addition

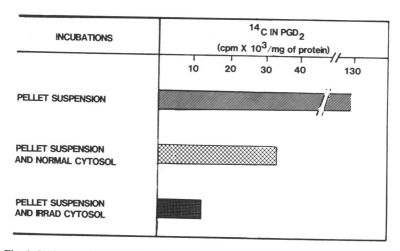

Fig. 4. Inhibition of [14C]-AA transformation into [14C]-PGD$_2$ by 100,000 x g supernatant fraction from 24-hr postirradiation skin

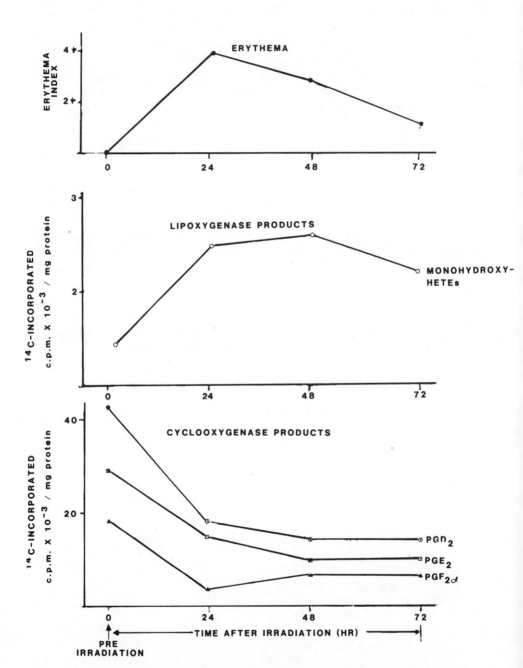

Fig. 5. Time-course relationship between development of cutaneous erythema and generation of eicosanoid

of the 105,000 x g supernatant fraction obtained from the 24-hr postirradiation epidermis to the particulate incubation mixture significantly inhibited [^{14}C]-PGD$_2$ biosynthesis from [^{14}C]-AA by unirradiated particulate fraction.

CONCLUSIONS

In the present studies, the incubation of AA with a 20,000 x g homogenate (containing both microsomal and cytoplasmic fractions) from UVB-irradiated guinea pig epidermis (24-72 hr) resulted in the decreased transformation of [^{14}C]-AA into the cyclooxygenase products (PGD$_2$, PGE$_2$, and PGF$_{2\alpha}$) whereas the incorporation of ^{14}C into lipoxygenase products (15-HETE and 12-HETE) increased. A relationship demonstrating the time course of development of erythema, release of arachidonic acid, and generation of cyclooxygenase and lipoxygenase products is shown in Fig. 5. The increased synthesis of lipoxygenase products was most prominent by incubations with epidermal enzymes prepared from 24-hr postirradiation skin, a time that paralleled maximal erythema in our guinea pigs. In light of our present studies, the increase in released PG's that have been reported by various investigators (2-4) after UVB irradiation seems not to have been generated by the UVB-disrupted epidermis per se but to have been contributed by the infiltrating inflammatory cells. The findings of Barr et al. (14) on the presence of lipoxygenase products in UVB-irradiated skin and our findings of *in vitro* enzymatic transformation of [^{14}C]-AA into HETE's (by epidermal preparations from UVB-induced erythematous skin) suggest the following: In addition to the role of the prostaglandins (cyclooxygenase products) as inflammatory mediators in the early response of the skin to UVB irradiation, the lipoxygenase products (particularly the mono-HETE's) seem more likely to contribute to the inflammatory reactions evident after 24 hr postirradiation. This observation might explain the often-reported ineffectiveness of indomethacin (a more selective inhibitor of the cyclooxygenase pathway) in suppressing the erythema evident beyond 24 hr after irradiation. Our findings support the view that locally, the epidermis is capable of the biosynthesis of cyclooxygenase products and the lipoxygenase products, and at the same time is capable of generating a soluble inhibitor(s) that acts to suppress the continued biosynthesis of inflammatory mediators of UVB-induced radiation. Since the 24-hr postirradiation slices are contaminated by the invading inflammatory cells, it is likely that some of the lipoxygenase products indicated in our studies may have been contributed by these cells. Nonetheless, the epidermis seems unique and possesses its own regulatory machinery to generate and arrest an ongoing UVB-induced inflammatory process. Data from the present studies underscore the novel finding of increased lipoxygenase activity in postirradiation epidermis as well as the generation of a soluble factor(s) that inhibits the biosynthesis of epidermal cyclooxygenase products.

ACKNOWLEDGMENTS

This work was supported in part by Research Grant AM-30679 of the U.S. Public Health Service.

REFERENCES

1. Snyder, D. S., and Eaglstein, W. Topical indomethacin and sunburn. Br. J. Dermatol. 90: 91-93, 1974.
2. Snyder, D. S. Cutaneous effects of topical indomethacin, an inhibitor of prostaglandin synthesis, on UV-damaged skin. J. Invest. Dermatol. 64: 322-325, 1975.
3. Black, A. K., Greaves, M. W., Hensby, C. N., Plummer, N. A., and Warin, A. P. The effects of indomethacin on arachidonic acid and prostaglandins E_2 and $F_{2\alpha}$ levels in human skin 24 hrs after UVB and UVC irradiation. Br. J. Clin. Pharmacol. 6: 261-266, 1978.
4. Blaok, A. K., Greaves, M. W., Hensby, C. N., and Plummer, N. A. Increased Prostaglandins E_2 and $F_{2\alpha}$ in human skin at 6 and 24 hours after ultraviolet B irradiation (290-320). Br. J. Clin. Pharmacol. 5: 431-436, 1978.
5. Black, A. K., Fincham, N., Greaves, M. W., and Hensby, C. N. Time course changes in levels of arachidonic acid and prostaglandins D_2, E_2, and $F_{2\alpha}$ in human skin following ultraviolet B irradiation. Br. J. Clin. Pharmacol. 10: 453-457, 1980.
6. Black, A. K., Fincham, N., Greaves, M. W., and Hensby, C. N. Time course changes in levels of arachidonic acid and prostaglandins D_2, E_2, $F_{2\alpha}$ and 6-OXO-PGF$_{1\alpha}$ in human skin following ultraviolet B irradiation. Br. J. Clin. Pharmacol. 11: 109P, 1980.
7. Black, A. K., Fineham, N., Greaves, M. W., and Hensby, C. N. Changes in levels of arachidonic acid and prostaglandins D_2, E_2, $F_{2\alpha}$, and 6-OXO-F$_{1\alpha}$ in human skin within 48 hours after ultraviolet B irradiation. Br. J. Dermatol. 105: 353-354, 1980.
8. Camp, R., Greaves, M. W., Hensby, C. N., Plummer, N. A., and Warin, A. P. Irradiation of human skin by short wavelength ultraviolet irradiation (100-290 nm) (UVC): Increased concentrations of arachidonic acid and prostaglandins E_2 and $F_{2\alpha}$. Br. J. Clin. Pharmacol. 6: 145-148, 1978.
9. Ruzicka, T., Walter, J. F., and Printz, M. P. Changes in arachidonic acid metabolism in UV-irradiated hairless mouse skin. J. Invest. Dermatol. 81: 300-303, 1983.
10. Lowry, O. H., Rosebrough, N. F., Farr, A. L., and Randall, R. J. Protein measurement with the Folin phenol reagent. J. Biol. Chem. 193: 265-275, 1951.
11. Morozowich, M., and Douglass, S. L. Resolution of prostaglandin p-nitrophenacylesters by liquid chromatography and conditions for rapid, quantitative p-nitrophenacylation. Prostaglandins 10: 19-40, 1975.
12. Sraer, J., Baud, L., Bens, M., Podjarny, E., Schlondorff, D., and Aradailleu, R. Glomeruli cooperate with macrophages in converting arachidonic acid to prostaglandins and hydroxyeicosatetraenoic acids. Prostaglandins Leukotrienes Med. 13: 67-74, 1984.
13. Ruzicka, T., and Printz, M. P. Arachidonic acid metabolism in guinea pig skin. Biochim. Biophys. Acta 711: 391-397, 1982.

14. Barr, R. M., Brain, S. D., Black, A. K., Camp, R. D., Greaves, M. W., Mallet, A. I., Wong, E., and Hensby, C. N. Lipoxygenase products of arachidonic acid in inflamed skin. J. Invest. Dermatol. 80(4): 345, 1983.

LEUKOTRIENE SYNTHESIS BY UV-IRRADIATED MACROPHAGE CELL CULTURES

J. E. Hardcastle[1] and S. Minoui[2]

[1]Chemistry Department and [2]Biology Department
Texas Woman's University
Denton, Texas 76204

ABSTRACT

Mouse peritoneal macrophage cells, prelabeled with ^{14}C-arachidonic acid, were treated with ultraviolet radiation (254 nm) or the Ca^{2+} ionophore A23187. After incubation, leukotrienes were extracted from the cultures and were analyzed by high-performance liquid chromatography and liquid scintillation counting. Five ^{14}C peaks were observed in both the UV-irradiated and Ca^{2+}-ionophore-treated culture extracts. Leukotriene C_4 and leukotriene B_4 were identified by coelution with standards. The UV-irradiated cell cultures produced twice as much LTC_4 and four times as much LTB_4 as did the Ca^{2+}-ionophore-treated cell cultures.

INTRODUCTION

Little is known about the effect of electromagnetic radiation on leukotriene production in cells. The synthesis of leukotrienes by macrophages *in vivo* occurs when the mammalian system is under trauma conditions (such as immediate hypersensitivity caused by allergic reactions) or inflammation caused by internal injuries (1,2). Irradiation of cells with ultraviolet (UV) radiation may mimic the conditions that occur during inflammatory events, and could result in the production of leukotrienes by the cells.

Ultraviolet radiant energy causes cellular damage, cell membrane alteration, and formation of free radicals in the cells (3,4). DeLeo et al. (4) found that UV radiation (280-329 nm) stimulates human keratinocytes in culture to produce certain prostaglandins. The synthesis of leukotrienes by cells subjected to UV radiation has not been reported. The objective of this research is to characterize, qualitatively and quantitatively, the leukotriene production by UV-irradiated macrophage cell cultures.

METHODS

Macrophage cells were collected from the peritoneal cavity of mice (BALB/C strain) and were cultured using standard techniques (5). The cells were suspended in minimum essential medium (MEM, GIBCO, Grand Island, NY) containing 10% fetal bovine serum (FBS, GIBCO). Then approximately 3×10^6 cells were seeded into 35-mm culture dishes and incubated for 2 hours at 37°C in 5% CO_2-95% air. The nonadherent cells were removed and the cultures were incubated overnight in fresh MEM-FBS. Then 1 ml of fresh MEM-FBS containing 0.5 μCi/ml ^{14}C-arachidonic acid (39 mCi/mmol, Research Products International Corp., Mount Prospect, IL) was added to each culture, and the cultures were incubated for 16 hours. About 80%-90% of the ^{14}C-arachidonic acid was absorbed by the macrophage cells. This technique labels the cell membrane with ^{14}C-arachidonic acid (5). Excess radioactivity was washed from the cultures with MEM after the 16-hour incubation period, and fresh MEM was added to each culture dish.

Some cell cultures were irradiated with UV light of 254 nm by placing the culture dishes with the lids removed under a UV lamp. The cultures were irradiated for 30 seconds, and the radiant flux was 0.14 W/m^2 as measured by a Blak-Ray (Ultraviolet Products, Inc., San Gabriel, CA) ultraviolet intensity meter. Calcium ionophore A23187 (Sigma Chemical Company, St. Louis, MO) was added to other macrophage cultures to give an ionophore concentration of 10^{-6} mmol per ml of culture medium. The Ca^{2+} ionophore was used for the non-immunological induction of leukotriene production by the macrophage cells (1). Macrophage cultures that had not been treated with UV radiation or ionophore were used as controls. After the cultures were incubated for 1 hour, they were extracted to obtain the leukotrienes. An alcohol-ether technique (6) was used to extract the leukotrienes from the treated cell cultures with an extraction efficiency of about 90%.

High-performance liquid chromatography (HPLC) and radiochemical assay were used to separate, quantify, and characterize the arachidonic acid metabolites produced by the macrophage cells. A Waters Associates (Milford, MA) Model 6000A pumping system with a U6K injector, and a Varian (Sunnyvale, CA) Vari-Chrom variable wavelength UV detector were used. A μ-Bondapak (Waters Associates) C-18 reverse-phase column (4.6 x 150 mm, ODS) was used. The mobile-phase solvent was methanol:water:acetic acid (65:35:0.02, v/v/v) adjusted to pH 5.65 with NH_4OH. The mobile phase was filtered and degassed before use. The flow rate was 1 ml/min, and the absorbance was monitored at 280 mm. One-ml fractions of the HPLC column eluent were collected and radioassayed by liquid scintillation counting. Before the experimental samples were injected onto the HPLC column, tritiated leukotriene B_4 (^{3}H-LTB$_4$) and tritiated leukotriene C_4 (^{3}H-LTC$_4$) standards were added to the samples. The radioassay was performed with a Beckman (Waldwick, NJ) Model LS9000 liquid scintillation counter using a program to count ^{14}C and ^{3}H double labeled samples.

Macroscopic examination of the UV-treated cultures showed cells that appeared to be lysed, disintegrated, or deformed compared to normal viable cells in the Ca^{2+}-ionophore-treated cultures. These cells were presumed to be dead. The dead cells were counted, and a survival rate of 50% was determined for the UV radiation dose used. Very few dead cells were observed in the Ca^{2+}-ionophore-treated and the control cell cultures.

RESULTS AND DISCUSSION

The HPLC analysis and radioassay of extracts from the UV-irradiated macrophage cell samples showed five peaks (Figure 1) that were ^{14}C-labeled and therefore were

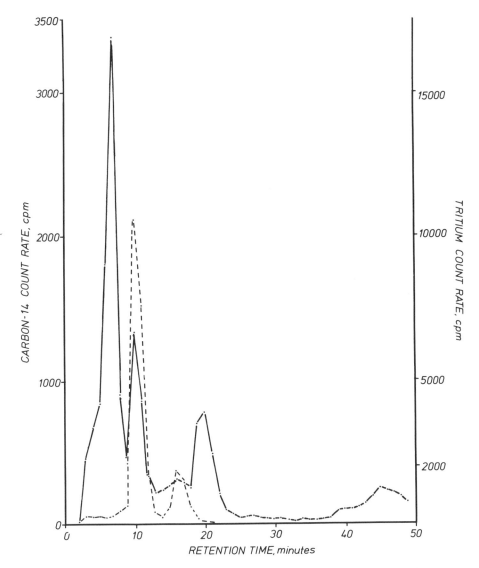

Fig. 1. Radioassay of 1-ml fractions from HPLC column eluent. Macrophage cells were treated with UV radiation after being labeled with ^{14}C-arachidonic acid. ——— , ^{14}C-arachidonic acid metabolites; · — · — · — · , ^{3}H-LTB$_4$ and ^{3}H-LTC$_4$ standards.

arachidonic acid metabolites. These peaks eluted from the column at 7, 10, 15, 20, and 45 minutes. The peaks at 10 and 15 minutes coeluted with the ^{3}H-LTB$_4$ and ^{3}H-LTC$_4$ standards, respectively. No standards were available to compare with the peaks that eluted at 7, 20, and 45 minutes. The peak that eluted at 20 minutes appears to be a major fraction of arachidonic acid metabolism in this system, and may be LTE$_4$ or 5,12-dihydroxyeicosatetraenoic acid (7). Other less polar hydroxyeicosatetraenoic acids may be eluted in the broad peak at 45 minutes (8). The peak that eluted at 7 minutes may be more polar cyclooxygenase metabolites of arachidonic acid (7). Samples from the Ca^{2+}-ionophore-treated macrophages showed the same five peaks (Figure 2). Analysis of the extracts of the control macrophage cell cultures showed no ^{14}C-labeled peaks.

It was found that ultraviolet radiation caused cellular damage and caused viable macrophages to produce leukotrienes and other arachidonic acid metabolites. The UV-treated macrophages produced three times as much arachidonic acid metabolites as did the Ca^{2+}-ionophore-treated cells. UV-irradiated cells produced four times as much LTB$_4$ and twice as much LTC$_4$ as did the Ca^{2+}-ionophore-treated cells. These results are presented in Table 1. Resting, or untreated, cells produced no measurable amount of leukotrienes.

Table 1. Amounts of LTB$_4$ and LTC$_4$ Produced by 10^6 Macrophage Cells Stimulated with UV Radiation and Ca^{2+} ionophore

A. UV-irradiated cells	LTC$_4$, ng	LTB$_4$, ng
1	13	2.8
2	22	6.9
3	31	7.9
4	19	6.0
Average	21 ± 7 SD	5.9 ± 2 SD

B. Ca^{2+} ionophore cells	LTC$_4$, ng	LTB$_4$, ng
1	7.4	1.1
2	9.2	1.6
3	11.5	1.1
Average	9.3 ± 2.1 SD	1.3 ± 0.3 SD

Amounts of leukotrienes were calculated from total disintegrations per minute under a given peak and specific activity of ^{14}C-arachidonic acid.

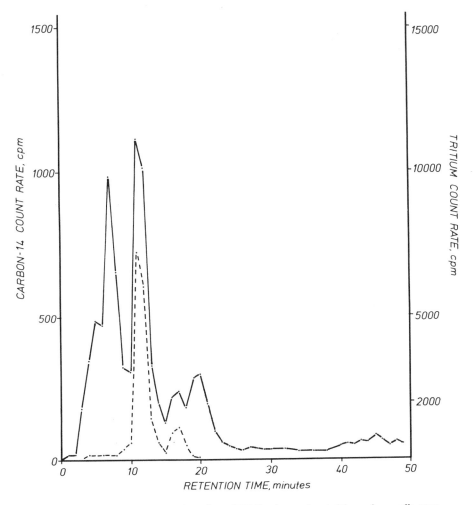

Fig. 2. Radioassay of 1-ml fractions from HPLC column eluent. Macrophage cells were treated with Ca ionophore after being labeled with ^{14}C-arachidonic acid. ———— , ^{14}C-arachidonic acid metabolites; · — · — · — · , ^{3}H-LTB$_4$ and ^{3}H-LTC$_4$ standards.

REFERENCES

1. Samuelsson, B. Leukotrienes: Mediators of immediate hypersensitivity reactions and inflammation. Science 220: 568-576, 1983.
2. Ford-Hutchinson, A. W. Leukotrienes: Their formation and role as inflammatory mediators. Fed. Proc. 44(1): 25-29, 1985.
3. Braun, G. A., and Buckner, A. C. Evidence of far ultraviolet light-mediated changes in plasma membrane structure and function. Biochem. Biophys. Acta 648: 263-266, 1981.

4. DeLeo, V. A., Horlick, H., Hanson, D., Eisinger, M., and Harber, L.C. Ultraviolet radiation induces changes in membrane metabolism of human keratinocytes in culture. J. Invest. Derm. 83: 323-326, 1984.
5. Scott, W. A., Pawlowski, N. A., Murray, H. W., Andreach, M., Zrike, J., and Cohn, Z. A. Regulation of arachidonic acid metabolism by macrophage activation. J. Exp. Med. 155: 1148-1160, 1982.
6. Clancy, R. M., and Hugli, T. E. The extraction of leukotrienes (LTC$_4$, LTD$_4$, LTE$_4$) from tissue fluids. Analyt. Biochem. 133: 30-39, 1983.
7. Henke, D. C., Kouzan, S., and Eling, T. E. Analysis of leukotrienes, prostaglandins and other oxygenated metabolites of arachidonic acid by HPLC. Analyt. Biochem. 140: 87-94, 1984.
8. Osborne, D. J., Peters, B. J., and Meade, C. J. The separation of leukotrienes and hydroxyeicosatetraenoic acid metabolites of arachidonic acid by HPLC. Prostaglandins 26: 817-832, 1982.

INFLUENCE OF PSORALEN AND ULTRAVIOLET THERAPY ON PLATELET FUNCTION AND ARACHIDONIC ACID METABOLISM IN PATIENTS WITH VITILIGO

G. H. R. Rao[1], M. Hordinsky[2], C. J. Witkop[3], and J. G. White[1]

[1]Department of Laboratory Medicine and Pathology

[2]Department of Dermatology

[3]Department of Oral Pathology
University of Minnesota Medical School
Minneapolis, Minnesota 55455

ABSTRACT

Shortwave ultraviolet light (UVC; 200-290 nm) can cause irreversible aggregation in stirred samples of human platelet-rich plasma (PRP). Patients with vitiligo are commonly treated with oral psoralen and longwave ultraviolet light (UVA; 320-400 nm) to induce repigmentation. The effect of this treatment on platelet function has not been evaluated. In the present study, we evaluated platelet function in eight patients with vitiligo, one of whom was receiving oral psoralen and UVA treatment (PUVA). Platelets from the seven nontreated patients with vitiligo aggregated normally with all agents. The platelets of the patient who was receiving PUVA therapy did not aggregate to the addition of arachidonate (0.45 mM; AA), but did aggregate after exposure to collagen (3 μg/ml) and a combination of epinephrine (5 μM) and AA. *In vitro* studies showed that exposure of platelets to 8-methoxsalen (0.5 mM) blocked arachidonate-induced aggregation. All patients' cells converted ^{14}C-arachidonic acid through the cyclooxygenase and lipoxygenase pathways to eicosanoid products as efficiently as control cells. Platelet prostaglandin synthesis, AA metabolism, and aggregation were also studied from two patients with non-treated vitiligo before ingestion of 20 mg of oxsoralen (8-MOP), 2 hours after psoralen ingestion, and immediately following the administration of UVA radiation. No significant changes in arachidonate-induced activation or prostaglandin synthesis were found. The results of our study demonstrate that (a) platelets from patients exposed to PUVA therapy do not show any alterations in arachidonate-induced activation or prostaglandin synthesis; and (b) PUVA therapy is associated with the induction of a specific refractory state to arachidonate by an unknown mechanism.

INTRODUCTION

Studies from our laboratory and that of others have demonstrated that the *in vitro* exposure of human and animal platelets to shortwave ultraviolet rays induces activation leading to irreversible aggregation (1). Oral psoralen and longwave ultraviolet light (PUVA) therapy is routinely prescribed for the treatment of vitiligo. In addition, in recent years there has been an increase in midrange wavelength ultraviolet light (UVB) (290-320 nm) and longwave ultraviolet light (UVA) administration for cosmetic purposes (suntan salons). However, little is known about the effect of UVA or UVB on platelets. In the present study, we evaluated platelet function and arachidonic acid metabolism in eight patients with vitiligo and followed the influence of PUVA. We also studied platelet function in three healthy controls who received UVA or UVB in increasing amounts for 4 consecutive days. The results of our study show that PUVA, UVA, or UVB therapy does not induce any significant alterations in platelet arachidonic acid metabolism. However, prolonged administration of PUVA is associated with the induction of a specific refractory state to arachidonate-induced stimulation.

MATERIALS AND METHODS

Arachidonic acid as the sodium salt was from NuChek Prep (Elysian, MN); injectable adrenalin and bovine thrombin were from the Parke-Davis Company (Detroit, MI). Radiolabeled (^{14}C) arachidonic acid (AA) was purchased from New England Nuclear (Boston, MA). Unless otherwise indicated, all other chemicals were purchased from the Sigma Chemical Company (St. Louis, MO).

Platelet Preparation

Approval for this study was obtained previously from the Human Research Committee at the University of Minnesota. Blood was obtained by venipuncture from seven nontreated patients with vitiligo; one patient with vitiligo who was receiving PUVA therapy; and two patients with vitiligo before methoxsalen ingestion (20 mg), 2 hours after ingestion, and again following UVA administration. Blood was also obtained from five control patients. Three of these also underwent platelet function studies after receiving increasing amounts of UVA (7.8 mW/cm^2 for a total of 10 joules; n = 1) or UVB (0.325 mW/cm^2 for a total of 0.16 joules; n = 2). No medications were allowed for at least 2 weeks before venipuncture. Blood was mixed immediately after drawing with citrate anticoagulant, and PRP was obtained by the methods of Rao et al. (2). In one case, platelet aggregation was also studied after AA (0.45 mM) and 8-methoxsalen solution (5 μl) were added to control platelets. The ability of platelets to convert radiolabeled arachidonic acid to thromboxane was followed using radiolabeled arachidonic acid.

Arachidonic Acid Metabolism

For the measurement of radiolabeled arachidonic acid conversion to thromboxane by intact platelets, each reaction mixture containing 1.5×10^9 cells, suspended in 1 ml of HBSS, was stirred on an aggregometer for 5 minutes at 37°C with 1 μg of labeled arachidonic acid. Extraction separation and quantitation were done according to the procedures described by Rao et al. (2).

RESULTS

Platelet Function Studies

Platelets from normal control donors aggregated irreversibly when stirred with agonists such as epinephrine (5 μM), adenosine diphosphate (3 μM), collagen (2-4 μg/ml), thrombin (0.2 U/ml), and arachidonate (0.45 mM). Platelets from patients with vitiligo did not show any dysfunction when tested with various agonists. UVA or UVB treatment of controls did not induce significant alterations in the response of platelets to the action of the various agonists tested. No platelet dysfunction was observed in the vitiligo patients who received one PUVA treatment. However, prolonged administration of PUVA therapy was associated with some dysfunction. Platelets from the patient who had been receiving PUVA therapy for 3 months did not aggregate when stirred with arachidonate. Moreover, the addition of 8-MOP plus AA to normal platelets caused delayed aggregation.

Effect of PUVA Therapy on Arachidonic Acid Metabolism

Normal control platelets converted 34% of the substrate arachidonic to thromboxane (Table 1). Platelets from patients also converted normal amounts of arachidonic acid (28%) to thromboxane. No significant difference in the ability of platelets to convert arachidonic acid to lipoxygenase and cyclooxygenase

Table 1. ^{14}C-Arachidonic Acid (AA) Conversion by Platelets From Patients With Vitiligo and Normal Control Donors

	Percent Conversion of AA[1]			
	AA	HETE	HHT	TXB$_2$
Control (n = 3)	2.9 ± 0.6	31.2 ± 5.1	29.2 ± 5.3	34.7 ± 0.9
Patients	9.6 ± 0.9	43.1 ± 4.3	18.9 ± 4.3	28.3 ± 2.1

[1]Mean and standard error
HETE = 12L-hydroxy-5,8,10,14-eicosatetraenoic acid
HHT = 12L-5,8,10-heptadecatrienoic acid
TXB$_2$ = thromboxane B$_2$

Table 2. Effect of Oral Psoralen on ^{14}C-Arachidonic Acid (AA) Conversion by Intact Human Platelets

	Percent Conversion of AA[1]		
	HETE	HHT	TXB$_2$
Pre-psoralen	34.4	28.9	26.0
2 hr post-psoralen	38.4	22.4	28.0
UVA following drug	30.0	28.8	26.8

[1]Average values from two different donors

metabolites could be found in the patients who received PUVA therapy (Table 2). In addition, the administration of psoralen alone did not alter platelet function.Platelets of control donors who had been subjected to either UVA or UVB treatment also converted normal amounts of arachidonic acid to thromboxane.

DISCUSSION

Vitiligo is a systemic disorder in which unknown factors destroy pigment cells in the skin. It is not well understood why PUVA therapy is stimulatory to pigment cell growth since psoralens, when excited by UVA radiation, covalently bind nucleic acids and effectively prevent replication, transcription, and translation (3). Except for the studies on the effect of UVC radiation on platelet function, no data are available on the *in vivo* effect of UVA or UVB light on *ex vivo* platelet arachidonic acid metabolism and function.

The data from this preliminary study on the effects of PUVA therapy on platelet function in vitiligo patients demonstrate that (a) platelet function in patients with vitiligo is normal; (b) PUVA therapy does not induce significant alterations in platelet arachidonic acid metabolism, but is associated with the development of minor platelet dysfunction; and (c) UVA or UVB treatment of normal individuals does not significantly alter platelet arachidonic acid metabolism and function. Furthermore, the addition of 8-MOP plus AA to normal platelets is associated with delayed aggregation. Why some platelet dysfunction is induced by psoralens remains to be elucidated. The finding of altered platelet function in PUVA-treated patients with vitiligo needs to be confirmed by studying a larger number of patients.

ACKNOWLEDGMENTS

Supported by USPHS grants HL-11880, CA-21737, GM-22167, HL-30217, NIH 5F32 AM06800-02, a grant from the March of Dimes (MOD 1-886), and the Minnesota Medical Foundation (HDRI-38-85).

REFERENCES

1. Dickson, R. C., Doery, J. C. G., and Lewis, A. T. Induction of platelet aggregation. Sci. 172: 1140-1141, 1971.
2. Rao, G. H. R., Peller, J. D., Semba, C. P., and White, J. G. Influence of the calcium sensitive fluorophore, Quin 2, on platelet function. Blood 67: 354-361, 1986.
3. Bredberg, A. Genetic toxicity of psoralen and ultraviolet radiation in human cells. Acta Dermato-venerelogica, Suppl. 104: 1-40, 1982.

SOURCES OF INCREASED *IN VIVO* CYCLOOXYGENASE PRODUCT RELEASE FOLLOWING WHOLE-BODY IRRADIATION OF RATS

M. J. Schneidkraut, P. A. Kot, and P. W. Ramwell

Department of Physiology and Biophysics
Georgetown University Medical Center
Washington, DC 20007

ABSTRACT

Ionizing radiation alters cyclooxygenase product synthesis in a time- and dose-related manner. Rats exposed to 10 or 20 Gy whole-body gamma radiation showed a significant increase (p $<.05$) in urine immunoreactive thromboxane B_2 (iTXB$_2$) 4-120 hours after 10 Gy as well as 4 and 12 hours after 20 Gy exposure. Irradiation with 2 Gy had no effect. The source(s) of this radiation-induced increase in iTXB$_2$ excretion was studied by regional shielding. Rats were anesthetized and exposed to sham radiation, 15 Gy whole-body radiation, or a dose of 20 Gy radiation with either the abdomen or thorax shielded. Four hours after exposure, the animals were re-anesthetized and urine samples collected. Unshielded animals exposed to a 15 Gy dose of radiation showed a 2.5-fold (p $<.05$) increase in iTXB$_2$ excretion. Abdominal shielding attenuated the radiation-induced increase in iTXB$_2$ excretion by 41%, but thoracic shielding prevented the increase in iTXB$_2$ excretion. Thus, the thoracic organs are an important source of the radiation-induced increase in iTXB$_2$ excretion, and the abdominal organs may also contribute to the increased *in vivo* release of iTXB$_2$. In the next series of experiments, the individual contribution of the kidneys and lungs to the increased excretion of cyclooxygenase products was studied. Rats were subjected to 20 Gy whole-body radiation and 4 hours after exposure, either the kidneys or lungs were isolated and perfused with a cell-free medium. Radiation did not alter urine acidification by the isolated perfused kidney. However, the concentration of urine from irradiated kidneys was 18.3% (p $<.05$) less than urine from control kidneys. Whole-body gamma radiation also elicited a 2.2-fold (p $<.05$) and 3.6-fold (p $<.05$) increase in the excretion of iPGE$_2$ and 6-keto prostaglandin $F_{1\alpha}$ (i6KPGF$_{1\alpha}$) from isolated perfused kidneys. The excretion rate of iTXB$_2$ from perfused kidneys after irradiation was not significantly different from sham-irradiated controls. On the other hand, isolated perfused rat lungs released 100% (p $<.05$) more iTXB$_2$ following irradiation than lungs from sham-irradiated

185

animals. The release of i6KPGF$_{1\alpha}$ was also significantly elevated. These studies showed a regional release of cyclooxygenase products following irradiation. The elevated excretion rate of iTXB$_2$ appears to be primarily due to an increased pulmonary release of this arachidonate metabolite.

INTRODUCTION

The cellular and tissue effects of ionizing radiation are well documented, but the molecular manifestations are poorly understood. Cyclooxygenase products, by virtue of their powerful vasoactive and platelet aggregatory properties (1-13), may play an important role in radiation-induced injury. Credence is lent to this hypothesis by the observations that radiation injury is associated with the formation of free radicals as well as lipid peroxides (14) and that peroxides play a pivotal role in the activation of cyclooxygenase (15-18). Since cyclooxygenase activity may be one of the rate-limiting steps in arachidonic acid metabolism, an increase in lipid peroxide formation could increase the release of these vasoactive agents.

This issue was addressed by Eisen and Walker (19), who showed that exposure of mice to 7.0 gray (Gy) X radiation increased pulmonary and splenic prostaglandin E-like activity. Other studies indicated increased hepatic and cerebral prostaglandin E-like activity and increased prostaglandin F$_{2\alpha}$ synthesis after irradiation with doses of 5.0-7.5 Gy (20). Exposure of mice to 9.0 Gy gamma radiation significantly increased *in vitro* synthesis of prostaglandin E$_2$ (PGE$_2$) and prostaglandin F$_{2\alpha}$ by hepatic microsomes and homogenates of brain and testis (21,22). These and other studies demonstrated that *in vitro* synthesis of the classical prostaglandins is increased following exposure to ionizing radiation (19-23).

The effect of ionizing radiation on thromboxane A$_2$ (TXA$_2$) synthesis is less clear. Maclouf et al. (24) demonstrated a significant increase in thromboxane B$_2$ (TXB$_2$) release by splenic microsomes isolated from rats exposed to 9.0 Gy whole-body radiation. Steel and Catravas (25) showed increased TXB$_2$ release from guinea pig lung parenchymal strips at 1-3 hours postirradiation. On the other hand, Steel et al. (26) did not show an increase in TXB$_2$ synthesis by guinea pig lung airway tissue following irradiation. Similarly, Allen et al. (27) found no change in TXB$_2$ release by platelets irradiated *in vitro*.

Although numerous studies addressed the *in vitro* release of cyclooxygenase products following radiation exposure (19-40), few investigations evaluated the effect of whole-body ionizing radiation exposure on the *in vivo* release of these products (41-45). The present series of studies are divided into three sections: time course of *in vivo* cyclooxygenase product release following irradiation, *in vivo* analysis of the source(s) of radiation-induced alterations in cyclooxygenase product release, and *in vitro* cyclooxygenase product release from selected isolated perfused organ systems.

TIME COURSE OF *IN VIVO* CYCLOOXYGENASE PRODUCT RELEASE FOLLOWING IRRADIATION

Male Sprague-Dawley rats (200-250 g) were used in all studies. The animals were anesthetized with sodium pentobarbital (30 mg/kg, i.p.) and exposed to either sham radiation or 2.0, 10.0, or 20.0 Gy whole-body gamma radiation in a ventrodorsal orientation from a 7.4×10^{13} becquerel cesium-137 radiation source (Best Industries small animal irradiator, Arlington, VA). The rate of delivery was calibrated biannually and the exposure times adjusted to the delivery rate.

The rats were re-anesthetized at 2, 4, 12, 24, 72, or 120 hours after irradiation or sham irradiation, and the *in vivo* cyclooxygenase product release was determined from urine samples removed directly from the urinary bladders. Each bladder was exposed by a 1-cm midline abdominal incision, and the urine was drained from the bladder via a syringe. The urine samples were frozen at -20°C and later thawed as a group for the determination of cyclooxygenase product levels by radioimmunoassay.

Animals exposed to 2.0 Gy showed no change in urine immunoassayable thromboxane B_2 (iTXB$_2$) levels at all observation times (Table 1). Rats exposed to 10.0 Gy demonstrated a significant increase in urine iTXB$_2$ at 4-120 hours postirradiation but not at 2 hours after exposure (Table 1). Increasing the exposure dose to 20.0 Gy resulted in a 233% and 105% increase in urine iTXB$_2$ at 4 and 12 hours, respectively (Table 1). Exposure to 20.0 Gy did not alter the release of iTXB$_2$ at 2, 24, or 72 hours. Urine samples could not be taken 120 hours after 20.0 Gy because this dose of radiation was 100% lethal by 4 days postirradiation. At each dose of radiation used and at each observation time, urine volumes in control and in irradiated animals were not significantly different.

IN VIVO ANALYSIS OF SOURCES OF RADIATION-INDUCED ALTERATIONS IN CYCLOOXYGENASE PRODUCT RELEASE

The studies presented above indicated that whole-body radiation exposure increases the *in vivo* release of iTXB$_2$. Since the kidney is a radiosensitive organ and urinary iTXB$_2$ was measured, studies were performed to determine if the increase in urine iTXB$_2$ reflected alterations in local renal synthesis. To examine this possibility, an isogravitometric cross-perfusion system was developed.

Isogravitometric Cross-Perfusion

Recipient rats were anesthetized with ether and injected with indomethacin (20 mg/kg, i.v.) 24 hours before cross-perfusion in order to suppress endogenous cyclooxygenase activity. The indomethacin-pretreated animals were cross-perfused with either indomethacin-pretreated rats (indomethacin), sham-irradiated rats (control), or rats exposed to 20.0 Gy whole-body radiation 4 hours before cross-perfusion (irradiated). Urine samples were removed only from the indomethacin-

Table 1. *In Vivo* Release of iTXB$_2$ Following Whole-Body Irradiation

| | Hours Postirradiation | | | | | |
	2	4	12	24	72	120
Control	3.21 ±0.96	2.75 ±0.85	2.25 ±0.41	2.61 ±1.26	2.50 ±0.58	3.13 ±1.26
2 Gy	2.94 ±0.82	3.16 ±0.71	2.36 ±0.58	3.74 ±1.10	1.98 ±0.58	3.27 ±1.21
Control	2.64 ±0.58	2.58 ±0.41	1.87 ±0.22	4.01 ±0.55	2.61 ±0.30	3.05 ±0.80
10 Gy	3.13 ±0.80	5.00[1] ±0.38	5.55[1] ±0.58	9.26[1] ±2.03	7.50[1] ±1.18	6.43[1] ±1.24
Control	3.32 ±0.60	1.51 ±0.30	1.70 ±0.50	2.69 ±0.47	2.25 ±0.58	ND
20 Gy	2.09 ±0.55	5.03[1] ±0.82	3.49[1] ±0.58	3.38 ±0.52	1.13 ±0.14	[2]

Values are expressed as mean ± standard error in nanomoles thromboxane B$_2$ per liter of urine for 6-10 animals per group. iTXB$_2$ = immunoreactive thromboxane B$_2$; ND = not done.
[1] p < .05 by unpaired Student's t test analysis
[2] Animals expired before samples removed
(Source: J. Appl. Physiol. 57: 833-838, 1984)

pretreated recipients of the cross-perfused pair prior to cross-perfusion and after 60 minutes of cross-perfusion.

The rats were anesthetized with sodium pentobarbital i.p. and the tracheas were intubated. The common carotid artery and jugular vein were catheterized and the urinary bladder exposed. The animals were then anticoagulated with sodium heparin (1000 USP units/kg, i.v.) prior to attachment to the cross-perfusion device.

The cross-perfusion system consisted of a 10-ml syringe reservoir, to which was attached low-resistance (PE240) polyethylene tubing. The reservoir was located 2-3 cm above a two-pan balance on which the anesthetized, catheterized rats were placed. The height of the reservoir above the animals resulted in a calculated venous pressure of 1.5-2.3 mm Hg and an estimated flow rate of 0.4 ml per minute. The tubing and reservoirs were filled with isotonic saline containing 3% bovine serum albumin to maintain a normal blood volume. The carotid artery catheter was attached to the outflow tubing running to the reservoir, and the jugular vein catheter was attached to the inflow tubing running from the reservoir. The tubing and reservoirs were allowed to fill with blood before the two circulations were linked.

Immediately before the circulations were connected, the bladder was drained and the urine discarded. The inflow line from the left reservoir was then attached

Table 2. Determination of Source of Urine by Isogravitometric Cross-Perfusion

	Pre-Cross-Perfusion TXB$_2$	Cross-Perfusion (60 min) TXB$_2$
Indomethacin	0.10 ±0.06	0.32 ±0.17
Control	0.03 ±0.03	2.14[1] ±0.89
20 Gy	0.002 ±0.002	0.52[1] ±0.19

Values are expressed as mean ± standard error in nanomoles thromboxane B$_2$ (iTXB$_2$) per liter of urine for seven pairs of animals per group.
[1]p <.01 by paired Student's t test analysis
(Source: J. Appl. Physiol. 57: 833-838, 1984)

to the jugular vein catheter of the rat on the right pan, and vice versa. The relative weights of the rats were balanced using the scale weights. After this time and until the end of the cross-perfusion, the relative weights of the cross-circulating rats were maintained by adjusting the flow of blood to the reservoir by the screw clamps on the outflow lines (42,46).

Indomethacin-pretreated recipients cross-perfused with a second group of indomethacin-pretreated rats showed no change in urine iTXB$_2$ levels (Table 2). In contrast, urine iTXB$_2$ from indomethacin-pretreated recipients was significantly increased (p <.05) when these animals were cross-perfused with either control or irradiated rats (Table 2). No significant difference was observed between the control group and irradiated cross-perfused group. These data indicate that urine iTXB$_2$ is primarily derived from extrarenal sources; however, the source(s) of the radiation-induced alteration in cyclooxygenase product release might still have been the kidney.

Regional Shielding

In order to assess the role that abdominal organs play in the radiation-induced alterations in cyclooxygenase product release, a regional shielding study was performed. Rats were anesthetized and divided into four groups: (a) unshielded, exposed to 15.0 Gy whole-body radiation, (b) abdomen shielded at the level of the kidneys and exposed to 20.0 Gy gamma radiation, (c) thorax shielded with the same shield used to protect the abdomen and exposed to 20.0 Gy ionizing radiation, and (d) sham-irradiated controls. The radiation shield was made from a lead alloy in the shape of a horseshoe. The shield had an internal diameter of 7.0 cm, width of 4.2 cm, and thickness of 4.1 cm. Postmortem examination was performed on all animals. The placement of the shield was evaluated by marking the location of the shield on the skin and by postmortem confirmation of the organs that were protected. Animals in which there was partial shielding of the lungs or the kidneys were omitted from the experimental groups. The effectiveness of the radiation shield was shown to be 92% using lithium iodide thermoluminescent crystals. In order to expose all groups to an equivalent dose of whole-body radiation, the unshielded group was exposed to a 25% smaller dose of radiation than were the shielded animals. This was done because in both the thorax-shielded and abdomen-shielded groups, 25% of the body was covered and not exposed to ionizing radiation.

The iTXB$_2$ excretion rate of rats exposed to 15.0 Gy whole-body gamma radiation was 269% (p <.05) greater than sham-irradiated controls (Table 3). Thorax shielding prevented the radiation-induced increase in iTXB$_2$ excretion. Thus, animals exposed to gamma radiation with the thorax shielded showed a significant decrease (p <.05) in iTXB$_2$ excretion compared to unshielded animals, and no significant difference in iTXB$_2$ excretion compared to sham-irradiated controls (Table 3).

Abdominal shielding reduced the radiation-induced increase in the iTXB$_2$ excretion rate by 42%. This attenuation of the iTXB$_2$ excretion rate was not significantly different from that measured in either the unshielded irradiated group or sham-irradiated group. The radiation-induced increase in iTXB$_2$ excretion was not due to altered rates of urine formation because the urine volume released into the urinary bladder over the 3- to 4-hour postirradiation period was not significantly different in any of the groups.

Table 3. Urine iTXB$_2$ Levels in Regionally Shielded or Unshielded Animals Compared to Control

	iTXB$_2$ (nmol/liter/hour)
Control	5.06 ± 0.87
15 Gy unshielded	18.66[1] ± 4.79
20 Gy abdomen-shielded	10.86 ± 2.34
20 Gy thorax-shielded	5.97 ± 1.28

Data are expressed as mean ± standard error for 13-19 animals per group.
[1]p <.05 compared to control by analysis of variance and a Dunnett's test.
(Source: J. Appl. Physiol. 61: 1264-1269, 1986)

IN VITRO RELEASE OF CYCLOOXYGENASE PRODUCTS FROM SELECTED ISOLATED PERFUSED ORGAN SYSTEMS

The data presented above demonstrated that urine iTXB$_2$ is primarily derived from extrarenal sources. The thoracic organs appear to be a major source of the increased iTXB$_2$ excretion observed following radiation exposure, but the upper abdominal organs also may contribute to this response. In order to analyze the extent to which the kidneys contribute to the altered iTXB$_2$ excretion seen following whole-body irradiation, an isolated perfused kidney system was developed.

Renal Perfusion

Rats were anesthetized and exposed to 20.0 Gy whole-body radiation or sham radiation. Three hours later, the animals were re-anesthetized and urine samples removed. These urine samples, obtained prior to the isolation and perfusion of the kidney, were designated as pre-perfusion urine samples. After removal of the pre-perfusion urine sample, an endotrachael tube was inserted. The vena cava cephalad to the renal veins and caudad to the diaphragm was isolated, and a loose ligature of umbilical tape was placed around the vessel. Three loose ligatures of 4-0 silk were then placed around both the inferior vena cava and the abdominal aorta proximal to the iliolumbar vessels and distal to the renal vessels. The superior mesenteric artery, coeliac artery, and spermatic arteries were then ligated. Finally, a loose ligature of umbilical tape was placed around the abdominal aorta proximal to the renal arteries and distal to the diaphragm. The animal was then anticoagulated with 1000 USP units heparin/kg i.v. (Elkin-Sinn Inc., Cherry Hill, NJ).

Ten minutes after the administration of heparin, the most distal tie around the abdominal aorta and vena cava was tied. Next the loose ligature of umbilical tape around the aorta proximal to the renal arteries was tightened, and a catheter (PE-60 tubing) was introduced into the abdominal aorta caudal to the renal arteries and cephalad to the iliolumbar vessels. The renal bed was then cleared of blood by perfusion with Kreb's-Ringer bicarbonate containing 3% bovine serum albumin and 5.6 mM glucose. The vena cava proximal to the renal veins was then ligated, and a second catheter (PE-50 tubing) was placed in the inferior vena cava distal to the renal veins. The bladder was then drained of urine, and the urine was discarded.

The perfusate was oxygenated with room air prior to perfusion through the kidneys. The warmed solution (37°C) was perfused through the renal bed for 10 minutes at a rate of 3.9 ml/min and a perfusion pressure of 90 mm Hg. Venous effluent samples were taken from the vena caval catheter after 5 minutes of perfusion. The urine formed during the 10-minute perfusion period was removed from the urinary bladder, and the volume was recorded. Aliquots (0.2 ml) of both the venous effluent and urine were removed from each sample, and the osmolality and pH were determined (Advanced Digimatic Osmometer, Needham, MA) (Fisher Accumet pH Meter Model 310, Pittsburgh, PA). The remaining samples were frozen at -20°C immediately after removal from the animal. Samples were later thawed as a group and assayed for cyclooxgenase products by radioimmunoassay. This renal perfusion system remained viable for periods of up to 20 minutes without evidence of formation of gross edema. The shorter perfusion period of 10 minutes was chosen to minimize the possibility of edema formation, which could alter cyclooxygenase product release. Immediately following perfusion, the kidneys were examined by visual inspection. The perfused kidneys were then injected with dye (trypan blue) to determine the effectiveness of the perfusion.

The renal function of perfused kidneys from irradiated and sham-irradiated animals was assessed by comparing the urine and venous effluent pH and osmolality. These isolated organs both acidified and concentrated the perfusate. Whole-body gamma irradiation did not alter the renal acidifying or concentrating mechanism, nor did it change renal venous effluent pH and osmolality.

The urine from both irradiated and sham-irradiated perfused kidneys contained significant amounts of iTXB$_2$, immunoassayable PGE$_2$ (iPGE$_2$), and immunoassayable 6-keto prostaglandin F$_{1\alpha}$ (i6KPGF$_{1\alpha}$) (Table 4). The renal venous effluent contained only iTXB$_2$ and i6KPGF$_{1\alpha}$. Urine output, venous flow, and perfusion pressure were not significantly altered in irradiated or sham-irradiated perfused kidneys.

The effect of ionizing radiation on the renal excretion of cyclooxygenase products was investigated. In the intact animal, whole-body gamma irradiation significantly increased ($p < .05$) iTXB$_2$ excretion and significantly decreased ($p < .05$) iPGE$_2$ release (Table 4). The excretion rate of i6KPGF$_{1\alpha}$ in the intact animal was not determined because it is an unreliable indicator of prostacyclin (PGI$_2$) release *in vivo* (47).

Table 4. Selected Cyclooxygenase Product Excretion Rates in Isolated Perfused Kidneys Exposed to Ionizing Radiation

	iTXB$_2$ (pmol/hr)	iPGE$_2$ (pmol/hr)	i6KPGF$_{1\alpha}$ (pmol/hr)
Sham-irradiated	5.18 ±0.95	3.05 ±0.99	4.66 ± 1.05
20 Gy Irradiated	6.81 ±1.31	9.79[1] ±2.70	21.40[1] ± 5.46

Data are expressed as mean ± standard error of mean for 6-13 animals per group.
[1]$p < .05$ compared to sham-irradiated controls by unpaired Student's t test analysis
iTXB$_2$, immunoreactive thromboxane B$_2$
iPGE$_2$, immunoreactive prostaglandin E$_2$
i6KPGF$_{1\alpha}$,immunoreactive 6-keto prostaglandin F$_{1\alpha}$
(Source: J. Appl. Physiol. 61: 1264-1269, 1986)

The $iTXB_2$ excretion rate of irradiated isolated perfused rat kidneys was comparable to that of the sham-irradiated perfused organs. The excretion of $iPGE_2$ and $i6KPGF_{1\alpha}$ from irradiated perfused kidneys was increased to 221% and 359% (p <.05), respectively, compared to perfused kidneys from nonirradiated animals (Table 4).

Finally, the effect of radiation exposure on the release of cyclooxygenase products into the renal venous effluent was assessed. Whole-body irradiation did not alter the levels of renal venous $iTXB_2$ or $i6KPGF_{1\alpha}$. Immunoreactive PGE_2 was not detectable in the venous effluent from irradiated or sham-irradiated kidneys.

The isolated perfused kidney studies showed that the kidney does contribute to urine $iTXB_2$ levels but that the radiation-induced increase in $iTXB_2$ excretion is not due to an altered endogenous renal synthesis. The renal cyclooxygenase pathway is affected by whole-body radiation exposure in that both $iPGE_2$ and $i6KPGF_{1\alpha}$ excretion from irradiated isolated perfused kidneys is significantly greater than from control isolated perfused kidneys.

The regional shielding studies indicated that the thoracic organs contributed to the increased $iTXB_2$ excretion seen following whole-body irradiation. The effect of radiation exposure on the pulmonary release of cyclooxygenase products was assessed using an isolated perfused rat lung model.

Lung Perfusion

Rats were anesthetized 4 hours after whole-body gamma irradiation. The trachea was cannulated and the lungs were ventilated with 95% air-5% CO_2 at a tidal volume of 2.5 ml and a frequency of 55 breaths per minute (Harvard Model Rodent Respirator, Boston, MA). Positive end expiratory pressure was 1 cm of water. The blood was removed from the lungs by perfusing the pulmonary vascular bed with Krebs-Ringer bicarbonate via the cannulated hepatic portal vein. Blood drained from the lungs through the transected abdominal aorta. After the lungs were cleared of blood, the thoracic cavity was opened and the pulmonary artery was cannulated via the right ventricle. Perfusion of the lungs was transferred from the hepatic portal vein to the pulmonary artery. The lungs were then removed from the animal and suspended in a water-jacketed *in vitro* perfusion chamber. The temperature of the chamber was held constant at 37°C.

Once the lungs were suspended in the *in vitro* perfusion chamber, the perfusate was allowed to recycle through the lungs. The perfusate was pumped through the pulmonary artery (Multistatic 4-channel pump, Haake Buchler, Saddle Brook, NY) and drained through the cut left ventricle into the reservoir at the bottom of the perfusion chamber. The perfusate was passed through a bubble trap (Travenol i.v. set) prior to entry in the pulmonary artery. The flow rate was held constant at 10 ml/min. Perfusion pressure and airway resistance were continuously monitored throughout the perfusion period. Changes in tracheal pressure were used as an indicator of pulmonary edema. Lungs were also evaluated for edema by dry weight/

Table 5. Effect of Whole-Body Ionizing Radiation Exposure on
Cyclooxygenase Product Release From Isolated Rat Lungs

	iTXB$_2$ (pmol/1)	i6KPGF$_{1\alpha}$ (pmol/1)
Sham-irradiated	142.2 ± 26.8	2,086.7 ± 177.5
20 Gy Irradiated	283.0[1] ± 45.0	3,563.0[1] ± 387.9

Data are expressed as mean ± standard error of mean for 9-10 animals per group.
[1]p <.02 compared to sham-irradiated controls by unpaired Student's t test analysis
(Life Sci. 41: 479-484, 1987)

wet weight ratios. Lungs with a dry weight/wet weight ratio of less than 16% or
with elevated tracheal pressures were excluded from the groups. Perfusate samples
(3 ml per sample) were withdrawn from the reservoir 20 minutes after the start
of *in vitro* pulmonary perfusion. Samples were frozen at -20°C and later thawed
as a group for assay of iTXB$_2$ and i6KPGF$_{1\alpha}$ concentrations.

Rats exposed to 20.0 Gy whole-body radiation at 4 hours before the lungs were
isolated showed a 99% increase (p <.02) in pulmonary effluent iTXB$_2$ levels compared
to control (Table 5). Similarly, i6KPGF$_{1\alpha}$ levels were significantly elevated (p <.02)
in the pulmonary effluent from irradiated lungs (Table 5). These data show that
ionizing radiation exposure will alter the pulmonary release of cyclooxygenase
products. These results, coupled with the regional shielding data, suggest that the
lungs are a major source of the increased iTXB$_2$ excretion observed following whole-
body irradiation.

DISCUSSION AND CONCLUSIONS

The data presented in this review show that whole-body ionizing radiation exposure
alters the *in vivo* excretion of iTXB$_2$ (41-43). The dose of radiation necessary to
increase iTXB$_2$ excretion is greater than 2.0 Gy and less than 10.0 Gy exposure
(41,42). These studies also showed that there is a delay (approximately 4 hours)
before this increased *in vivo* synthesis is observed (41,42).

The source of the observed increase in iTXB$_2$ release *in vivo* was investigated.
The data presented indicate that the upper abdominal organs, but not the kidney,
contribute to the radiation-induced increase in iTXB$_2$ excretion (43). The lungs
were implicated as a major source of the increased *in vivo* release of iTXB$_2$ following
whole-body irradiation (30,43).

The mechanism by which radiation exposure alters cyclooxygenase product release can be divided into three broad categories: increased arachidonic acid release, increased activity of the cyclooxygenase pathway enzymes, and decreased catabolism of the cyclooxygenase products.

Evidence for increased release of arachidonic acid following radiation exposure is supported by the work of Hahn et al. (31). That study showed that arachidonic acid release from pulmonary endothelial cells exposed to 20.0 Gy radiation and challenged with bradykinin was 66% greater than the release of this fatty acid from sham-irradiated cells exposed to the same concentration of this agent (31).

The increased release of arachidonate metabolites seen following irradiation also may be due to an increased activity of the cyclooxygenase. This supposition is based on the findings that radiation exposure increases the formation of free radicals and peroxides (14), and that this free radical formation may activate the cyclooxygenase. Seregi et al. (17) reported a stimulation of brain cyclooxygenase activity *in vitro* with the addition of hydrogen peroxide. Hemler et al. (15) extended these findings by demonstrating a dose-dependent *in vitro* activation of cyclooxygenase by both hydrogen peroxide and lipid peroxides. On the other hand, Egan et al. (16) showed that higher concentrations of free radicals inactivated the cyclooxygenase and that this inactivation was prevented by the addition of free radical scavengers. These studies suggest that radiation-induced free radical and peroxide formation may play a role in the elevated *in vivo* and *in vitro* release of cyclooxygenase products observed in the present study.

The elevated concentrations of i6KPGF$_{1\alpha}$ and iPGE$_2$ in the urine from irradiated isolated perfused kidneys or the increased release of i6KPGF$_{1\alpha}$ from irradiated isolated perfused lungs may be due to a decrease in the catabolism of these cyclooxygenase products. This proposed mechanism is based on the observations of Eisen and Walker (19), who demonstrated that splenic tissue from 7.0-Gy irradiated mice inactivated prostaglandin E$_1$ to a lesser extent than the same tissue taken from sham-irradiated controls. Subsequent studies showed that the radiation-induced decrease in 15-hydroxyprostaglandin dehydrogenase activity in splenic tissue was present within 4 hours of radiation exposure (32). Previous studies have demonstrated the presence of 15-hydroxyprostaglandin dehydrogenase activity within the kidney and lung (48-58). Other studies have concluded that 15-hydroxyprostaglandin dehydrogenase is a major catabolic enzyme for both PGI$_2$ and PGE$_2$ (47,52-54,57-59). Therefore, the radiation-induced elevation in the release of iPGE$_2$ and i6KPGF$_{1\alpha}$ by these isolated organs may be due, in part, to the reduced catabolism of these two cyclooxygenase products.

The preliminary investigations reviewed here indicate that doses of radiation that induced tissue damage also increased the release of cyclooxygenase products. The mechanism by which radiation exposure alters cyclooxygenase product release and the role that the altered arachidonic acid metabolite levels play in radiation-induced tissue injury remain to be determined. If an association between cyclooxygenase product concentrations and tissue injury is demonstrated, novel treatments for radiation injury would be possible.

ACKNOWLEDGMENTS

The authors wish to thank Ms. Ellen Costello for her help with the preparation of this manuscript. The authors also wish to thank Dr. Ken Mossman for fabrication of the radiation shield and determination of its efficiency.

REFERENCES

1. Armstrong, J. M., Boura, A. L. A., Hamberg, M., and Samuelsson, B. A comparison of the vasodepressor effects of the cyclic endoperoxides PGG2 and PGH2 with those of PGD2 and PGE2 in hypertensive and normotensive rats. Eur. J. Pharmacol. 39: 251-258, 1976.
2. Angerio, A. D., Fitzpatrick, T. M., Kot, P. A., Ramwell, P. W., and Rose, J. C. Effect of verapamil on the pulmonary vasoconstrictor action of prostaglandin F2α and a synthetic PGH2 analogue. Brit. J. Pharmacol. 73: 101-103, 1981.
3. Charo, I. F., Feinmann, R. D., Detwiler, T. C., Smith, J. B., Ingerman, C. M., and Silver, M. J. Prostaglandin endoperoxides and thromboxane A2 an induce platelet aggregation in the absence of secretion. Nature 269: 66-69, 1977.
4. Cowan, D. H. Platelet adherence to collagen: Role of prostaglandin-thromboxane synthesis. Brit. J. Haematol. 49: 425-434, 1981.
5. Dusting, G. J., Chapple, D. J., Hughes, R., Moncada, S., and Vane, J. R. Prostacyclin (PGI2) induces coronary vasodilatation in anaesthetized dogs. Cardiovasc. Res. 12: 720-730, 1978.
6. Dusting, G. J., Moncada, S., and Vane, J. R. Vascular actions of arachidonic acid and its metabolites in perfused mesenteric and femoral beds of the dog. Eur. J. Pharmacol. 49: 65-72, 1978.
7. Ellis, E. F., Nies, A. S., and Oates, J. A. Cerebral arterial smooth muscle contraction by thromboxane A2. Stroke 8: 480-483, 1977.
8. Ellis, E. F., Oelz, O., Roberts, L. J., II, Payne, N. A., Sweetman, B. J., Nies, A. S., and Oates, J. A. Coronary arterial smooth muscle contraction by a substance released from platelets: Evidence that it is thromboxane A2. Science 193: 1135-1137, 1976.
9. Fletcher, J. R., and Ramwell, P. W. Hemodynamic evaluation of prostaglandin D2 in the conscious baboon. Adv. Prostaglandin Thromboxane Res. 7: 723-725, 1980.
10. Hamberg, M., Svensson, J., and Samuelsson, B. Thromboxanes: A new group of biologically active compounds derived from prostaglandin endoperoxides. Proc. Natl. Acad. Sci. USA 72: 2994-2998, 1975.
11. Kot, P. A., Johnson, M., Ramwell, P. W., and Rose, J. C. Effects of ganglionic and β-adrenergic blockade on cardiovascular responses to the bisenoic

prostaglandins and their precursor arachidonic acid. Proc. Soc. Exp. Biol. Med. 149: 953-957, 1975.

12. Rose, J. C., Johnson, M., Ramwell, P. W., and Kot, P. A.   Effects of arachidonic acid on systemic arterial blood pressure, myocardial contractility and platelets in the dog. Proc. Soc. Exp. Biol. Med. 147: 652-655, 1974.

13. Rose, J. C., Kot, P. A., Ramwell, P. W., Doykos, M., and O'Neill, W. P. Cardiovascular responses to the prostaglandin endoperoxide analogs in the dog. Proc. Soc. Exp. Biol. Med. 153: 209-212, 1976.

14. Petakau, A. Radiation carcinogenesis from a membrane perspective. Acta Physiol. Scand. Suppl. 492: 81-90, 1980.

15. Hemler, M. E., Cook, H. W., and Lands, W. E. M. Prostaglandin biosynthesis can be triggered by lipid peroxides. Arch. Biochem. Biophys. 193: 340-345, 1979.

16. Egan, R. W., Paxton, J., and Kuehl, F. A., Jr. Mechanism for irreversible self-destruction of prostaglandin synthetase. J. Biol. Chem. 251: 7329-7335, 1976.

17. Seregi, A., Serfozo, P., and Mergl, Z. Evidence for the localization of hydrogen peroxide-stimulated cyclooxygenase activity in rat brain mitochondria: A possible coupling with monoamine oxidase. J. Neurochem. 40: 407-413, 1983.

18. Taylor, L., Menconi, M. J., and Polgar, P. The participation of hydroperoxides and oxygen radicals in the control of prostaglandin synthesis. J. Biol. Chem. 258: 6855-6857, 1983.

19. Eisen, V., and Walker, D. I. Effect of ionizing radiation on prostaglandin-like activity in tissues. Brit. J. Pharmacol. 57: 527-532, 1976.

20. Pausescu, E. J., Teodosiu, T., and Chirvasie, R. Effects of total exposure to ^{60}Co gamma radiation on cerebral nicotinamide nucleotides and glutathione in dogs. Radiat. Res. 51: 302-309, 1972.

21. Nikandrova, T. I., Zhulanova, Z. I., and Romanstev, E. F. Prostaglandin-synthetase activity in the liver, brain, and testis of gamma irradiated F1 (CBA X C57 B1) mice. Radiobiologiia 21: 265-269, 1981.

22. Romantsev, E. F., Zhulanova, Z. I., and Nikandrova, T. I. Prostaglandin-synthetase activity of brain tissues in experimental animals with radiation sickness. Vestn. Akad. Med. Nauk. SSSR. 9: 86-89, 1982.

23. Trocha, P. J., and Catravas, G. N. Prostaglandins, lysosomes, and radiation injury. Adv. Prostaglandins Thromboxane Res. 7: 851-856, 1980.

24. Maclouf, J., Bernard, P., Rigaud, M., Rocquet, G., and Breton, J. C. Alteration of arachidonic acid metabolism with spleen microsomes of irradiated rats. Biochem. Biophys. Res. Comm. 79: 585-591, 1977.

25. Steel, L. K., and Catravas, G. N. Radiation-induced changes in production of prostaglandins F2α, E, and thromboxane B2 in guinea pig parenchymal lung tissue. Int. J. Radiat. Biol. 42: 517-530, 1982.

26. Steel, L. K., Swedler, I. K., and Catravas, G. N. Effects of ^{60}Co radiation on synthesis of prostaglandins F2α, E, and thromboxane B2 in lung airways of guinea pigs. Radiat. Res. 94: 156-165, 1983.

27. Allen, J. B., Sagerman, R. H., and Stuart, M. J. Irradiation decreases vascular prostacyclin formation with no concomitant effect on platelet thromboxane production. Lancet 2: 1193-1196, 1981.

28. Trocha, P. J., and Catravas, G. N. Prostaglandin levels and lysosomal enzyme activities in irradiated rats. Int. J. Radiat. Biol. 38: 503-511, 1980.

29. Trocha, P. J., and Catravas, G. N. Effect of radioprotectant WR2721 on cyclic nucleotides, prostaglandins, and lysosomes. Radiat. Res. 94: 239-251, 1983.

30. Heinz, T. R., Schneidkraut, M. J., Kot, P. A., Ramwell, P. W., and Rose, J. C. Radiation-induced alterations in cyclooxygenase product synthesis by isolated perfused rat lungs. Prog. Biochem. Pharmacol. 20: 74-83, 1985.

31. Hahn, G. L., Menconi, M. J., Cahill, M., and Polgar, P. The influence of gamma radiation on arachidonic acid release and prostacyclin synthesis. Prostaglandins 25: 783-791, 1983.

32. Walker, D. I., and Eisen, V. Effect of ionizing radiation on 15-hydroxy prostaglandin dehydrogenase (PGDH) activity in tissues. Int. J. Radiat. Biol. 36: 399-407, 1979.

33. Baluda, V. P., Sushkevich, N., Parshkov, E. M., and Lukoyanova, T. I. Influence of gamma rays ^{60}Co and fast neutrons on intravascular platelet aggregation and prostacyclin-like activity of the vascular wall. Buill. Eksp. Biol. Med. 91: 559-562, 1981.

34. Borowska, A., Sierakowski, S., Mackowiak, J., and Wisniewski, K. A prostaglandin-like activity in small intestine and postirradiation gastrointestinal syndrome. Experientia 35: 1368-1370, 1979.

35. Eldor, A., Vlodavsky, I., HyAm, E., Atzman, R., and Fuks, Z. The effect of radiation on prostacyclin (PGI2) production by cultured endothelial cells. Prostaglandins 25: 263-279, 1983.

36. Rubin, D. B., Drab, E. A., Ts'ao, C. H., Gardner, D., and Ward, W. F. Prostacyclin snythesis in irradiated endothelial cells cultured from bovine aorta. J. Appl. Physiol. 58: 592-597, 1985.

37. Sinzinger, H., Cromwell, M., and Firbas, W. Long-lasting depression of rabbit aortic prostacyclin formation by single-dose irradiation. Radiat. Res. 97: 533-536, 1984.

38. Sinzinger, R., Firbas, W., and Cromwell, M. Radiation induced alterations in rabbit aortic prostacyclin formation. Prostaglandins 24: 323-329, 1982.

39. Ts'ao, C. H., Ward, W. F., and Port, C. D. Radiation injury in rat lung. I. Prostacyclin (PGI2) production, arterial perfusion, and ultrastructure. Radiat. Res. 96: 284-293, 1983.

40. Ziboh, V. A., Mallia, C., Morhart, E., and Taylor, J. R. Induced biosynthesis of cutaneous prostaglandins by ionizing irradiation. Proc. Soc. Exp. Biol. Med. 169: 386-391, 1982.

41. Schneidkraut, M. J., Kot, P. A., Ramwell, P. W., and Rose, J. C. Urinary prostacyclin and thromboxane levels after whole body gamma-irradiation in the rat. Adv. Prostaglandin Thromboxane Leukotriene Res. 12: 107-111, 1983.

42. Schneidkraut, M. J., Kot, P. A., Ramwell, P. W., and Rose, J. C. Thromboxane and prostacyclin synthesis following whole body irradiation in rats. J. Appl. Physiol. 57: 833-838, 1984.

43. Schneidkraut, M. J., Kot, P. A., Ramwell, P. W., and Rose, J. C. Regional release of cyclooxygenase products after radiation exposure of the rat. J. Appl. Physiol. 61: 1264-1269, 1986.

44. Donlon, M., Steel, L., Helgeson, E. A., Shipp, A., and Catravas, G. N. Radiation-induced alterations in prostaglandin excretion in the rat. Life Sci. 32: 2631-2639, 1983.

45. Donlon, M., Steel, L., Helgeson, E. A., Wolfe, W. W., and Catravas, G. N. WR2721 inhibition of radiation-induced prostaglandin excretion in rats. Int. J. Radiat. Biol. 47: 205-212, 1985.

46. Pearce, J. W., Sonnenberg, H., Veress, A. T., and Ackermann, U. Evidence for a humoral factor modifying the renal response to blood volume expansion in the rat. Can. J. Physiol. Pharmacol. 47: 377-386, 1969.

47. Sun, F. F., Taylor, B. M., Sutter, D. M., and Weeks, J. R. Metabolism of prostacyclin. III. Urinary metabolite profile of 6-keto PGF1α in the rat. Prostaglandins 17: 753-759, 1979.

48. Forstermann, U., and Neufang, B. The role of the kidney in the metabolism of prostacyclin by the 15-hydroxyprostaglandin dehydrogenase pathway *in vivo*. Biochim. Biophys. Acta 793: 338-345, 1984.

49. Bakhle, Y. S. Inhibition by clinically used dyes of prostaglandin inactivation in rat and human lung. Brit. J. Pharmacol. 72: 715-721, 1981.

50. Hellewell, P. G., and Pearson, J. D. Effect of sulphasalazine on pulmonary inactivation of prostaglandin F2α in the pig. Brit. J. Pharmacol. 76: 319-326, 1982.

51. Hoult, J. R. S., and Robinson, C. Selective inhibition of thromboxane B2 accumulation and metabolism in perfused guinea-pig lung. Brit. J. Pharmacol. 78: 85-88, 1983.

52. Levenson, D. J., Simmons, C. E., and Brenner, B. M. Arachidonic acid metabolism, prostaglandins and the kidney. Am. J. Med. 72: 354-374, 1982.

53. McGuire, J. C., and Sun, F. F. Metabolism of prostacyclin. Oxidation by rhesus monkey lung 15-hydroxyl prostaglandin dehydrogenase. Arch. Biochem. Biophys. 189: 92-96, 1978.

54. Myatt, L., Jogee, M., Lewis, P. J., and Elder, M. G. Measurement of 13,14 dihydro-6,15 dioxo PGF1α by radioimmunoassay: Application to the study of prostacyclin metabolism. Prog. Lipid Res. 20: 807-810, 1981.

55. Robinson, C., and Hoult, J. R. S. Inactivation of prostaglandins in the perfused rat lung. Biochem. Pharmacol. 31: 633-638, 1982.

56. Robinson, C., Hoult, J. R. S., Waddell, K. A., Blair, I. A., and Dollery, C. T. Total profiling by GC/NICIMS of the major cyclooxygenase products from antigen and leukotriene-challenged guinea-pig lung. Biochem. Pharmacol. 33: 395-400, 1984.

57. Wong, P. Y. K., McGiff, J. C., Cagen, L., Malik, K. U., and Sun, F. F. Metabolism of prostacyclin in the rabbit kidney. J. Biol. Chem. 254: 12-14, 1979.

58. Miller, M. J. S., Spokas, E. G., and McGiff, J. C. Metabolism of prostglandin E2 in the isolated perfused kidney of the rabbit. Biochem. Pharmacol. 31: 2955-2960, 1982.

59. Brash, A. R., Jackson, E. K., Saggesse, C. A., Lawson, J. A., Oates, J. A., and Fitzgerald, G. A. Metabolite disposition of prostacyclin in humans. J. Pharmacol. Exp. Therap. 226: 78-87, 1983.

THERMAL INJURY CAUSES STIMULATION OF PHOSPHOLIPASE A₂ ACTIVITY IN MAMMALIAN CELLS

S. K. Calderwood[1], E. K. Farnum[1,2], and M. A. Stevenson[1,3]

[1]Joint Center for Radiation Therapy
Harvard Medical School
Boston, Massachusetts 02115

[2]George Washington University Medical School
Washington, DC

[3]Department of Medicine
Massachussetts General Hospital
Boston, Massachusetts

ABSTRACT

Heat stress (43°C, 45°C) stimulated phopholipase A₂ activity in the two mammalian cell lines investigated: HA-1 hamster fibroblasts and PC-12 rat pheochromocytoma cells. The resultant arachidonic acid (AA) release was of a similar order to that induced by thrombin and bradykinin. Dexamethasone (5 x 10^{-8} M) inhibited by 80%-90% the effect of heat stress on AA release. Heat-induced stimulation of phospholipase A₂ activity may be involved both in the direct cytotoxic effects of heat stress and in the inflammatory responses induced by thermal injury.

INTRODUCTION

Exposure of cells and tissues to high temperatures leads to a large number of cellular and biochemical changes and may result ultimately in cell death and tissue injury (1). It is still not clear how the biological effects of heat stress are mediated, although cell membranes seem to be implicated in many of the responses (1). In the present study we investigated the role of arachidonic acid (AA) release and eicosanoid production in thermal injury.

MATERIALS AND METHODS

Experiments were carried out on HA-1 hamster fibroblast cells and on PC-12 rat pheochromocytoma cells maintained in monolayer culture at 37°C. AA release was measured after preincubation of cells with ³H-AA for 2-24 hr. The isotope appeared to equilibrate with unlabeled AA in the main phospholipid species by

90 min at 37°C in both cell types. AA metabolites were extracted and fractionated by thin layer chromatography as described by Hong and Deykin (2).

In all cases investigated, over 95% of AA was released as unmodified arachidonate (as determined by comparing the rF values of released lipids, fractionated by thin layer chromatography, with those of verified lipid standards). Lysophospholipids were also analyzed as in reference 1.

RESULTS AND DISCUSSIONS

Temperature shock increased the rate of AA release in HA-1 fibroblasts (Figure 1A). Increasing the temperature of cells to 43°C doubled the rate of AA release and at 45°C, AA release was five times that in controls (Figure 1A). Heat (45°C) also induced a major twofold to threefold elevation in the rate of AA release by PC-12 cells (not shown). AA release has been shown to take place either directly through phospholipase A_2 (PLA_2) activity or indirectly by phospholipase C (PLC) cleavage of phospholipids to diacylglycerol (dAG) and secondary release of AA by the activity of diglyceride lipase (3). Two pieces of evidence suggest that heat induces AA release through PLA_2 action. First, α_1 adrenergic agonist phenylephrine, an agent that activates PLC in HA-1 fibroblasts (4), does not increase AA release (Figure 1). This indicates that PLC activation is not necessarily coupled to AA release and suggests, by implication, that heat-induced AA release may occur by a PLC-independent mechanism such as PLA_2 induction. Second, accumulation of dAG in heated cells does not occur until at least 30 min at 45°C (Figure 1B), in contrast to AA release, which is elevated immediately on heating (Figure 1A). Further support for heat-induced AA release through PLA_2 activation is provided by our preliminary findings (not shown) of heat-induced accumulation of lysophospholipids, the products of PLA_2 activity (2).

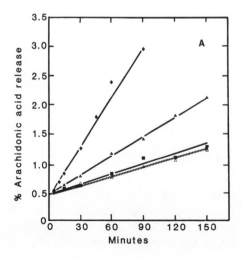

Fig. 1A. Effect of heat and phenylephrine on AA release in HA-1 fibroblasts. AA release at 37°C (■), 43°C (▲), 45°C (♦), and in 10^{-4} M phenylephrine (Δ) is shown. Data are expressed as % of total cell AA released into medium. Datum points are means of 10-12 determinations. When lipid fractions from heated cells were analyzed by TLC, we were able to detect increased levels of major prostaglandin species, PGE_2 and PGF_2, in addition to several as-yet-unidentified AA metabolites. Unmodified AA constituted over 95% of released ^{3}H-AA activity in all samples.

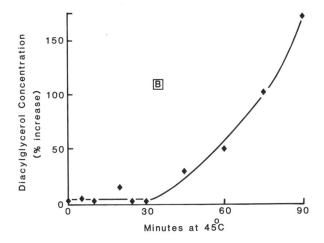

Fig. 1B. Diacylglycerol levels in HA-1 cells heated at 45°C. Means of six measurements are shown.

Stimulation of AA release by heat was of a similar order to the effects of bradykinin and thrombin (Figure 2). It may be significant that these two compounds play a key role in the acute inflammatory response (3,5). Thus, heat might induce secondary inflammatory changes in addition to causing direct injury. Ca^{2+} ionophore A23187 was a more effective inducer of AA release than were the other agents (Figure 2). Release of AA by A23187 treatment was not observed under conditions of low internal Ca^{2+} concentration (6), in the presence of EGTA (not shown), implicating Ca^{2+} in the activity of PLA₂ in HA-1 cells.

The effects of heat and ionophore A23187 on AA release were additive at each concentration tested (Figure 3). This suggests that the PLA₂-stimulating action of

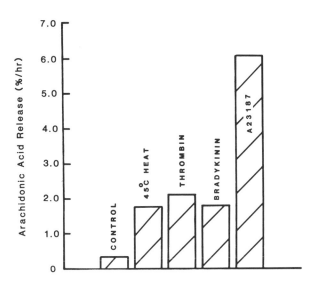

Fig. 2. Effect of heat (45°C), thrombin (10 U/ml), bradykinin (5 x 10^{-5} M), and Ca^{2+} ionophore A23187 (5 x 10^{-5} M) on AA release. Experiments were carried out for 60 min on HA-1 cells in serum-free medium. Means of 8-10 estimations are shown.

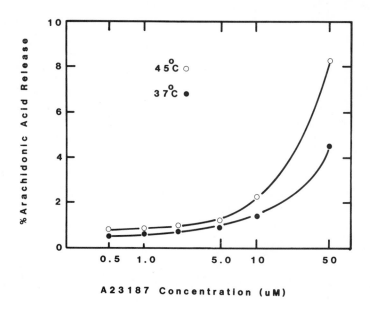

Fig. 3. Effect of 45°C on AA release by A23187 (60 min in 5 x 10^{-5} M ionophore) in HA-1 cells. Mean values from six assays are shown.

heat is not exclusively mediated through the increase in Ca^{2+} (6) that occurs in heated cells. Similarly, the failure of Ca^{2+}-directed agonist phenylephrine to induce AA release seems to preclude an exclusive regulatory role for Ca^{2+} in heat-induced AA release. We have shown also that two napthalene sulphonamide inhibitors of calmodulin action, W7 and W13 (7), fail to inhibit AA release in HA-1 fibroblasts (S. K. Calderwood, in preparation). Thus, heat-induced AA release through activation of Ca^{2+}/calmodulin is not indicated. Ca^{2+}-mediated AA release may occur only at the supraphysiological Ca^{2+} concentrations induced by A23187.

Pretreatment with dexamethasone (Figure 4) or hydrocortisone (not shown) blocked basal AA release in addition to AA release induced by heat (45°C), thrombin, and ionophore A23187. AA release induced by A23187 was least sensitive to inhibition by glucocorticoids (Figure 4). The mechanisms involved in inhibition of heat-induced AA release by glucocorticoids are under investigation.

CONCLUSIONS

Heat stress induces a major increase in PLA_2 activity in the two cell types investigated. The degrees of PLA_2 activation by heat and inflammatory agonists bradykinin and thrombin are similar (Figures 1A and 2). Activation of PLA_2 in heat-stressed cells appears to be independent (Figure 3) of temperature-induced rises in Ca^{2+} levels (6). However, PLA_2 has been shown to be subject to regulation by multiple factors, including Ca^{2+}, H^+, G proteins, and protein kinases (3), and more work is required to identify the mechanism of heat-induced AA release.

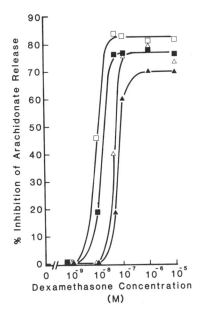

Fig. 4. Inhibition of AA release by dexamethasone. HA-1 cells were pretreated with dexamethasone for 24 hr. AA release was measured over 60 min in controls (■), at 45°C (□), in thrombin (10 U/ml; Δ) and in A23187 (5 x 10^{-5}; ▲). Data are means of five determinations.

The data suggest that stimulation of PLA₂ may be involved in cell and tissue responses to heat stress. Glucocorticoids inhibit both PLA₂ activity (Figure 4) and cell killing by heat (8). AA and its metabolites also may be involved in vascular and inflammatory changes in heated tissues as observed in hyperthermic treatment of cancer (9) and burn trauma (6).

REFERENCES

1. Hahn, G. M. "Hyperthermia and Cancer." Plenum Press, New York, 1982.
2. Hong, S. L., and Deykin, D. The activation of phosphatidyl inositol hydrolyzing phospholipase A₂ during prostaglandin synthesis in transformed mouse Balb-C 3T3 cells. J. Biol. Chem. 256: 5215-5219, 1981.
3. Johnson, M., Carey, F., and McMillen, R. M. Alternative pathways of arachidonate metabolism: Prostaglandins, thromboxanes, and leukatrienes. Essays Biochem. 15: 40-139, 1983.
4. Calderwood, S. K., Stevenson, M. A., and Hahn, G. M. Heat stress stimulates inositol trisphosphate release and phosphorylation of phosphoinositides in CHO and Balb-C 3T3 cells. J. Cell Physiol. (in press).
5. Movat, H. Z., ed. The acute inflammatory reaction. In: "Inflammation, Immunity, and Hypersensitivity." Harper and Row, New York, 1979, pp. 1-162.
6. Stevenson, M. A., Calderwood, S. K., and Hahn, G. M. Rapid increases in inositol trisphosphate and intracellular Ca^{2+} after heat shock. Biochem. Biophys. Res. Commun. 137: 826-833, 1986.

7. Hidaka, H., Sasaki, Y., Tamaka, T., Toyoshi, E., Fujii, Y., and Nagato, T. N-(6-amino hexyl)-5-chloro-1-napthalenesulphonamide, a calmodulin antagonist, inhibits cell proliferation. Proc. Natl. Acad. Sci. USA 78: 4354-4357, 1981.

8. Fisher, G. A., Anderson, R. L., and Hahn, G. M. Glucocorticoid-induced heat resistance in mammalian cells. J. Cell Physiol. 128: 127-132, 1986.

9. Dickson, J. A., and Calderwood, S. K. Thermosensitivity of neoplastic tissues *in vivo*. In: "Hyperthermia in Cancer Therapy." F. K. Storm, ed. G. K. Hall, Boston, 1983, pp. 63-140.

EFFECTS OF NEUTRON IRRADIATION ON PGE_2 AND TxB_2 LEVELS IN BIOLOGICAL FLUIDS: MODIFICATION BY WR-2721

L. K. Steel[1] and G. D. Ledney[2]

Departments of [1]Radiation Biochemistry and [2]Experimental Hematology
Armed Forces Radiobiology Research Institute
Bethesda, Maryland 20814-5145

ABSTRACT

Whole-body fission neutron irradiation of mice results in changes in plasma and urinary prostaglandin levels. No relationship was found between urinary TxB_2 excretion and the neutron dose. Elevated plasma PGE_2 at 2-4 days postexposure coincides with increased PGE_2 in the urine. A second increase in urinary PGE_2 occurred 6-10 days postirradiation, and was dose dependent up to 3.85 Gy. Plasma PGE_2 levels at this time were normal. WR-2721 administration markedly increased (24 hr) and then reduced (2-4 days) PGE_2 excretion in nonirradiated animals. The urinary PGE_2 of WR-2721-pretreated, irradiated mice paralleled those of drug-treated, nonirradiated animals for the first 5 days. However, WR-2721 injection did not modify the elevation of PGE_2 occurring 6-10 days postexposure. Urinary TxB_2 from nonirradiated groups was not altered by WR-2721, but the combination of drug and radiation resulted in increased levels 1-7 days postexposure. These results further implicate prostaglandins in the biological response to radiation exposure and suggest another mechanism of WR-2721 radioprotection.

INTRODUCTION

A role for prostaglandins (PG) as modulators of radiation-induced tissue injury is becoming increasingly evident. Antagonistic as well as cytoprotective effects of PG have been demonstrated in radiation injuries (reference 1 and references therein). To date most reports have focused on alterations in arachidonic acid metabolites as a consequence of low-LET (linear energy transfer, i.e., X or gamma) radiation exposure (2). The higher density of ionization produced by neutrons, compared to that of gamma rays, may intensify the interaction between free radicals and unsaturated fatty acids in membrane lipids. This could result in greater membrane damage from lipid peroxidation than observed from gamma or X rays. The effect of fission neutrons on the synthesis of these compounds has recently been addressed (1).

The present study reviews the *in vivo* effect of whole-body exposure to fission neutrons on plasma PGE_2, urinary TxB_2, and urinary PGE_2 in mice. In addition, the phosphorothioate compound WR-2721 was examined for its radioprotective efficacy against fission neutron irradiation, and its effect on urinary PG synthesis/ excretion. The results further implicate PG's in the biological response to ionizing radiation, and suggest a role for PG's in the mechanism of radioprotection by WR-2721.

METHODS

The radioprotective compound WR-2721 was provided by COL David E. Davidson, Jr., Division of Experimental Therapeutics, Walter Reed Army Institute of Research. The drug was dissolved in normal saline (pH 7.6) immediately before use and administered intraperitoneally (i.p.) in a single 0.5-ml dose of 453 mg/ kg of body weight. Control mice received saline only.

Animal Irradiation

Neutron irradiations were conducted using the AFRRI TRIGA Mark-F reactor as previously described (1,3). A minimum of 27 mice were used at each dose (2.0-5.0 Gy). Whole-body radiation was delivered at a midline tissue (MLT) dose rate of 0.4 Gy/min. The gamma component of the MLT dose was approximately 4.5% of the total dose (neutron-to-gamma ratio of approximately 21). The average neutron and gamma energies are estimated to be 0.68 MeV and 1.8 MeV, respectively (4).

Urine and Plasma

B6D2F1 (C57BL♀ x DBA/2J♂) female mice (Jackson Labs, Bar Harbor, ME) were housed in lucite metabolic cages, three animals per cage, for 7 days before initiating urine collection. Urine was collected from each cage at 24-hr intervals beginning 7 days before exposure and continuing to 30 days postirradiation. Urine samples were quantitated, aliquoted, organically extracted, and chromatographed on individual silica columns as described (1). Immunoreactive PGE_2 and TxB_2 in the column extracts were measured by radioimmunoassay as described (5,6).

Urine was collected daily for 7 days prior to drug injection. On the seventh day, mice were administered either WR-2721 or saline, and urine was collected daily for 7 days postinjection. In radiation experiments, WR-2721 or saline was administered 30 min before whole-body irradiation (4 Gy). Following exposure, animals were returned to their metabolic cages, and urine was collected up to 30 days postexposure.

Blood was obtained via cardiac puncture on anesthetized animals and plasma isolated by centrifugation at 4°C (7). PGE_2 was extracted from individual mouse plasma samples using solid phase C-18 extraction columns (8), and quantitated by radioimmunoassay (1,8). Radioactivity was measured in an LKB gamma counter, and PGE_2 concentrations were determined by extrapolation from standard curves (1,5,6,8).

Statistical Analyses

Student's t test for paired samples (9) was applied to determine significance between postirradiation and/or postinjection values and baseline (postirradiation and/or preinjection) values. Thirty-day and 100-day survival results were analyzed by probit analysis (10) to determine the radiation $LD_{50/30}$ or $LD_{50/100}$. Survival data were compared by the method of Finney (10). Correlations across postirradiation time, within each radiation dose [Spearman's rank correlation (11)] or across radiation doses (12), were examined for each sampling day postirradiation. Other statistical analyses and comparison tests were performed as described previously (1).

RESULTS

Radiation-Induced Changes in Plasma and Urinary PG

Urine excretion rate during the 7 days prior to irradiation averaged 1.17 ± 0.05 ml/24 hr/mouse (M $\pm$ SE; range 0.8 to 1.35 ml). Urinary PGE_2 and TxB_2 averaged 311.79 ± 25.65 and 472.86 ± 29.70 pg/24 hr/mouse, respectively (range, 211.12-416.67 pg PGE_2 and 390.67-643.17 pg TxB_2). Preirradiation excretion rates were relatively constant within each radiation group (see legends to Figures 1 and 2). No correlation was found between PGE_2, TxB_2, and urine volume excretion rates (p >0.05).

Total-body exposure to fission neutrons resulted in biphasic elevations in PGE_2 excretion (Figure 1). Urinary PGE_2 levels at days 2-3 were 1.5-fold to threefold

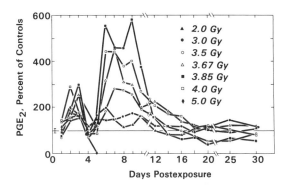

Fig. 1. Urinary excretion of PGE_2 from mice exposed to fission neutrons. Data are % of preirradiation baseline excretion rates (pg/24 hr/mouse: 2.0 Gy, 426.7 ± 25.6; 3 Gy, 253.0 ± 22.8; 3.5 Gy, 389.9 ± 31.2; 3.67 Gy, 277.0 ± 16.6; 3.85 Gy, 334.5 ± 33.5; 4.0 Gy, 211.1 ± 23.2; 5.0 Gy, 312.1 ± 15.6). $\mathbf{I}$, overall preirradiation excretion at day 0.

higher than controls, but this elevation was not dose dependent. A second increase in urinary PGE_2 occurred days 6-10 postexposure. Increases in radiation dose up to 3.85 Gy were associated with increasingly elevated PGE_2 excretion. At day 10 postirradiation, urinary PGE_2 was still elevated, but there was no significant difference in excretions between radiation doses. Days 12-16 were characterized by declining urinary PGE_2; by 18 days postexposure, most groups were within control levels.

Plasma levels of PGE_2 were examined on days 1-5, 7, 9 and 11 following exposure to 3.0-Gy fission neutrons (Figure 2). Nearly a threefold increase in PGE_2 concentrations occurred 3-4 days after exposure, declining to control levels on the 5th day. Thereafter, the levels of PGE_2 in plasma from irradiated mice were similar to those of controls.

The levels of thromboxane B_2 positively correlated with urine excretion only in those animals exposed to 2.0 Gy (p $<$0.01). Urinary TxB_2 concentrations were unchanged or reduced (3.67 or 3.85 Gy) during the first 3 days postexposure (Figure 3). Groups that received 3.67-3.86 Gy had reduced TxB_2 levels between days 4 and 8, while 4.0-Gy-exposed animals exhibited elevated TxB_2 excretion. The second week postexposure, urinary TxB_2 was transiently elevated in animals exposed to 2.0 Gy, while higher doses resulted in unchanged or transiently reduced levels. On days 14-16 postexposure, patterns of excretion were not statistically different across doses. Variable transient increases in urinary TxB_2 occurred 18-30 days postexposure in mice exposed to $\leq$3.0 Gy. Those that received $\geq$ 3.5 Gy had unchanged or reduced (3.85 Gy) urinary TxB_2.

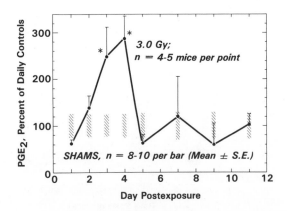

Fig. 2. PGE_2 in mouse plasma after whole-body fission neutron irradiation (3.0 Gy). Data are presented as percentage of control (nonirradiated) levels (▨). Mean control plasma levels were 30 $\pm$ 5 pg PGE_2/ml.

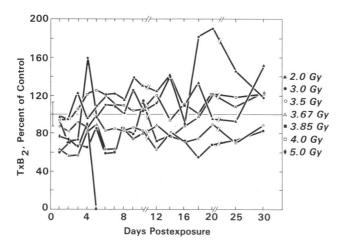

Fig. 3. Urinary excretion of TxB_2 from mice exposed to fission neutrons. Data are presented as percentage of preirradiation baseline excretion rates (pg/24 hr/mouse: 2.0 Gy, 390.7 ± 27.4; 3.0 Gy, 422.4 ± 54.9; 3.5 Gy, 463.0 ± 32.4; 3.67 Gy, 643.2 ± 45.0; 3.85 Gy, 571.8 ± 45.7; 4.0 Gy, 523.9 ± 21.0; 5.0 Gy, 399.9 ± 20.0). I, overall preirradiation excretion illustrated at day 0.

WR-2721 Administration

Under the conditions of these studies, the $LD_{50/30}$ value for female mice exposed to fission neutrons is 3.97 Gy (95% C.I., 3.92, 4.02 Gy) (1). Administration of WR-2721 (453 mg/kg, i.p.) 30 min prior to irradiation extends the $LD_{50/30}$ to 4.87 Gy (4.50, 5.83 Gy), with a DMF of 1.23 (1.17, 1.29) (39).

Administration of WR-2721 (453 mg/kg, i.p.) to nonirradiated mice resulted in nearly a threefold increase in urinary PGE_2 excretion during the first 24 hr (Figure 4). On the second day postinjection, PGE_2 levels were significantly reduced; PGE_2 excretion was 38 ± 5% of control (saline-treated) values. WR-2721-treated animals also exhibited reduced urinary PGE_2 concentrations 3 and 4 days postadministration (61% and 65% of controls, respectively), which returned to control levels by the 5th day (Figure 4). Urine volume and urinary TxB_2 excretion were not altered by WR-2721 administration 1-7 days postinjection, compared to saline-treated controls (data not shown). In the present studies, no toxic deaths occurred following a single i.p. injection of 453 mg/kg WR-2721.

Urinary excretion of PGE_2 from mice administered WR-2721 30 min prior to whole-body fission neutron irradiation (4.0 Gy) is shown in Figure 5. Pretreated, irradiated animals exhibited increased PGE_2 excretion at 24 hr and reduced excretion 48 hr postirradiation to the same extent as nonirradiated mice injected with WR-2721 (Figure 4). WR-2721 eliminated the expected elevation in urinary PGE_2 between 2 and 3 days postirradiation (Figures 1,5), but not the second PGE_2 elevation at days 6-10 postexposure (Figure 5).

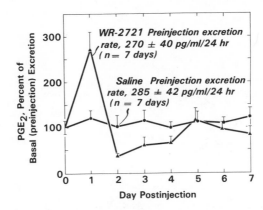

Fig. 4. Urinary PGE_2 from mice after i.p. injection of saline or WR-2721 (453 mg/kg). Data are presented as percent of preinjection excretion rates. ▲—▲, WR-2721 (preinjection excretion, 270 ± 40 pg PGE_2/ml/24 hr). ●—●, saline (preinjection excretion, 285 ± 42 pg PGE_2/ml/24 hr).

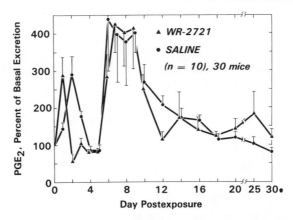

Fig. 5. Urinary PGE_2 from mice after i.p. injection of WR-2721 (453 mg/kg) and whole-body fission neutron irradiation (4 Gy). Data are presented as a percentage of preirradiation excretion rates. ▲—▲, WR-2721; ●—●, control.

Although WR-2721 did not modify urinary TxB_2 levels in nonirradiated animals, the combination of drug and fission neutron irradiation (4 Gy) resulted in increased urinary TxB_2 1-7 days postexposure (Figure 6). Mice that received 4 Gy but no drug pretreatment exhibited a modest elevation (20%-26%) in TxB_2 excretion on days 4-6 only. On these same days, urinary TxB_2 levels from WR-2721-injected, irradiated animals were 80%-121% higher than control (preirradiation) excretion. TxB_2 levels were within normal values in both WR-2721-treated and nontreated groups 8-30 days postexposure.

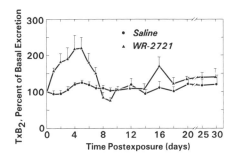

Fig. 6. Urinary TxB_2 from mice after i.p. injection of WR-2721 (453 mg/kg) and whole-body fission neutron irradiation (4 Gy). Data are presented as a percentage of preirradiation excretion rates. ▲—▲, WR-2721; ●—●, control.

DISCUSSION

Biphasic increases in mouse urinary excretion of PGE_2 occurred following whole-body exposure to fission neutrons (1). The initial transient elevation 2-3 days postexposure was not correlated with radiation dose. The second elevation occurring 6-10 days postexposure was dose dependent up to 3.85 Gy (Figure 7). These results suggest that urine PGE_2 may be a useful indicator of radiation exposure.

PGE_2 in urine is normally a reflection of renal production (13). Increases in urine PGE_2 levels may indicate a direct effect of radiation on the kidney. The fact that plasma PGE_2 was normal 6-10 days postexposure suggests that the increase in urine is primarily of renal origin. However, radiation exposure alters PG levels *in vivo* and *in vitro* in a variety of tissues and fluids (1,2,5,6,14,15-23), and the elevated plasma PGE_2 2-4 days postexposure (3 Gy) doses coincide with the initial increase observed in the urine. Thus, the extra-renal contribution of elevated circulating PGE_2 to urinary levels is a distinct possibility. Tanner and co-workers (14) have reported an increase in plasma PG-like material in patients treated with synchronous radiotherapy and chemotherapy for head and neck cancer. Conversely, Lifshitz et al. (24) found no elevation in plasma PG's and no correlation between radiation dosage and PG levels in patients undergoing pelvic irradiation for gynecologic cancer. Neither of these studies examined plasma PG concentrations within the first 7 days following radiotherapy, and urinary PG's were not measured. Several investigations (15-17) have reported increases in PG activity in rat and mouse tissues after exposure to ionizing radiation. These increases were found to occur 3-4 and 7-8 days postexposure, markedly similar to the biphasic elevations seen in urinary PGE_2 excretion. Conversely, Eisen and Walker examined PG-like activity (bioassay, reference 25) in whole blood and kidneys of mice 1-7 days after irradiation with 7 Gy X rays, and found no significant changes (15).

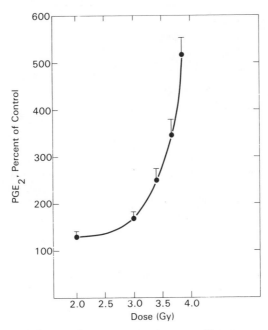

Fig. 7. Urinary excretion of PGE$_2$ 6-9 days postexposure. Data are presented as means ± S.E.M. (% of controls) of mean urinary levels of days 6, 7, 8, and 9 postexposure, for each radiation dose.

Sublethal neutron irradiation (2-3 Gy) resulted in unchanged or transiently elevated TxB$_2$ excretion, while intermediate doses (3.5-3.85 Gy) resulted in transient reductions. We and others have found dose-related increases in guinea pig lung tissue (5) and rat urine (18,19) TxB$_2$ concentrations after whole-body gamma irradiation. These differences may be due to species and radiation quality.

Intraperitoneal administration of WR-2721 (453 mg/kg) to nonirradiated mice results in nearly a threefold increase in PGE$_2$ excretion the first 24 hr, reduced excretion 2-4 days postinjection, and normal levels on the 5th day. Urinary TxB$_2$ levels in these animals were not altered by WR-2721 treatment. Trocha and Catravas (20) have reported no change in PGE$_2$ and PGF$_2$ content in rat spleen or liver tissues 1 hr to 3 days after WR-2721 administration. Similarly, Donlon et al. (21) did not observe a change in PG excretion rates in WR-2721-treated rats. However, Pryanishnikova and co-workers (22) demonstrated increased PGE$_2$ and PGF$_2$ synthesis in mouse cerebral and testicular cystolic fractions and hepatic microsomes incubated *in vitro* with WR-2721. Furthermore, these increases could be inhibited by the addition of indomethacin.

Biodistribution studies have shown that WR-2721 has a short half-life in the body, being rapidly distributed to the kidneys and excreted in the urine (26). High levels of the compound (or its metabolites) and long retention times (>2 hr) are seen in kidney tissues (27). The most frequent adverse effects of WR-2721 include nausea, vomiting, hypotension, and hypocalcemia (reviewed in reference 28). These same symptoms also occur with PG administration. The observed increase in PGE_2 synthesis/excretion following WR-2721 administration suggests that PG's may contribute to these side effects. Further work is required to determine whether PGE_2 acts as a modulator or a necessary component. One logical approach is premedication with a PG inhibitor (i.e., indomethacin) before WR-2721 injection.

The combination of WR-2721 (453 mg/kg, i.p.) and radiation (4 Gy) resulted in 100% animal survival. Urinary PGE_2 levels in WR-2721-pretreated and irradiated mice paralleled those of drug-treated nonirradiated animals for 5 days postinjection/irradiation. The radioprotectant did not modify the elevation of PGE_2 occurring 6-10 days postexposure. In comparison, neither drug nor fission neutron exposure (4 Gy) markedly altered TxB_2 excretion, but in combination, TxB_2 synthesis/excretion was elevated for 7 days. These results further implicate PG's in the biological response to ionizing radiation and suggest another mechanism of phosphorothioate radioprotection. Both natural and synthetic PGE_2 (16,16-dimethyl PGE_2) have been shown to provide radioprotection when given before irradiation, but are ineffective when given after irradiation (29,30). Simultaneous administration of synthetic PGE_2 (5 μg, subcutaneously) and WR-2721 (200 mg/kg, i.p.) to mice 15 min prior to exposure to gamma radiation from a ^{60}Co source additively increases the DMF (T. Walden, personal communication). Although acetylsalicylate has been shown to effectively control many gastrointestinal symptoms occurring in patients postradiotherapy (31), neither dexamethasone nor indomethacin premedication (400 μg or 100 μg, intramuscularly, respectively, every 12 hr for 3 days prior to 3.9-5.0 Gy fission neutron irradiation) modified animal weight loss or mortality (Steel, unpublished observations). These findings suggest that elevated PGE_2 levels at the time of radiation exposure (either by endogenous stimulation with WR-2721 or by exogenous addition) contribute to radioresistance. The potential mechanism(s) of PG-induced radioprotection (i.e., through modulation of intermediary substances such as cAMP, erythropoietin, glutathione, superoxide dismutase, nonprotein sulfhydryls, or other physiological changes) is currently under examination.

Renal production of Tx is normally low, but many pathological and inflammatory conditions are associated with increased biosynthesis and excretion (reviewed in reference 32). Free radicals formed by radiation exposure (33) may potentiate changes in arachidonate metabolism. The extent of radical and peroxide formation may determine the magnitude and direction of arachidonate metabolism. Lipid peroxides and H_2O_2 activate cyclooxygenase in a dose-dependent manner (34-37); however, high concentrations of hydroperoxide or free radicals inactivate cyclooxygenase (35,38). The addition of free radical scavengers has been shown to prevent this inactivation (37). Elevated urinary TxB_2 following WR-2721 administration and fission neutron irradiation may reflect the ability of the radioprotectant to effectively trap free radicals and thereby inhibit cyclooxygenase inactivation. Enhanced Tx synthesis/release may also be the result of direct or indirect stimulation of one or more of the enzyme systems participating in endoperoxide intermediate conversions.

ACKNOWLEDGMENTS

Supported by the Armed Forces Radiobiology Research Institute, Defense Nuclear Agency, under Research Work Units MJ 00064 and B2152. The views presented in this paper are those of the authors; no endorsement by the Defense Nuclear Agency has been given or should be inferred. Research was conducted according to the principles enunciated in the "Guide for the Care and Use of Laboratory Animals," prepared by the Institute of Laboratory Animal Resources, National Research Council. The authors gratefully acknowledge the secretarial assistance of Mrs. M. J. Waldbillig; the AFRRI Reactor Facilities staff for the performance of animal irradiations; Cpt M. Dooley, Radiological Physics Division, for performance of dosimetry techniques; and the technical expertise of Mr. W. Wolfe, Ms. M. Rafferty, and Ms. D. A. Kennedy. The authors also thank LCDR H. N. Hughes and T. L. Walden, Jr., for their critical evaluation and advice in the preparation of this manuscript.

REFERENCES

1. Steel, L. K., Rafferty, M. A., Wolfe, W. W., Egan, J. E., Kennedy, D. A., Catravas, G. N., Jackson, W. E., III, and Dooley, M. A. Urinary excretion of cyclic nucleotides, creatinine, prostaglandin E_2 and thromboxane B_2 from mice exposed to whole-body irradiation from an enhanced neutron field. Int. J. Radiat. Biol. 50: 695-715, 1986.

2. Polgar, P., Hahn, G., and Taylor, L. Radiation effects on eicosanoid formation. In: "Biochemistry of Arachidonic Acid Metabolism." W. E. M. Lands, ed. Martin Nijhoff Publishing, Boston, 1985, pp. 161-173.

3. Stewart, D. A., Ledney, G. D., Baker, W. H., Daxon, E. G., and Sheehy, P. A. Bone marrow transplantation of mice exposed to a modified fission neutron (N/G-30:1) field. Radiat. Res. 92: 268-279, 1982.

4. Zeman, G. H., and Ferlic, K. P. Paired ion chamber constants for fission gamma-neutron fields. Technical Report TR84-8. Armed Forces Radiobiology Research Institute, Bethesda, Maryland, 1984.

5. Steel, L. K., and Catravas, G. N. Radiation-induced changes in production of prostaglandins $F_{2\alpha}$, E, and TxB_2 in guinea pig parenchymal lung tissues. Int. J. Radiat. Biol. 42: 517-530, 1982.

6. Steel, L. K., Sweedler, I. K., and Catravas, G. N. Effects of ^{60}Co radiation on synthesis of prostaglandins $F_{2\alpha}$, E and thromboxane B_2 in lung airways of guinea pigs. Radiat. Res. 94: 156-165, 1983.

7. Ledney, G. D., Steel, L. K., Exum, E. D., and Gelston, H. M. Pathophysiologic responses in mice after neutron irradiation combined with either wound or burn trauma. In: "The Pathophysiology of Combined Injury and Trauma: Management of Infectious Complications in Mass Casualty Situations." D. Gruber, R. I. Walker, T. J. MacVittie, and J. J. Conklin, eds. Academic Press, Inc., Orlando, Florida, 1987.

8. El-Bayer, H., Steel, L., Montcalm, E., Danquechin-Dorval, E., Dubois, A., and Shea-Donohue, T. The role of endogenous prostaglandins in the regulation of gastric secretion in Rhesus monkeys. Prostaglandins 30: 401-419, 1985.

9. Snedecor, G. W., and Cochran, W. G. The comparison of two samples. In: "Statistical Methods." Iowa State University Press, Ames, Iowa, 1980, pp. 85-86.

10. Finney, D. J. In: "Statistical Method in Biological Assay." Third Edition. Macmillan Publishing Co., Inc., New York, 1978, pp. 349-390.

11. Conover, W. J. In: "Practical Nonparametric Statistics." Second Edition. John Wiley and Sons, Inc., New York, 1980, pp. 144-151.

12. Snedecor, G. W., and Cochran, W. G. Analysis of variance. In: "Statistical Methods." Iowa State University Press, Ames, Iowa, 1980, pp. 235-237.

13. Frölich, J. C., Wilson, T. W., Sweetman, B., Smigel, M., Nies, A. S., Carr, K., Watson, J. T., and Oates, J. A. Urinary prostaglandins. Identification and origin. J. Clin. Invest. 55: 763-770, 1975.

14. Tanner, N. S. B., Stamford, I. F., and Bennett, A. Plasma prostaglandins in mucositis due to radiotherapy and chemotherapy for head and neck cancer. Brit. J. Cancer 43: 767-771, 1981.

15. Eisen, V., and Walker, D. I. Effect of ionizing radiation on prostaglandin-like activity in tissues. Br. J. Pharmacol. 57: 527-532, 1976.

16. Trocha, P. J., and Catravas, G. N. Prostaglandin levels and lysosomal activities in irradiated rats. Int. J. Radiat. Biol. 38: 503-522, 1980.

17. Päusescu, E., Chirvasie, R., Teodosiu, T., and Päun, C. Effect of ^{60}Co γ-radiation on the hepatic and cerebral levels of some prostaglandins. Radiat. Res. 65: 163-171, 1976.

18. Donlon, M., Steel, L., Helgeson, E. A., Shipp, A., and Catravas, G. N. Radiation-induced alterations in prostaglandin excretion in the rat. Life Sci. 32: 2631-2639, 1983.

19. Schneidkraut, M. J., Kot, P. A., Ramwell, P. W., and Rose, J. C. Urinary prostacyclin and thromboxane levels after whole-body gamma irradiation in the rat. Adv. in Prostaglandin, Thromboxane, and Leukotriene Res. 12: 107-112, 1983.

20. Trocha, P. J., and Catravas, G. N. Effects of WR-2721 on cyclic nucleotides, prostaglandins and liposomes. Radiat. Res. 94: 239-251, 1983.

21. Donlon, M. A., Steel, L. K., Helgeson, E. A., Wolfe, W. W., and Catravas, G. N. WR-2721 inhibition of radiation-induced prostaglandin excretion in rats. Int. J. Radiat. Res. 47: 205-212, 1985.

22. Pryanishnikova, E. N., Zhulanova, Z. I., and Romantsev, E. F. Effect of S-[N-(3)aminopropyl)-2-aminoethyl]thiophosphoric acid on prostaglandin synthetase activity in mouse tissues (in Russian). Vopr. Med. Khim. 26: 685-688, 1980.

23. Dubois, L. A., Dorvall, E. D., Steel, L., O'Connell, L., and Durakovic, A. Do prostaglandins mediate radiation-induced suppression of gastric acid output? 33rd Annual Meeting of Radiation Research Society, Los Angeles, CA. May 5-9, 1985. Abstr. Jf-4, p. 133.

24. Lifshitz, S., Savage, J. E., Taylor, K. A., Tewfik, H. H., and VanOrden, D. E. Plasma prostaglandin levels in radiation-induced enteritis. Int. J. Radiat. Oncol. Biol. Phys. 8: 275-277, 1982.

25. Jaffe, B. M., Behrman, H. R., and Parker, C. W. Radioimmunoassay measurement of Prostaglandins E, A and F in human plasma. J. Clin. Invest. 52: 398-405, 1972.

26. Utley, J. F., Marlowe, C., and Waddell, W. J. Distribution of ^{35}S-labeled WR-2721 in normal and malignant tissues of the mouse. Radiat. Res. 68: 284-291, 1976.

27. Rasey, J. S., Nelson, N. J., Mahler, P., Anderson, K., Krohn, K. A., and Menard, T. Radioprotection of normal tissues against gamma rays and cyclotron neutrons with WR-2721: LD_{50} studies and ^{35}S-WR-2721 biodistribution. Radiat. Res. 68: 284-291, 1976.

28. Kligerman, M. M., Glover, D. J., Turrisi, A. T., Norfleet, A. L., Yuhas, J. M., Coia, L. R., Simone, C., Glick, J. H., and Goodman, R. L. Toxicity of WR-2721 administered in single and multiple doses. Int. J. Radiat. Oncol. Biol. Phys. 10: 1733-1776, 1984.

29. Hanson, W. R., and Thomas, C. 16,16-dimethyl prostaglandin E_2 increases survival of murine intestinal stem cells when given before photon radiation. Radiat. Res. 96: 393-398, 1983.

30. Walden, T. L., Snyder, S. L., Patchen, M. L., and Steel, L. K. Enhanced LD 50/30 survival in mice by 16,16 dimethyl prostaglandin E_2. 34th Annual Meeting of Radiation Research Society, Las Vegas, Nevada, April 12-17, 1986. Abstract Cl-11, p. 41.

31. Mennie, A. T., Dalley, V. M., Dinneen, L. C., and Collier, H. O. J. Treatment of radiation-induced gastrointestinal distress with acetylsalicylate. Lancet 2: 942-943, 1975.

32. Granström, E., Diczfalusy, U., and Hamberg, M. The thromboxanes. In: "Prostaglandins and Related Substances." C. Pace-Asciak and E. Granström, eds. Elsevier Science Publishers, Amsterdam, The Netherlands, 1983, pp. 45-93.

33. Granger, D. N., and Parks, D. A. Role of oxygen radicals in the pathogenesis of intestinal ischemia. Physiologist 26: 159-164, 1983.

34. Taylor, L., and Polgar, P. Stimulation of prostaglandin synthesis by ascorbic acid via hydrogen peroxide formation. Prostaglandins 19: 693-700, 1980.

35. Taylor, L., Menconi, M. J., and Polgar, P. The participation of hydroperoxides and oxygen radicals in the control of prostaglandin synthesis. J. Biol. Chem. 258: 6855-6857, 1983.

36. Seregi, A., Serfozo, P., and Mergl, Z. Evidence for the localization of hydrogen peroxide-stimulated cyclooxygenase activity in rat brain mitochondria: A possible coupling with monoamine oxidase. J. Neurochem. 40: 407-413, 1983.

37. Hemler, M. E., Cook, H. W., and Lands, W. E. M. Prostaglandin biosynthesis can be triggered by lipid peroxides. Arch. Biochem. Biophys. 193: 340-345, 1979.

38. Egan, R. W., Paxton, J., and Kuehl, F. A., Jr. Mechanism for irreversible self-destruction of prostaglandin synthetase. J. Biol. Chem. 251: 7329-7335, 1976.

39. Steel, L. K., Jacobs, A. J., Giambarresi, L. I., and Jackson, W. E., III. Protection of mice against fission neutron irradiation by WR-2721 or WR-151327. Radiat. Res. 109: 469-478, 1987.

THROMBOXANE AND PROSTACYCLIN PRODUCTION BY IRRADIATED AND PERFUSED RAT KIDNEY

Z. Weshler[1], A. Raz[4], E. Rosenmann[2], S. Biran[1], Z. Fuks[1], and A. Eldor[3]

[1]Department of Radiation and Clinical Oncology
[2]Department of Pathology
[3]Department of Hematology
Hadassah University Hospital
Jerusalem, Israel

[4]Department of Biochemistry
Tel-Aviv University
Tel-Aviv, Israel

ABSTRACT

The effect of ionizing radiation on renal prostaglandin synthesis was investigated in rats subjected to a single radiation fraction of 15 Gy administered to the left kidney. Seven days later, irradiated kidneys and control kidneys from sham-irradiated animals were excised and perfused with a buffer to allow the evaluation of prostaglandin synthesis by the tubular and the capillary systems. Effluent fluids collected from the renal vein and ureter were assayed by radioimmunoassay for thromboxane A_2 (TXB_2) and prostacyclin (6-keto-$PGF_{1\alpha}$). Angiotensin II stimulation was used to determine the capacity of the kidneys to synthesize the above prostaglandins. Morphological changes observed 1 week following irradiation were minimal and consisted of interstitial edema; however, changes in prostaglandin synthesis were profound. A significant increase in the basal and stimulated releases of thromboxane A_2 was observed in the ureteral and venous effluents of irradiated kidneys whereas the basal and stimulated releases of prostacyclin were increased only in the ureteral effluents. These changes may be involved in the pathogenesis of the postradiation capillary and tubular injuries.

INTRODUCTION

The effects of ionizing radiation on the synthesis of prostaglandins by various organs and tissues have recently attracted much attention. Tissue injury is usually associated with the release of prostaglandins, which are measurable in body fluids or secretions, in perfusates of isolated organs, or in a growth medium of cultured cells (1-4).

The kidney is a relatively radiosensitive organ, a fact that limits the therapeutic radiation doses given to treat abdominal tumors (6,7). The role of prostaglandins in normal kidney function as well as under pathological conditions including the effects of ionizing radiation have been investigated (3-5). Exposure to whole-body radiation resulted in rapid changes in the urinary excretion of thromboxane A2 (TXA$_2$) (3,4), while no changes were observed in prostacyclin (PGE$_2$) production (3).

The present study was undertaken to investigate the effects of ionizing radiation on the basal release and stimulated production of these prostaglandins in an isolated and perfused rat kidney. This model allows the evaluation of early postradiation effects on prostaglandin production by the renal capillary system (assayed in fluids collected from the venous effluent) and by the renal tubular system (assayed in fluids collected from the ureteral effluent).

MATERIALS AND METHODS

Twelve-week-old male (Wistar-Sabra) rats weighing 250-300 g were anesthetized by intraperitoneal injection of 30 mg/kg pentobarbital sodium. Each animal was placed in the supine position and the left kidney was identified by palpation and simulation. The surrounding tissues were shielded by 7 mm lead, and the kidney was irradiated with an Orthovoltage machine (127 kV, 20 mA, HVL 0.2 mm Cu, target distance 50 cm) with a single fraction of 15 Gy. Control rats were also anesthetized, fixed in a supine position, and not irradiated. One week later, the irradiated and sham-irradiated rats were again anesthetized. The left kidney was exposed and the left renal artery, vein, and ureter were cannulated. The kidney was excised and perfused through the renal artery for 15 min (equilibration time) with Kreb's-Henseleit buffer (pH 7.4), at a temperature of 37°C, gassed with 95% O$_2$ and 5% CO$_2$ at 8-10 ml/min.

Effluent fluids were collected from the renal vein (flow rate of 8-10 ml/min) and the ureter (flow rate 0.2-0.5 ml/min) for a period of 10 min following the equilibration period. This served for the estimation of the basal release of TXA$_2$ and prostacyclin (PGI$_2$). Twenty min later, angiotensin II (0.1 mg) (CIBA Pharmaceutical Company, Summit, NJ) was injected into the arterial line to stimulate a rapid production of prostaglandins, which was estimated in samples of effluent fluids collected for 5 min. These fluids were assayed by radioimmunoassay for thromboxane B$_2$ (TXB$_2$) and 6-keto-PGF$_{1\alpha}$ (6KF) as previously reported (1).

The kidneys were fixed in 4% formalin and prepared for histological examination. Samples were viewed by light microscopy and by transmission electron microscopy. Mean quantities of prostaglandins synthesized and released by the control and irradiated kidneys were compared by Student's unpaired t test and by the one-way analysis of variance (Kruskal-Wallis test).

RESULTS

Light and electron microscopy examinations of the perfused kidneys of the control animals were unremarkable. In the irradiated kidneys, an interstitial edema was evident with no other cellular changes (Fig. 1). The glomeruli and tubular system showed no pathological changes despite the high radiation dose, and transmission electron microscopy also did not reveal any changes.

While the histological changes were minimal, the effects of radiation on the basal and stimulated releases of the prostaglandins were profound. Table 1 shows that following irradiation, a significant increase (p <0.01) of the basal release of TXB_2 was observed in both venous and the ureteral effluents. Stimulation with the vasoconstrictor hormone angiotensin II induced increased synthesis of TXB_2 in both the normal and irradiated kidneys (Table 1). However, following irradiation, the release of TXB_2 in response to angiotensin II provocation was significantly

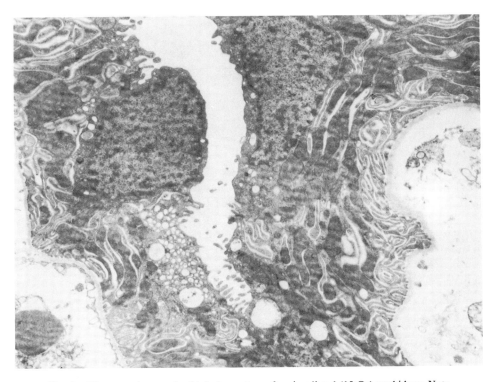

Fig. 1. Electron micrograph of tubular system of an irradiated (15 Gy) rat kidney. Note minimal interstitial edema without cellular damage.

Table 1. Effect of Irradiation on Renal Production of TXA_2 and PGI_2 Measured as TXB_2 and 6-keto $PGF_{1\alpha}$ (6KF) (pg/ml)[1]

Method of Release	Control	Irradiated	P
	Venous Effluent		
	TXB_2:		
Basal Release	50 ± 90	800 ± 160	<0.01
Stimulated Release[2]	310 ± 80	7200 ± 960	<0.01
P	<0.01	<0.01	
	6KF:		
Basal Release	105 ± 58	177 ± 62	N.S.[3]
Stimulated Release[2]	236 ± 49	315 ± 88	N.S.
P	<0.01	<0.01	
	Ureteral Effluent		
	TXB_2:		
Basal Release	300 ± 65	2330 ± 800	<0.01
Stimulated Release[2]	615 ± 88	23100 ± 11200	<0.0
P	<0.01	<0.01	
	6KF:		
Basal Release	126 ± 49	343 ± 78	<0.05
Stimulated Release[2]	263 ± 64	531 ± 138	<0.05
P	<0.01	<0.01	

[1]Mean ± S.D. of 10 rats
[2]With angiotensin II
[3]Not significant

higher (p <0.01) in both the tubular and capillary systems, compared with release in the nonirradiated kidney.

PGI$_2$ production was also altered in irradiated kidneys. The basal release of 6KF in venous and ureteral effluents was higher in the irradiated kidneys; however, the differences were smaller than the corresponding differences of TXB$_2$ levels (Table 1). Only the differences between the levels of 6KF in the ureteral effluents of the irradiated and control kidneys were significant (p <0.05). Angiotensin II stimulation was associated with the production of increased amounts of 6KF by the capillary and tubular systems of the nonirradiated and irradiated kidneys; however, the differences were significant (p <0.05) only in the ureteral fluids.

DISCUSSION

In the present study, we found that irradiation of rat kidneys with a dose that produced minimal histological changes was associated with profound changes in the release of TXA$_2$, which is a potent vasoconstrictor and a strong stimulator of platelet activation and aggregation. Interestingly, radiation had only a moderate effect on the production of PGI_2, which antagonizes the biological effects of TXA_2 and is a strong vasodilator and inhibitor of platelet activation.

The results of our study are very similar to those reported by Schneidkraut et al. (3), who exposed rats to whole-body radiation (10 Gy) and found a marked increase in the urinary excretion of TXA_2 while PGI_2 excretion was unchanged. In our study, radiation was administered only to the kidney (with shielding of other organs), and its effect on basal and angiotensin II-stimulated renal prostaglandin synthesis was measured directly in fluids perfused or filtered through the isolated renal tissue. With this system, it was possible to differentiate between prostaglandin production by the renal capillary system and the tubular system. We found that irradiation of the kidney was associated with increased TXA_2 production in both the venous and ureteral effluents, while PGI_2 production by the vascular system was only slightly affected and a significant increase was observed only in the ureteral fluids. Perfusion of the kidney with angiotensin II stimulates cells that line the lumen of the capillaries and tubular system to produce prostaglandins. Hence our findings imply a disturbed balance between TXA_2 and PGI_2 production capacities of the vascular and tubular systems following irradiation. Renal prostaglandins participate in the maintenance of the renal perfusion during "stress" conditions, such as congestive heart failure, liver disease, and hemorrhage (5). Radiation is associated with significant damage to renal tissue (6,7). Such damage may be affected by the above disturbed balance of TXA_2 and PGI_2 since TXA_2 is a potent vasoconstrictor and antagonizes the vasodilatory effect of PGI_2 (5).

The effects of radiation on the TXA_2-PGI_2 balance observed in the kidney are different from those observed in other irradiated organs or tissues. Heinz et al. (2) found that irradiation of the rat lung with low radiation doses was associated with an increase in both TXA_2 and PGI_2 production, while with higher radiation doses (10-20 Gy) the production of PGI_2 decreased. We have investigated the effects of radiation on PGI_2 production by cultured bovine aortic endothelial cells (1). Metabolically active cells that survived the radiation damage and remained confluent and firmly attached to the culture dish exhibited a marked decrease in their capacity to synthesize PGI_2 upon exposure to various stimuli of the arachidonic acid cascade (1). The increased capacity to produce TXA_2 observed in irradiated kidney contradicts our findings with endothelial cell cultures, and this may be explained by inherent differences between the endothelial cells and the large vessels and capillaries.

In summary, prostaglandin synthesis in an organ may serve as a readily quantifiable marker of tissue injury. Characterization of the changes in renal prostaglandin production following irradiation are important, since these substances have a major role in maintaining renal blood flow and the glomerular filtration rate (5). Measurement of prostaglandin release and prostaglandin production capacity can be a useful tool in radiobiology for studies of radiosensitizing and radioprotective agents.

ACKNOWLEDGMENTS

We wish to express our gratitude to S. Jackson and E. Hy-Am for their expert technical assistance. This study was supported by a grant from the Israel Atomic Energy Commission and the Council for Higher Education.

REFERENCES

1. Eldor, A., Vlodavsky, I., Hy-Am, E., Atzmon, R., and Fuks, Z. The effect of radiation on prostacyclin (PGI_2) production by cultured endothelial cells. Prostaglandins 25: 263-279, 1983.
2. Heinz, T. R., Schneidkraut, M. J., Kot, P.A., Ramwell, P.W., and Rose, J.C. Radiation-induced alteration in cycloxygenase synthesis by isolated perfused rat lungs. Symposium on cyclooxygenase and lipoxygenase modulators in lung reactivity. In: "Progress in Biochemical Pharmacology," Vol. 20. F. Barett, S. Hurd, and R. J. Hegyeli, eds. S. Karger, Basel, Switzerland, 1985, pp. 74-83.
3. Schneidkraut, M. J., Kot, P. A., Ramwell, P. W., and Rose, J. C. Urinary prostacyclin and thromboxane levels after whole body gamma irradiation in rat. In: "Advances in Prostaglandin Thromboxane and Leukotriene Research." R. Paoletti and P. Ramwell, eds. Raven Press, New York, 1983, pp. 107-111.
4. Donlon, M., Steel, L., Helgeson, E. A., Shipp, A. and Catravas, G. N. Radiation-induced alterations in prostaglandin excretion in the rat. Life Sciences 32: 2631-2639, 1983.
5. Levenson, D. J., Simmons, C. E., and Brenner, B. M. Arachidonic acid metabolism, prostaglandins and the kidney. Am. J. Med. 72: 354-374, 1982.
6. Rubin, P., and Casarett, G. W. The kidney. In: "Clinical Radiation Pathology." W. B. Saunders, Philadelphia, 1968, pp. 293-327.
7. Mostofi, F. K., and Berdjis, C. C. Radiopathology of kidney. In: "Pathology of Irradiation." C. C. Berdjis, ed. Williams and Wilkins, Baltimore, 1971, pp. 635-697.

GENERATION OF A NEUTROPHIL CHEMOATTRACTANT OF THE LIPOXYGENASE PATHWAY BY IRRADIATED BOVINE AORTIC ENDOTHELIAL CELLS

A. Eldor[1], M. Cohn[1], I. Vlodavsky[2], E. Razin[3],
E. Hy-Am[1], Z. Fuks[2], and Y. Matzner[4]

[1]Department of Hematology
[2]Department of Radiation and Clinical Oncology
Hadassah University Hospital
Jerusalem, Israel

[3]Department of Biochemistry
Hadassah-Hebrew University Medical School
Jerusalem, Israel

[4]Hematology Unit
Hadassah University
Mount Scopus
Jerusalem, Israel

ABSTRACT

We investigated the capacity of cultured bovine aortic endothelial cells (BAEC) to generate chemotactic factor(s) following ionizing radiation injury. BAEC in serum-free medium were irradiated, and the conditioned medium was assayed for chemotactic activity using normal human neutrophils in Boyden chambers. A rapid release of chemotactic activity dependent on the dose of radiation and the time between irradiation and collection of the medium was found. Damage of BAEC by freezing and thawing was not associated with the release of chemotactic activity. The chemotactic activity produced by irradiated BAEC was not destroyed by boiling or by treatment with trypsin, and its release was inhibited by pretreatment of the cells with a lipoxygenase inhibitor but not with a cyclooxygenase inhibitor. The release of chemotactic activity by vascular endothelium may account for the acute inflammatory response observed following ionizing irradiation.

INTRODUCTION

Radiation injury to blood vessels is associated with an acute inflammatory process (1). The vascular damage induced by ionizing irradiation has been attributed to its effect on the endothelial cell lining of the intima (2,3). Radiation-induced vascular injury consists of endothelial cell nuclear and cytoplasmic swelling and sloughing of endothelial cells (2,3). These changes lead to the deposition of platelets and white blood cells on the injured intima (1,3). Although impairment of one of the vascular defense mechanisms against thrombosis (namely, the capacity to produce prostacyclin) was described following irradiation (3), the mechanism of the radiation-induced acute inflammatory reaction has not been elucidated yet. Recently, the capacity of endothelial cells to secrete inflammatory mediators, either spontaneously (4) or following irradiation (5), was described. In this communication, we present evidence that, shortly following irradiation, cultured endothelial cells produce a potent neutrophil chemotactic factor that is probably an arachidonate metabolite of the lipoxygenase pathway.

METHODS

Irradiation of Endothelial Cells

Clonal populations of bovine aortic endothelial cells (BAEC) derived from the adult aortic arch were obtained as previously described (3). Cells from different cell passages were irradiated after having formed a contact inhibited cell monolayer (3). The culture medium was always replaced with serum-free medium prior to irradiation. The cells were irradiated as previously reported with either a Phillips 250 keV X-ray machine (half value layer 0.5 μg of Cu at surface-to-surface distance of 50 cm) or with a cobalt-60 Gammacell 220 (Atomic Energy of Canada) at a dose rate of 50 Gy per min. Nonirradiated cells, which served as controls, were subjected to the same conditions as were the irradiated cells. The conditioned media were harvested following different periods of incubation at 37°C, centrifuged briefly to remove any detached cells, and diluted with Dulbecco's phosphate buffer (PBS) to 25% (v/v).

In some experiments, nordihydroguaiaretic acid (NDGA, 10 μg/ml) was added to the endothelial cell monolayers after replacement of the medium, 60 min before irradiation. NDGA itself had no effect on neutrophil migration. Endothelial cultures were also treated with acetylsalicylic acid (ASA, 100 μM) for 60 min and then washed and incubated with fresh medium before irradiation. The removal of ASA was indicated because of its antichemotactic activity (6). Boiling and trypsin digestion of postirradiation conditioned media were carried out as previously described (7).

Chemotaxis

Neutrophils were isolated from the blood of healthy donors as decribed (6) and suspended at 10^6 cell/ml in PBS containing 1 mg/ml glucose and 6 mg/ml bovine serum albumin. Directed migration of neutrophils was assayed using the "leading front" technique (8). The neutrophil suspension was placed in the upper compartment of a Boyden chamber (Nucleopore Corp., Bethesda, MD) and the endothelial cell-conditioned medium in the lower compartment. The depth of cell migration was expressed in microns (μm), after a 45-min incubation at 37°C (7). Control assays were conducted using 1% zymosan-activated serum. Chemotaxis and inhibition of chemotaxis were calculated as reported (7). The statistical significance was determined by the paired t test.

RESULTS

The generation of chemotactic activity by irradiated endothelial cells depended on the duration of incubation following irradiation and on the dose of irradiation. Significant chemotactic activity was detected in the conditioned medium of endothelial cells as early as 10 min after irradiation, and maximal activity was observed after 1 hour. No further increase in the chemotactic activity was observed in the 24 hours following irradiation (Table 1A). The release of chemotactic activity depended also on the dose of radiation. Significant chemotactic activity was detected at 1 hour postirradiation with a dose as low as 5 Gy, and it increased with higher radiation doses (Table 1B). Chemotactic activity is also released by nonirradiated endothelial cell cultures (4). However, in our experiments, it was not detected in the culture medium that was incubated over nonirradiated cells for 1 hour. Longer incubation periods of the control cultures revealed a slowly increasing release of chemotactic activity which was always significantly lower than that released from comparable irradiated cultures (results not shown). The release of chemotactic activity from irradiated cells was not due to cell death or lysis, because suspensions of endothelial cells that were subjected to quick freezing (liquid N_2) and thawing (37°C) did not release any chemotactic activity into their culture medium. This result suggested that the release of chemotactic activity by irradiated cells was due to *de novo* synthesis of a chemotactic mediator triggered by irradiation and not due to leakage from disrupted cells. In an attempt to characterize this rapidly produced chemotactic mediator, we treated the supernatants obtained from irradiated endothelial cells with either trypsin or boiling. These treatments did not destroy the chemotactic activity (Table 1C), indicating that the chemotactic factor produced by the irradiated cells is not a protein and is likely to be a lipid. Hence, in the following experiments, the cultures were treated before irradiation with either lipoxygenase or cyclooxygenase inhibitors. The addition of the lipoxygenase inhibitor NDGA significantly inhibited production of the chemotactic activity, whereas addition of ASA had no effect on the chemotactic activity of the conditioned medium, and in many experiments increased activity was observed.

DISCUSSION

Our findings indicate that, following irradiation, cultured endothelial cells release increasing amounts of chemotactic mediator(s) for human neutrophils. Generation of this chemotactic activity is dependent on the dose of radiation and the duration of incubation following irradiation. Maximal amounts of chemotactic activity were observed in the culture supernatants after 1 hour of incubation, and the release of chemotactic activity was induced with radiation doses as low as 5 Gy.

Table 1. Effects of Radiation on Release of Chemotactic Activity by BAEC

A. Time Response (10 Gy Irradiation):

Time (min)	Chemotaxis (μm)
0	12
10	20
20	24
30	33
45	43
60	42
24 hours	40

B. Dose Response (Medium Collected at 1 Hour):

Radiation dose (Gy)	Chemotaxis (μm)
0	12
5	31
10	41
20	49
40	60

C. Effect of Different Treatments and Inhibitors:

Treatment	Inhibition of chemotaxis (%)
Boiling	0.4
Trypsin	6.5
NDGA	64.0
ASA	-18.0

Results represent eight experiments performed with endothelial cultures of different passages.
Each data point is mean of six culture dishes. Differences in chemotactic activity in these
dishes did not exceed 20%.
BAEC, bovine aortic endothelial cells
NDGA, nordihydroguaiaretic acid
ASA, acetylsalicylic acid

The capacity of endothelial cells to produce neutrophil chemoattractants was recently reported (4,9,10). O'Brien et al. (9) showed that BAEC generate a heat-stable, lipid-extractable neutrophil chemoattractant upon stimulation with thiourea. Farber et al. (10) reported that hypoxic damage to BAEC is associated with the release of a chemotactic lipoxygenase product. Mercandetti et al. (4) reported that cultured human endothelial cells in medium that contained serum generated at least two neutrophil chemoattractants with relative mass of 35,000 dalton and 1500 dalton. The latter was a lipid-extractable compound that coeluted with leukotriene B$_4$ on high-performance liquid chomatography. Our experiments were performed in serum-free medium in order to exclude the possibility of activation of serum components (i.e., C5a), and we treated the conditioned media after irradiation with either boiling or trypsin. These treatments did not destroy the chemotactic activity, suggesting that it is not a protein. Pretreatment of the cultures with lipoxygenase inhibitor, but not with cyclooxygenase inhibitor, resulted in inhibition of release of the chemotactic activity, suggesting that it may be an arachidonate product.

Recently, Dunn et al. (5) reported the elaboration of chemotactic activity by irradiated BAEC. They found a significant increase in the chemotactic activity in the conditioned medium of irradiated cells only 72 hours after irradiation, and this activity was dependent on protein synthesis and was destroyed by trypsin. As mentioned before, endothelial cells can produce both a chemotactic lipid and a chemotactic protein (4). Our findings imply that radiation injury induces a rapid release (within 1 hour) of a lipid-extractable chemotactic activity, and the findings of Dunn et al. suggest a late enhanced release (72 hours) of a trypsin-sensitive chemotactic factor (probably a protein). Further investigations are being conducted in our laboratory to characterize the chemotactic lipid.

The mechanism for generation of a chemotactic lipid induced by irradiation of endothelial cells is at present not known. A possible mechanism may be the generation of toxic oxygen metabolites such as superoxide anions, hydroxyl radicals, and singlet oxygen by ionizing irradiation. Oxygen-derived free radicals were shown to stimulate the arachidonic acid metabolism and to yield products with chemotactic activity (11). Following different types of injury, endothelial cells can generate chemotactic mediators that are metabolites of the lipoxygenase pathway (4,9,10). These chemotactic agents may also play a role in the inflammatory reaction that follows ionizing irradiation.

ACKNOWLEDGMENTS

The authors wish to thank Ms. Ruth Drexler and Ms. Sima Calko for their excellent technical assistance. This work was supported by the Israel Atomic Energy Commission and the Council for Higher Education, by the Kovshar Foundation, and by an NIH grant CA 30289 awarded to I. V.

REFERENCES

1. Hopewell, J. W., Foster, J. L., Young, C. A. M., and Wernik, G. Late radiation damage to pig skin. Radiology 130: 783-788, 1979.
2. Law, M. P. Radiation induced vascular injury and its relation to late effects of normal tissues. Adv. in Radiation Biol. 9: 37-93, 1981.
3. Eldor, A., Vlodavsky, I., Hy-Am E., Atzmon, R., and Fuks, Z. The effect of radiation on prostacyclin (PG12) production by cultured endothelial cells. Prostaglandins 25: 263-279, 1983.
4. Mercandetti, A. J., Lane, T. A., and Colmerauer, M. E. M. Cultured human endothelial cells elaborate neutrophil chemoattractants. J. Lab. Clin. Med. 104: 370-380, 1984.

5. Dunn, M. M., Drab, E. A., and Rubin D. B. Effects of irradiation on endothelial cell-polymorphonuclear leukocyte interaction. J. Appl. Physiol. 60: 1932-1937, 1986.

6. Matzner, Y., Drexler, R., and Levy, M. Effect of dipyrone, acetylsalicylic acid and acetaminophen on human neutrophil chemotaxis. Eur. J. Clin. Invest. 14: 440-443, 1984.

7. Matzner, Y., Partridge R. E. H., and Babior, B. M. Chemotactic inhibitor in synovial fluid. Immunol. 49: 131-138, 1983.

8. Zigmud, S. H., and Hirsch, J. G. Leukocyte locomotion and chemotaxis. New methods for evaluation and demonstration of cell-derived chemotactic factor. J. Exp. Med. 137: 387-410, 1973.

9. O'Brien, R., Seton, M., Makarski, J., Center, D., and Rounds, S. Endothelial cells treated with thiourea release a chemoattractant for neutrophils (abstr). Am. Rev. Respir. Dis. 125: 229, 1982.

10. Faber, H. W., Center, D. N., and Rounds, S. Hypoxic endothelial cells release a neutrophil chemoattractant (abstr). Circulation 68, Supp III-7: 292, 1983.

11. Perez, H. D., Weksler, B. B., and Goldstein, I. M. Generation of a chemotactic lipid from arachidonic acid by exposure to a superoxide-generating system. Inflammation 4: 313-328, 1980.

PROSTAGLANDINS IN RADIATION
PROTECTION AND RECOVERY

RADIATION PROTECTION BY EXOGENOUS ARACHIDONIC ACID AND SEVERAL METABOLITES

W. R. Hanson

Department of Therapeutic Radiology
Rush-Presbyterian-St. Luke's Medical Center
Chicago, Illinois 60612

ABSTRACT

Based on the observations that 16,16-dimethyl prostaglandin E_2 (dm PGE_2) protected intestinal clonogenic cells (ICC) from some degree of radiation injury, studies were done to see if exogenous arachidonic acid (AA) and its products were radioprotectants. One mg AA was required to measure protection that increased ICC survival from a single dose of 15.0 Gy cesium-137 from about 9 to 21 cells per jejunal circumference ($P < 0.001$). With phospholipase A_2, lower doses of AA were protective. All prostaglandins (PG's) were protective with the exception of E_1. PGI_2 was the most protective natural PG (300% increase in survival), and 15-deoxy,16-hydroxy,16-methyl PGE_1 (misoprostol) was the most protective to date (600% over controls). 16,16-dm PGE_2 also protected ICC from some degree of injury by the high-energy neutrons of Fermilab. Of the leukotrienes (LT's) tested, LTC_4 was the most protective (about 250% above controls) and LTD_4 was next in order (150% increase in survival). These results suggest that AA and products of the AA cascade can modify the response of tissue to radiation, which may have both mechanistic and clinical implications.

INTRODUCTION

The first report that prostaglandins (PG's) protected cells *in vitro* from photon radiation was by Prasad in 1972 (1), who showed that survival of Chinese hamster ovary (CHO) cells was increased by PGE_1. Subsequently, Lehnert (2), published in 1975 a series of complex survival curves showing that PGE_1 protected V-79 cells on the shoulder portion of the survival curve but increased the slope (a decreased D_0) of the terminal portion of the curve. Both of these authors suggested that the PG-induced increase in cAMP was the mechanism whereby PG modified the response of cells to radiation.

In 1983, it was reported (3) that 16,16-dimethyl (dm) PGE_2 protected murine intestinal clonogenic cells (ICC) *in vivo* from cesium-137 injury. Both the shoulder and the D_0 of the clonogenic cell survival curve were increased. The ICC were protected if radiation delivery was begun 5 min after a subcutaneous (s.c.) injection; however, if 3 to 4 hr elapsed between the PG administration and the initiation of cesium-137 irradiation, no PG-induced protection was seen. No protection was seen if the PG was given following irradiation. Subsequently it was reported that the shoulder and the D_0 of the bone marrow stem cell survival curve were also increased by 16,16-dm PGE_2 (4). The same PGE_2 analogue increased the LD50/6 (4) and the LD50/30 of irradiated mice (5). To further investigate the role of arachidonic acid (AA) and its products in modifying the response of tissue to radiation, a series of experiments was done to measure radiation protection by exogenous AA, several PG's, and several leukotrienes (LT's) from gamma injury. In addition, the most widely investigated radioprotective PG used to date, 16,16-dm PGE_2, was used to explore radiation protection from the high-energy neutrons at Fermilab.

MATERIALS AND METHODS

B6D2F1 male mice from Jackson Laboratories (Bar Harbor, ME) were used for these studies. Upon arrival at 70 days of age, several mice were checked and found to be free of *Pseudomonas*. All animals were fed standard Purina chow and acid water (pH = 2.7) to prevent the possible infection and spread of *Pseudomonas*. The animals were aged to 100-130 days before use. The care and use of these mice were overseen by the Institutional Animal Care and Use Committee of Rush University to insure that animals were kept in accordance with the recommendations as stipulated in DHEW publication No. NIH 85-23.

Arachidonic Acid-Induced Radiation Protection

Arachidonic acid (sodium salt, Sigma, St. Louis, MO) was dissolved in phosphate buffer and injected s.c. in the dorsal neck region of mice 2.5 hr before a single dose of 15.0 Gy cesium-137 radiation. The following concentrations of AA were given to groups of five mice each: 0 (phosphate buffer), 0.05, 0.1, 0.5, and 1.0 mg/mouse. Each in another group of mice that received 0.5 mg of AA 2.5 hr before irradiation was given two units of phospholipase A_2 (PLA_2) 1.5 hr before irradiation. As a control for this group, each in another group of mice was given PLA_2 1.5 hr before irradiation. These time intervals between drug treatment and radiation exposure have been shown to be effective for radiation protection by PG's. At 4 days after 15.0 Gy cesium-137, the mice were killed by cervical fracture and the upper jejunum was removed; fixed in alcohol, formalin, and acetic acid (AFA, 20:2:1); and embedded in paraffin. Five-micron cross sections were prepared and stained with hematoxylin and eosin. The number of regenerative epithelial colonies in the mucosa of each cross section [microcolony assay of Withers and Elkind (6)] was counted.

PG-Induced Radiation Protection

The following PG's were purchased from Upjohn: E_1, E_2, A_1, A_2, D_2, $F_{2\alpha}$, I_2, and 16,16-dm PGE_2. An E_1 analogue, 15-deoxy,15-hydroxy,16-methyl PGE_1 methyl ester (misoprostol), was a gift from G. D. Searle. Both the time course and the PG dose response for protection were measured for a single dose of 15.0-Gy cesium-137 radiation. To measure the time course of PG-induced radiation protection, a dose of 25 μg of each PG was given s.c. in the dorsal neck to groups of five mice at the following times before irradiation: 15 min, 30 min, 45 min, 1 hr, 2 hr, and 3 hr. The mice were killed by cervical fracture, and the microcolony assay was used to measure surviving ICC per circumference of intestine as described above.

To measure the PG dose response for radiation protection, the time of PG administration for maximum protection was selected from above, and the following doses of each PG were given to groups of five mice: 0 (the vehicle for each PG), 5, 10, 25, 50, and 100 μg/mouse. The microcolony assay was used to measure ICC survival.

Leukotriene-Induced Radiation Protection

Four LT's and 5-hydroxy, eicosatetraenoic acid (5-HETE) were kindly provided by Dr. Rokach, Merck-Frosst, Canada. Due to demand and the limited amount of these pure LT's, a single dose of 10 μg of each LT was given to groups of five mice 1 hr before 15.0 Gy cesium-137 irradiation. Only 5 μg of the 5-HETE could be given to each of five mice before exposure. The LT's C_4, D_4, and E_4 were shipped in aqueous solution whereas LTB_4 and 5-HETE were shipped in methanol. Therefore, the same aqueous solution was given to one control group and 0.1 ml methanol to another control group of five mice. The microcolony technique was used to assay for ICC survival as described.

16,16-dm PGE_2-Induced Radiation Protection From Fermilab Neutrons

B6D2F1 mice were transported from Rush to the Fermi National Accelerator Laboratory 35 miles west of Chicago in Batavia, IL. Groups of five mice each were given 10 μg 16,16-dm PGE_2 1 hr before graded doses of high-energy neutrons. Control groups of five mice each were injected with a 5% ethanol-saline solution, the vehicle for this PG. The Fermilab beam characteristics and the irradiation procedure have been described in detail previously (7). Briefly, unanesthetized mice were placed in perforated plastic tubes, which were fixed by elastic bands between two tissue-equivalent disks of Shonka A-150 material (0.854 g/cm^2) sufficiently thick to improve the homogeneity of the dose distribution. This assembly was aligned perpendicularly to the port 150 cm from the beryllium target. An Exradin air-filled ionization chamber, made with Shonka A-150 tissue-equivalent plastic walls, was used for neutron dosimetry. With this configuration, the estimated gamma component of the beam was about 5%, which is included in the reported total dose. Ten animals (five PG-treated and five controls) were irradiated at a dose rate of 0.70 Gy/min. Following irradiation, the mice were transported back to Rush and killed 4 days later for the microcolony assay as described.

RESULTS

A dose of 15.0 Gy cesium-137 radiation resulted in the survival of about 9 ICC per jejunal circumference (Figure 1). This value is on the exponential terminal slope of the ICC radiation survival curve. Therefore, radioprotection at this dose could be seen as an increased number of microcolonies without interference of the broad shoulder region that is a characteristic of the intestinal microcolony assay (6). No protection was seen when doses of AA below 1.0 mg/mouse were given 2.5 hr before irradiation. A dose of 1.0 mg AA per mouse increased ICC survival significantly (P < 0.001) to about 21 colonies per circumference. PLA$_2$ alone had no effect on ICC survival. However, when PLA$_2$ was combined with 0.5 mg AA (a dose of AA that had no protective effect by itself), ICC survival was increased to about 24 for a dose modification factor of about 2.4 at 15.0 Gy.

In contrast to the mg quantity of AA (the parent compound of the AA cascade) necessary to demonstrate radiation protection, μg amounts of the PG's were protective. The time course for the most effective PG-induced protection of ICC from an s.c. injection is shown in Figure 2. Data for the PG's (A$_1$, A$_2$, and F$_{2\alpha}$) are not shown since they were the same as PGE$_2$ in their time course for protection. No radiation protection was seen at any time for 25 μg PGE$_1$. For other PG's, an increase in survival was seen 5-10 min after injection and reached a maximum

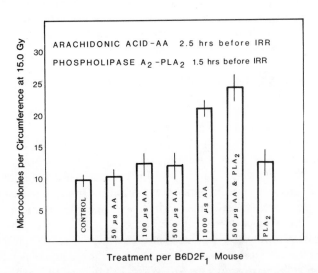

Fig. 1. Microcolonies of regenerating jejunal mucosal epithelium 4 days after 15.0 Gy cesium-137 irradiation in controls or in mice given increasing doses of arachidonic acid (AA), or phospholipase A$_2$ (PLA$_2$), or combination of AA and PLA$_2$. Bars represent mean of five mice per group (n = 5) ± 1 SEM.

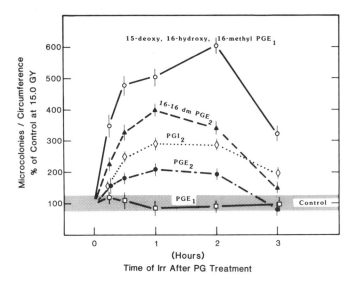

Fig. 2. Time course of radiation protection of ICC by several PG's as reflected by increase in number of microcolonies of regenerating jejunal mucosal epithelium 4 days after 15.0 Gy cesium-137. Data of cell survival are expressed as percent of control values at 15.0 Gy, which in these experiments were about 6 microcolonies per circumference. Values are the mean from five animals $\pm$ 1 SEM.

by about 1 hr. For misoprostol, the greatest proection was seen at 2 hr after injection (8), shown here for comparison. By 3-4 hr following PG administration, the protective activity of most of the PG's was no longer seen. Protection by misoprostol was declining at 3 hr post-injection; however, there was still a significant degree of protection at this time interval. The shape of the time course for all protective PG's investigated was essentially the same except for PGD_2 (Figure 3). In this case, there was a biphasic response with a rapid increase in survival followed by a sharp drop and then a second broader peak.

The pretreatment time interval for maximum protection from radiation derived from Figure 2 for each PG was used to investigate the PG dose response shown in Figure 4. The shape of the PG dose-response curve was similar for all the PG's investigated with the continued exception of E_1, which did not protect ICC at doses up to 100 μg/mouse. There was rapid rise in cell survival, followed by a plateau above 10-25 μg up to 100 μg/mouse. Although all the PG's shown in Figure 4 were investigated at all the PG doses listed, the 5-μg, 10-μg, and 25-μg/mouse data for some PG's are left out for clarity.

In contrast to the similarity in the shape of the PG dose-response curve, there was a marked difference in the degree to which the PG's afforded protection. Of the natural PG's $I_2$was the most protective. The E_1 analogue, misoprostol, was the most protective PG investigated to date, increasing ICC survival to 600% of controls.

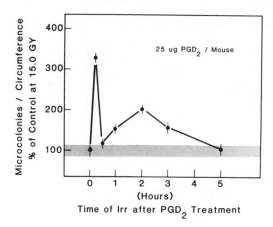

Fig. 3. Time course of radiation protection of ICC by PGD₂. Microcolonies per jejunal circumference are plotted versus time of administration of 10 μg PGD₂ per mouse before 15.0 Gy cesium-137 irradiation. Values are mean from five mice ± 1 SEM.

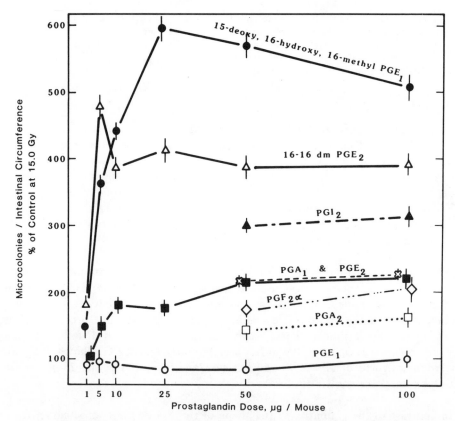

Fig. 4. PG dose response for radiation protection of ICC by several PG's. Data are number of intestinal microcolonies per circumference at 15.0 Gy cesium-137 expressed as percent of control values versus dose of PG given s.c. at most protective time interval before irradiation derived from data in Figure 2. Each symbol is mean from five animals ± 1 SEM.

Radiation protection of ICC by several LT's, 5-HETE, and their appropriate controls is shown in Figure 5. LTC$_4$ was the most protective followed by LTD$_4$. LTE$_4$ showed no protective activity at this time interval or dose. Although an increased number of surviving ICC was seen in LTB$_4$-treated and 5-HETE-treated mice, there is no apparent protection since it appeared that the small amount of methanol (the vehicle for these LT's) also induced some radiation protection.

16,16-dm PGE$_2$ protected ICC from Fermilab high-energy neutrons as shown in Figure 6. In this case, a change in the shoulder portion of the curve was less

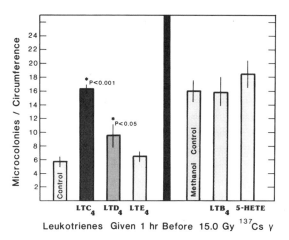

Fig. 5. Microcolonies of regenerating intestinal epithelium 4 days after 15.0 Gy cesium-137 in two groups of controls [aqueous (a) or methanol (m)] and in groups given LTB$_4$ (m), LTC$_4$ (a), LTD$_4$ (a), LTE$_4$ (a), or 5-HETE (m). Each bar represents mean of five animals ± 1 SEM.

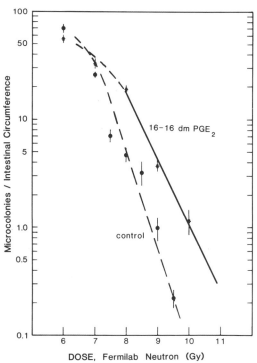

Fig. 6. Microcolonies per jejunal circumference 4 days after graded doses of Fermilab neutrons in controls or in animals given 10 μg 16,16-dm PGE$_2$ 1 hr before irradiation. Each symbol represents mean value from five mice ± 1 SEM.

apparent than with 16,16-dm PGE$_2$-induced protection from gamma irradiation. However, there was a distinct increase in the slope of the ICC survival curve from 0.95 ± 0.11 to 1.37 ± 0.14 Gy.

DISCUSSION

Arachidonic acid, a dietary constituent, is bound in small quantities to phosphatidylinositol and to phosphatidylcholine in the membranes of cells. Following its release from the membrane, which is mediated in part by phospholipases, it is metabolized through at least two major pathways: the cyclooxygenase or the 5-lipoxygenase pathways (9,10). The studies reported here suggest that AA, the parent compound, and products of both pathways protected ICC from photon radiation injury. In addition, 16,16-dm PGE$_2$ protected ICC from neutron injury. A large dose (1 mg/mouse) or AA was needed to protect cells from irradiation; however, half that dose with the addition of PLA$_2$ was protective. PLA$_2$ alone had no effect. These results suggest that AA itself may not induce protection, but if given in excessive amounts (1.0 mg) or if given in smaller amounts and then released by PLA$_2$, the products of the cascade may be produced, which in turn induce protection.

This contention is strengthened by the results showing that both PG's and LT's are radiation protectors. ICC cell survival was increased by all the PG's tested other than PGE$_1$. This is noteworthy since E$_1$ was reported to protect CHO cells (1) and V-79 cells (2) in vitro. PGE$_1$ is not found in abundance (if it is present at all) in the intestine of rodents, and it is possible that ICC do not have receptors for E$_1$. This may explain the discrepancy since PG's appear to function through receptor-mediated events (11), and the general shape of the PG dose-response curve for protection of ICC (Figure 2) also suggests that a receptor-mediated mechanism is involved. Although the shape for the PG dose-reponse curve was similar for the various PG's, the degree to which the PG's protected was markedly different. Of the natural PG's, the order of increasing protection was E$_1$ (no protection), A$_2$, F$_{2\alpha}$, E$_2$, A$_1$, and I$_2$ (about a 300% increase in survival). PGD$_2$ was the only PG investigated to date that showed a biphasic time course for protection. It appears that D$_2$ (protective by itself) was rapidly metabolized into products that were also protective. Both analogues tested were more protective than the natural PG's. 16,16-dm PGE$_2$ increased survival to 400% of controls, and misoprostol increased survival to 600% of controls. Comparison of protection by the E-series PG's (8) suggests that the alpha and beta side chains are important in determining protective function; however, the studies reported here show that A$_1$ is more protective than A$_2$, which is reversed compared to the effects of E$_1$ and E$_2$. Therefore, the structure of the cyclopentane ring may also determine in part the degree of radiation protection by PG's.

Although less extensively studied, it appears that at least two of the 5-lipoxygenase products of AA are also radioprotective. The LTC$_4$ and LTD$_4$ dose response and time course from radiation protection should now be investigated. These results agree with those of Walden et al. (12), who showed some radiation protection of V-79A03 cells *in vitro* by LTC_4.

The E_2 analogue, 16,16-dm PGE_2, protected from the high-energy neutrons of Fermilab. The significance of the reduced effect of this PG on the shoulder of the ICC survival curve in neutron-irradiated compared to gamma-irradiated mice is unknown. However, a similar effect of other protectors such as WR-2721 has been seen (13). The D_0 increase was similar for both cesium-137 and this source of neutrons.

There are several implications of the experimental observations that AA and several of the products of the AA cascade protect cells *in vivo* from radiation injury: (a) Nearly all cells in humans and animals can participate in the AA cascade; however, the extent of AA metabolism and the specific PG's and perhaps LT's produced appear to vary from tissue to tissue. These differences may, in part, account for some variation in normal tissue radiation sensitivity. (b) A variety of injurious agents, including radiation (14-17), causes an increase in the production of PG's. Our results suggest that these PG's may in turn protect tissue from subsequent injury. Therefore, PG's synthesized in response to injury from the initial doses in a normal 6-week course of radiation therapy of a tumor may alter tissue radiation sensitivity to subsequent fractions. (c) The evidence showing a large difference in the structure-function relationship for radiation protection by different PG's suggests that untested or newly synthesized PG analogues may be more protective than those investigated to date. These PG's could potentially be given alone or in combination with the thiophosphoroate WR-2721 before each fraction of radiation during radiation therapy to protect normal tissue. The success of this approach to the differential protection of normal tissue is based on the assumption that the protectors will concentrate in well-vascularized normal tissue whereas there will be less protector in areas of poorly vascularized tumor tissue (18). (d) There is substantial evidence that many, if not all, human tumors produce excess PG's (19-23), perhaps in response to tissue injury associated with invasion or tumor necrosis. Tumors with excessive PG's may be protected from radiation therapy to some degree by their PG's or possibly LT's. The potential exists that by blocking either the synthesis or the binding sites of PG's and LT's, tumors may be made more sensitive to irradiation; however, normal tissues may respond similarly. Continued studies are needed to investigate a possible gain in the therapeutic ratio by manipulating the AA cascade.

ACKNOWLEDGMENTS

The author is grateful to Ms. Kristi DeLaurentiis (RPSLMC) for her excellent technical assistance, and to Drs. R. Ten Haken, I. Rosenberg, M. Awschalom, and the staff of Fermilab for their kind help in facilitating the neutron investigations. These studies were supported in part by the Department of Therapeutic Radiology, Rush-Presbyterian St. Luke's Medical Center, and Contract No. DNA 001-86-0038 from the Defense Nuclear Agency.

REFERENCES

1. Prasad, K. N. Radioprotective effect of prostaglandin and an inhibitor of cyclic nucleotide phosphodiesterase on mammalian cells in culture. Int. J. Radiat. Biol. 22: 187-189, 1972.

2. Lehnert, S. Modification of postirradiation survival of mammalian cells by intracellular cyclic AMP. Radiat. Res. 62: 107-116, 1975.

3. Hanson, W. R., and Thomas, C. 16-16 dimethyl prostaglandin E_2 increases survival of murine intestinal stem cells when given before photon radiation. Radiat. Res. 96: 393-398, 1983.

4. Hanson, W. R., and Ainsworth, E. J. 16-16 dimethyl prostaglandin E_2 induces radioprotection in murine intestinal and hematopoietic stem cells. Radiat. Res. 103: 196-203, 1985.

5. Walden, T. L., Jr., Patchen M., and Snyder, S. L. 16,16-dimethyl prostaglandin E_2 increases survival in mice following irradiation. Rad. Res. (in press).

6. Withers, H. R., and Elkind, M. M. Microcolony survival assay for cells of mouse intestinal mucosa exposed to radiation. Int. J. Radiat. Biol. 17: 261-267, 1970.

7. Cohen, L., and Awschalom, M. The cancer therapy facility at Fermi National Acclerator Laboratory, Batavia, Illinois: A preliminary report. Appl. Radiol. 5: 51-60, 1976.

8. Hanson, W. R., and DeLaruentiis, K. Comparison of *in vivo* murine intestinal radiation protection by the E-series prostaglandins; E_1, E_2, 16-16 dm E_2, and Misoprostol. (Submitted to Prostaglandins).

9. Lands, W. E. M. The biosynthesis and metabolism of prostaglandins. Ann. Rev. Physiol. 41: 633-652, 1979.

10. Samuelsson, B. Prostaglandins, thromboxanes and leukotrienes: Biochemical pathways. In: "Prostaglandins and Cancer: First International Conference." T. J. Powles, R. S. Bockman, K. V. Honn, and P. Ramwell, eds. Alan R. Liss, Inc., New York, 1982, pp. 1-19.

11. Kuehl, F. A., Jr., and Hunes, J. L. Direct evidence for a prostaglandin receptor and its application to prostaglandin measurements. Proc. Nat. Acad. Sci., U.S.A. 69: 480-491, 1972.

12. Walden, T. L., Jr., Holahan, E. V., Jr., and Catravas, G. N. Development of a model system to study leukotriene-induced modification of radiation sensitivity in mammalian cells. Progr. Lipid Res. 25: 587-590, 1986.

13. Sigdestad, C. P., Grdina, D. J., Connor, A. M., and Hanson, W. R. A comparison of radioprotection from three neutron sources and Cobalt-60 by WR-2721 and WR-151327. Radiat. Res. 106: 224-233, 1986.

14. Eisen, V., and Walker, D. I. Effect of ionizing radiation on prostaglandin-like activity in tissues. Br. J. Pharmac. 57: 527-532, 1976.

15. Pausescu, E., Chirvasie, R., Teodosiu, T., and Paun, C. Effects of 60-Co gamma radiation of the hepatic and cerebral levels of some prostaglandins. Radiat. Res. 65: 163-171, 1976.

16. Trocha, P. J., and Catravas, G. N. Prostaglandin levels and lysosomal enzyme activities in irradiated rats. Int. J. Radiat. Biol. 38: 503-515, 1980.

17. Steel, L. K., and Catravas, G. N. Radiation-induced changes in production of prostaglandins $F_{2\alpha}$, E, and thromboxane B_2 in guinea pig parenchymal lung tissues. Int. J. Radiat. Biol. 42: 517-530, 1982.

18. Yuhas, J. M., and Storer, J. B. Differential chemoprotection of normal and malignant tissues. J. Natl. Cancer Inst. 42: 331-342, 1969.

19. Seyberth, H. W., Segre, G. V., Morgan, J. L., Sweetman, B. J., Potts, J. T., Jr., and Oates, J. A. Prostaglandins as mediators of hypercalcemia associated with certain types of cancer. N. Engl. J. Med. 293: 1278-1283, 1975.

20. Cummings, K. B., and Robertson, R. P. Prostaglandin: Increased production by renal cell carcinoma. J. Urol. 118: 720-723, 1977.

21. Prowles, T. J., Coombes, R. C., Neville, A. M., Ford, H. T., Gazet, J. C., and Levine, L. 15-Keto-13,14-dihydroprostaglandin E_2 concentrations in serum of patients with breast cancer. Lancet 2: 138-142, 1977.

22. Bennett, A., Charlier, E. M., McDonald, A. M., Simpson, J. S., Stamford, I. F., and Zebro, T. Prostaglandins and breast cancer. Lancet 2: 624-626, 1977.

23. Bennett, A., Carroll, M. A., Stamford, I. F., Whimster W. F., and Williams, F. Prostaglandins and human lung carcinomas. Br. J. Cancer 46: 888-893, 1982.

ALTERATIONS IN LOCOMOTOR ACTIVITY INDUCED BY RADIOPROTECTIVE DOSES OF 16,16-DIMETHYL PROSTAGLANDIN E_2

M. R. Landauer[1], T. L. Walden[2], H. D. Davis[1], and J. A. Dominitz[1]

[1]Behavioral Sciences Department
[2]Radiation Biochemistry Department
Armed Forces Radiobiology Research Institute
Bethesda, Maryland 20814-5145

ABSTRACT

16,16-Dimethyl prostaglandin E_2 (DiPGE2) is an effective radioprotectant when administered before irradiation. A notable side effect of this compound is sedation. In separate experiments, we investigated the dose-response determinations of the time course of locomotor activity and 30-day survival after 10 Gy gamma irradiation (LD_{100}). Adult male CD2F1 mice were injected subcutaneously with vehicle or DiPGE2 in doses ranging from 0.01 to 40 μg/mouse. A dose of 0.01 μg did not result in alterations in locomotor behavior or enhance survival. Doses greater than 1 μg produced ataxia and enhanced radiation survival in a dose-dependent fashion. Full recovery of locomotor activity did not occur until 6 and 30 hr after injection for the 10 μg and 40-μg groups, respectively. Radioprotection was observed when DiPGE2 was administered preirradiation but not postirradiation. Doses of 1 and 10 μg were maximally effective as a radioprotectant if injected 5 min prior to irradiation (80%-90% survival). A dose of 40 μg resulted in 100% survival when injected 5-30 min before irradiation. Therefore, increasing doses of DiPGE2 resulted in enhanced effectiveness as a radioprotectant. However, the doses that were the most radioprotective were also the most debilitating to the animal.

INTRODUCTION

The ideal radioprotective agent is one that maintains both high efficacy and mental alertness. An agent having both of these characteristics would be useful in the event of a radiation accident. Compounds providing a high degree of radioprotection but temporarily impairing mental alertness may still prove to be beneficial in radiotherapy. 16,16-Dimethyl prostaglandin E_2 (DiPGE2) has been shown to be effective in enhancing the radiation survival of hematopoietic and intestinal stem cells *in vivo* (1) and also for whole-animal survival (2). Although the mechanism(s) for the radioprotection is unknown, the induction of hypoxia through prostaglandin-mediated vascular changes remains a viable hypothesis. Radioprotective doses of $DiPGE_2$ result in significant increases in the hematocrit (2) and a concomitant reduction in the respiration rate (T. Walden, unpublished).

Exogenously administered prostaglandins, including those of the E series, have been demonstrated to produce sedative or tranquilizing effects in a variety of species (3-8). A decrease in gross locomotor movements has been observed following the administration of radioprotective doses of $DiPGE_2$, but the magnitude and duration of the response have yet to be quantified (2). Sedative effects might limit the usefulness of $DiPGE_2$ as a radioprotectant. We have examined the reduction in locomotor activity induced by $DiPGE_2$ over a dose range associated with radioprotection. In this paper we report that the doses that result in the greatest disruption of locomotor activity are the doses that offer the most radioprotection.

METHODS

Subjects

CD2F1 male mice, 3-5 months of age and obtained from Harlan Sprague-Dawley (Indianapolis, IN), served as subjects. They were housed under a 12-hour light-dark cycle. Immediately following arrival, animals were quarantined for 2 weeks to ensure that they were free from *Pseudomonas* and other common murine diseases, based on serological and histological examination. Mice were maintained in groups of 9-10 in clear plastic cages with filter tops, and were provided with Wayne Rodent Blox diet and acidified water (pH 2.5) ad libitum.

Drug

$DiPGE_2$ was obtained from the UpJohn Co. (Kalamazoo, MI). The drug was dissolved in a solution containing 4% ethanol in saline (0.9%). The mice received a single 100-μl subcutaneous injection in the skin at the back of the neck.

Locomotor Activity

Locomotor behavior was measured using a computerized Animal Activity Monitor (Model RXYZCM-16, Omnitech Electronics, Columbus, OH), which recorded horizontal activity (ambulation) by means of infrared photodetectors. Animals received subcutaneous injections of the vehicle or 0.01, 0.1, 1.0, 10.0, or 40.0 μg/mouse $DiPGE_2$ (N = 9/group) and were placed immediately into the test apparatus. Locomotor activity was monitored at 1-min intervals for the first 60 min to determine the latency of onset of the drug. Thereafter, activity was recorded at 1-hour intervals for the next 5 hours and again from hours 25 to 30 postinjection. After that time, all groups had returned to control values. All testing took place during the dark portion of the light-dark cycle.

Radioprotection

Additional mice were bilaterally irradiated with 10 Gy cobalt-60 radiation at a dose rate of 1 Gy/min, as previously described (9). Prior to irradiation, mice were injected subcutaneously with doses of vehicle or 0.1, 1.0, 10.0, or 40.0 μg/mouse DiPGE$_2$. The drug was administered at 5, 20, 30, or 60 min before irradiation. Separate groups of mice received these same doses postirradiation. Control animals received vehicle alone. Ten mice were used for each dose at each time point. Irradiated animals were monitored for the fraction surviving 30 days postirradiation.

Statistical Analysis

A one-way analysis of variance with repeated measures was performed on the locomotor activity data. Post hoc comparisons were made using Dunnett's test. Significant differences in percent 30-day survival were determined by using the method of Fleiss to compare proportions (10).

RESULTS

16,16-Dimethyl prostaglandin E$_2$ produced a dose-dependent decrease in locomotor activity. Analysis of variance indicated that the effect of dose on locomotor activity was significant over time (F $=$ 8.36, df $=$ 44/440, p $<$.0001). A dose of 0.01 μg did not significantly alter locomotor activity from control values. For doses between 0.1 and 40 μg/mouse, the latency to onset of the drug occurred within 2-3 min. Only those animals receiving a dose of 0.1 μg fully recovered during the first hour following injection (Figure 1). Increasing the concentration of the DiPGE$_2$ treatment extended the duration of the locomotor decrement (Figure 2). A dose of 1.0 μg resulted in full recovery of locomotor function by 2 hours postinjection, whereas mice administered doses of 10 and 40 μg did not return to control levels until 6 and 30 hours, respectively.

The time intervals for radioprotective activity of DiPGE$_2$ at concentrations tested for locomotor activity are shown in Table 1. The 40-μg dose provided the greatest level of radioprotection. At this dose, 100% survival from 10 Gy radiation was observed when DiPGE$_2$ was administered during the first 30 min before irradiation. Eighty percent of the animals survived when this dose was administered at 60 min preirradiation. When injected 5 and 20 min preirradiation, doses of 1 and 10 μg, respectively, were not significantly different from the 40-μg dose. However, at 30 and 60 min prior to irradiation, a dose of 40 μg was significantly more effective than doses of 0.1-10 μg DiPGE$_2$. Pretreatment with doses of less than 1 μg DiPGE$_2$ or administration of doses of 0.1-40 μg postirradiation did not yield any radioprotection. DiPGE$_2$ did not affect the rate of radiation-induced mortality. Radiation-induced deaths in both vehicle- and DiPGE$_2$-treated mice occurred between 9 and 14 days postirradiation. None of the vehicle-treated animals survived the 10-Gy radiation dose.

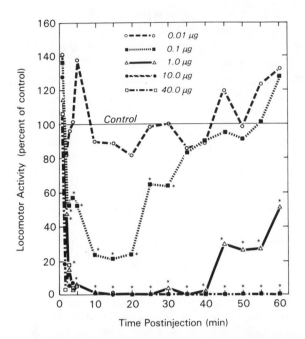

Fig. 1. Time course of DiPGE$_2$ as a function of dose on locomotor activity (ambulation) during first 60 min after injection. DiPGE$_2$ administered subcutaneously immediately before testing. Data are expressed as percent of vehicle control group (N = 9/group). *Significantly (p < 0.01) different from vehicle control group.

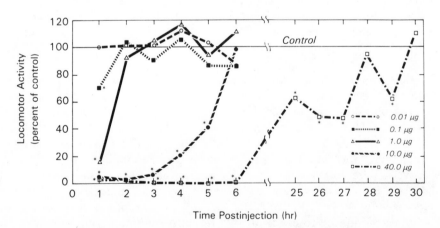

Fig. 2. Duration of action of effects of DiPGE$_2$ on locomotor behavior. Locomotor activity (ambulation) is expressed as percent of control group (N = 9/group). *Significantly (p < 0.01) different from vehicle control group.

Table 1. Effect of Dose and Time of PGE$_2$ on Percent 30-day Survival

Dose[1]	Min Preirradiation				Min Postirradiation			
	5	20	30	60	5	10	20	60
0.1	0	0	0[3]	0[3]	0	0	0	0
1.0	80[2]	60[2]	10[3]	0[3]	0	0	0	0
10.0	90[2]	70[2]	20[3]	10[3]	0	0	0	0
40.0	100[2]	100[2]	100	80	0	0	0	0

[1]Dose in μg/mouse subcutaneously (N = 10/group). All mice received 10 Gy gamma radiation at a dose rate of 1 Gy/min.

[2]These groups did not differ from each other and are significantly (p <0.05) different from 0.1-μg group.

[3]These groups did not differ from each other and are significantly different (p <0.05) from 40-μg group.

DISCUSSION

Radioprotective doses of DiPGE$_2$ (1-40 μg/mouse) resulted in rapid reduction of locomotor activity. Almost total cessation of ambulation occurred within 2 min after subcutaneous administration. The duration of the loss of activity was dose-dependent over a range of 0.1 to 40 μg. Activity recovered within 35 min following a dose of 0.1 μg, but remained depressed for up to 30 hours after the administration of 40 μg DiPGE$_2$. A dose of 0.01 μg did not disrupt locomotor activity. Prostaglandins of the E series administered to rodents have previously been shown to result in reduction of spontaneous locomotor activity (4,6-8), although the present study is the first to detail these effects for DiPGE$_2$.

Increasing the dose of DiPGE$_2$ enhanced the radioprotection but also extended the duration of the disruption of locomotor activity. A similar relationship between radioprotective efficacy and locomotor performance decrement in mice has been reported for other radioprotective compounds (11-13). The duration of the radioprotective activity of DiPGE$_2$ was much shorter than the disruptive effects on locomotor activity. For example, a dose of 10 μg resulted in 90% survival if injected 5 min preirradiation but only 10% survival wben administered 60 min before 10 Gy radiation. However, locomotor activity was significantly depressed from 2 min following injection, with total recovery delayed until 6 hours after drug administration. The differences in the duration of the radioprotective and behavioral effects may indicate that two separate processes are involved.

No significant difference was seen in the radioprotection induced by 1, 10, or 40 μg of DiPGE$_2$/mouse when administered 5 to 20 min prior to 10 Gy cobalt-60 radiation. However, at a higher dose of radiation (14.75 Gy), 40 μg DiPGE$_2$ has been shown to have greater radioprotective efficacy than a 10-μg dose (2).

The rapid onset of the reduction in locomotor activity may reflect a direct effect of DiPGE$_2$ on the central nervous system. Prostaglandins are capable of passing the blood-brain barrier (14), and have been shown to stimulate or modify neuronal activity (15-17) and disrupt behavior (3-8). Indirect effects on the central nervous system may result from a decrease in cerebral blood flow following the administration of DiPGE$_2$. PGE$_2$-induced sedation is believed to occur through this mechanism (4). Forty μg of DiPGE$_2$/mouse has been shown to increase the hematocrit from 61% to 71% within 10 min postadministration (2), and the cardiovascular response may also play a role in the suppression of activity.

The extended duration of depressed locomotor activity produced by radioprotective doses of DiPGE$_2$ may prevent the use of this compound as a general radioprotectant, particularly in those situations where operator activity must be maintained. The radioprotectant and the behavioral effects of DiPGE$_2$ appear to be the result of different mechanisms. Therefore, it may be possible to eliminate the deleterious behavioral side effects without the loss of radioprotective efficacy.

ACKNOWLEDGMENTS

We thank Dr. Douglas Morton of The UpJohn Company for very kindly supplying the prostaglandin used in this investigation. This research was supported by the Armed Forces Radiobiology Research Institute, Defense Nuclear Agency, under Research Work Units 4320-B4160 and 4230-B2152. The views presented in this paper are those of the authors; no endorsement by the Defense Nuclear Agency has been given or should be inferred. Research was conducted according to the principles enunciated in the "Guide for the Care and Use of Laboratory Animals" prepared by the Institute of Animal Laboratory Resources, National Research Council.

REFERENCES

1. Hanson, W. R., and Ainsworth, E. J. 16,16-Dimethyl prostaglandin E$_2$ induces radioprotection in murine intestinal and hematopoietic stem cells. Radiat. Res. 103: 196-203, 1985.
2. Walden, T. L., Jr., Patchen, M., and Snyder, S. L. 16,16-dimethyl prostaglandin E$_2$ increases survival in mice following irradiation. Radiat. Res. 109: 440-448, 1987.
3. Desiraju, T. Effect of intraventricularly administered prostaglandin E1 on the electrical activity of cerebral cortex and behavior in the unanesthetized monkey. Prostaglandins 3: 859-870, 1973.
4. Gilmore, D. P., and Shakh, A. A. The effect of prostaglandin E$_2$ in inducing sedation in the rat. Prostaglandins 2: 143-151, 1972.
5. Horton, E. W. Actions of prostaglandins E$_1$, E$_2$, and E$_3$ on the central nervous system. Brit. J. Pharmacol. 22: 189-192, 1964.

6. Weiner, M., and Olson, J. W. Comparative behavioral effects of dibutyryl cyclic AMP and prostaglandin E₁ in mice. Prostaglandins 9: 927-943, 1975.

7. Chiu, E. K. Y., and Richardson, J. S. Behavioral and neurochemical aspects of prostaglandins in brain function. Gen. Pharmac. 16: 163-175, 1985.

8. Potts, W. J., East, P. F., and Mueller, R. A. Behavioral effects. In: "The Prostaglandins," Vol. 2. P. W. Ramwell, ed. Plenum Press, New York, 1974, pp. 157-173.

9. Snyder, S. L., Walden, T. L., Jr., MacVittie, T. J., Patchen, M. L., and Fuchs, P. Radioprotective properties of detoxified lipid A from *Salmonella minnesota* R595. Radiat. Res. 107: 107-114, 1986.

10. Fleiss, J. L. "Statistical Methods for Rates and Proportions." John Wiley, New York, 1981, pp. 138-143.

11. Landauer, M. R., Davis, H. D., Dominitz, J. D., and Weiss, J. F. Dose and time relationships of the radioprotector WR-2721 on locomotor activity in mice. Pharmacol. Biochem. Behav. 27: 573-576, 1987.

12. Landauer, M. R., Davis, H. D., Dominitz, J. D., and Weiss, J. F. Comparative behavioral toxicity of four sulfhydryl radioprotective compounds in mice. WR-2721, cysteamine, diethyldithiocarbamate, and N-acetylcysteamine. Parmac. Ther., in press.

13. Landauer, M. R., Davis, H. D., Dominitz, J. D., and Weiss, J. F. Long-term effects of radioprotector WR-2721 on locomotor activity and body weight of mice following exposure to ionizing radiation. Toxicology, in press.

14. Bito, L. Z., Davson, H., and Hollingsworth, J. R. Facilitated transport of prostaglandins across the blood-cerebrospinal fluid and blood-brain barriers. J. Physiol. 253: 273-285, 1976.

15. Bergstrom, S., Farnebo, L. O., and Fuxe, K. Effect of prostaglandin E₂ on central and peripheral catecholamine neurons. Eur. J. Pharmacol. 21: 362-368, 1973.

16. Hayaishi, O. Prostaglandin D₂: A neuromodulator. In: "Advances in Prostaglandin, Thromboxane, and Leukotriene Research," Vol. 12. B. Samuelsson, R. Paoletti, and P. Ramwell, eds. Raven Press, New York, 1983, pp. 333-337.

17. Wolfe, L. S. Eicosanoids: Prostaglandins, thromboxanes, leukotrienes, and other derivatives of carbon-20 unsaturated fatty acids. J. Neurochem. 38: 1-14, 1982.

EFFECT OF PGE$_2$ ON RADIATION RESPONSE OF CHINESE HAMSTER V79 CELLS *IN VITRO*

E. V. Holahan[1], W. F. Blakely[1], and T. L. Walden[2]

[1]Radiation Sciences Department
[2]Radiation Biochemistry Department
Armed Forces Radiobiology Research Institute
Bethesda, Maryland 20814-5145

ABSTRACT

Several recent investigations have reported that 16,16-dimethyl prostaglandin E$_2$ (DiPGE$_2$) can protect murine intestinal epithelial cells and hematopoietic stem cells (CFU-S) *in vivo* from ionizing radiation. It has been postulated that PGE$_2$ may also increase radiation resistance *in vitro* by stimulating free radical scavenging or repair systems for oxidative damage. This study reports on the effect of PGE$_2$ in modifying radiation sensitivity in an *in vitro* mammalian cell line.

Chinese hamster V79A03 cells were cultured as monolayers in 6 ml of α-MEM medium supplemented with Earle's salts, 25 mM HEPES buffer, and 10% fetal bovine serum. Exponentially growing cells were incubated in medium containing 14 μM (5 μg/ml) PGE$_2$ for either 2 hr (acute) or >3 weeks (chronic) before exposure to graded doses of 250 kVp X rays. Cells were assayed for variations in intracellular levels of cyclic 3′,5′-adenosine monophosphate (cAMP), total protein, and glutathione (GSH), and radiation sensitivity was measured by cell survival before and after PGE$_2$ treatment.

An acute (2-hr) exposure induced a 25% increase in cAMP content with no significant change in intracellular GSH or protein and no effect on cell survival after exposure to radiation. Chronic exposure to PGE$_2$ increased intracellular GSH, protein, and cAMP levels by 82%, 3%, and 74%, respectively. However, no increase in radiation resistance was apparent following chronic exposure to PGE$_2$. Chronic PGE$_2$ exposures marginally increased the doubling time of the cells (10 versus 11 hr), although this perturbation was insufficient to alter radioresponse as a result of cell cycle perturbations. Consequently, the increase in the *in vivo* radiation response associated with PGE$_2$ treatment may not be the result of an intracellular response. Instead, the increased radiation resistance observed *in vivo* may be due to modifications such as localized tissue or organ system hypoxia.

INTRODUCTION

Prostaglandins are synthesized from arachidonic acid by most vertebrate tissues and certain invertebrate tissues. These prostaglandins modulate a number of humoral functions and possess a number of pharmacological actions whose physiological significance is only recently being deduced (for reviews, see references 1 and 2). Prostaglandin E_1 stimulates adenyl cyclase activity in a number of fibroblastic tissues, resulting in elevated cAMP levels (3-6). These increases can occur within 10 min of prostaglandin administration, and they remain elevated for up to 5 hr (4). Prostaglandin E_1 also is an effective radioprotective agent for cells irradiated *in vitro* (6,7). Prostaglandin E_1, when administered before irradiation, can stimulate a four- to fivefold increase in cAMP levels and decrease radiation sensitivity by increasing the shoulder of the radiation survival curve (7). Furthermore, Hanson et al. have demonstrated that a prostaglandin E_2 analogue, $DiPGE_2$, can also radioprotect murine intestinal epithelial cells (8,9) and hematopoietic stem cells (CFU-S) *in vivo* (9) by increasing the shoulder and decreasing the slope of the radiation survival curve if the analogue is injected before and not after irradiation.

The PGE-stimulated increase in adenyl cyclase and cAMP might be responsible for the decrease in mammalian cell radiosensitivity (6). Increased glutathione synthesis, which is coupled to the gamma-glutamyl amino acid transport system and whose biosynthesis is ATP dependent (10), may also contribute to the increased radioprotection observed *in vivo* and *in vitro*. We have postulated that the prostaglandin-induced increase in cAMP may also stimulate glutathione synthesis and thus confer a measurable amount of resistance to ionizing radiation. To test this hypothesis, we examined the effects of PGE_2 on the radiosensitivity of exponentially growing Chinese hamster fibroblasts.

MATERIALS AND METHODS

Chinese hamster V79A03 cells were cultured as monolayers on plastic in alpha MEM medium with Earle's salts and 25 mM HEPES buffer supplemented with 10% fetal bovine serum (Armour Pharmaceuticals, Tarrytown, NY), 14 mM sodium bicarbonate, 2 mM L-glutamine, and antibiotics. The cells were maintained at 37°C in a humidified incubator with a 3%-CO_2 and 97%-air mixture. For experiments in which asynchronous cells were treated, exponentially growing cells were trypsinized from monolayer culture, plated into 25-cm² flasks containing medium (6 ml at 37°C and pH 7.4), and incubated for an additional 18 to 24 hr before treatment (microcolony size of 1.5-2.2 cells per colony). The PGE_2 content of the medium supplemented with 10% fetal bovine serum was 38 pM (13.56 pg/ml). The number of cells inoculated into six replicate flasks was varied to yield 50-100 colonies per flask after treatment.

Except where indicated, the cells were incubated at 37°C with 14 μM (5 μg/ml) PGE_2 (Upjohn Co., Kalamazoo, MI) in medium for 2 hr. For chronic exposure to PGE_2, the cells were continuously subcultured every 48-72 hr for 3-5 weeks in medium containing 5 μg/ml PGE$_2$ without inducing any morphological changes. The medium was exchanged with fresh medium containing 5 μg/ml PGE$_2$ 2 hr before either irradiation or cell harvesting for cAMP or GSH assays.

The cells were irradiated on a rotating Plexiglas holder with a Phillips industrial X-ray machine (Philips GMBH, Hamburg, Germany) at room temperature (17°- 21°C). The irradiation conditions were 1.65 Gy/min with an effective energy of 80 keV (SSD = 43 cm, 250 kVp, 1 HVL = 1.0 mm Cu, 15 mA).

After drug and/or irradiation treatment, the cells were incubated for 8-10 days to allow macrocolony formation. The colonies were fixed to the flasks with methanol and stained with 0.5% crystal violet. Cell survival was adjusted for the plating efficiency and the average microcolony size at the time of treatment (11). The standard error of the mean is indicated when it is larger than the datum point. All curves have been fit by eye.

For the determination of intracellular glutathione content, a number of additional 100-mm petri dishes were inoculated with 2-5 x 10⁶ cells as described above. Following the various drug and/or irradiation treatments, the cells were rinsed with cold phosphate-buffered saline (PBS) and removed from the dishes by scraping the growth surface with a rubber policeman. The harvested cells were rinsed twice with cold PBS and then frozen in 0.5 ml of 0.6% sulfosalicylic acid. The cells frozen in sulfosalicylic acid were later thawed and assayed for total GSH content (oxidized and reduced) by the GSH reductase procedure (12). GSH content was expressed as nmoles/million cells and as nmoles/mg protein. Control values measured for 12 experiments ranged from 1.0 to 7.55 nmoles/million cells (0.30-2.32 μg GSH/ million cells).

For cAMP determinations, cells (4-5 x 10⁶) were harvested as described, and the cellular proteins were precipitated using cold trichloroacetic acid. The trichloroacetic acid was removed with ether, and the cAMP content was measured using a cAMP (¹²⁵I) radioimmunoassay kit (Dupont-New England Nuclear, Boston, MA). All samples were measured in triplicate with control values varying between 1.6 and 3.5 pmole cAMP/million cells for three separate experiments.

The stability of the PGE$_2$ in the tissue culture medium was determined by monitoring the radiolabeled products formed in the medium containing tritiated PGE$_2$. Briefly, the pH of a sample was adjusted to 3.0 with acetic acid followed by the addition of two volumes of ethyl acetate. The upper organic phase containing PGE$_2$ metabolites was removed, evaporated, and then resuspended in buffer containing 36% acetonitrile (v/v) and 64% phosphoric acid (pH 3.0). The PGE$_2$ samples were analyzed by reverse-phase high-performance liquid chromatography using a 5-micron Ultrasphere C-18 column (Beckman, San Ramon, CA), 4.6 x 250 mm. The samples were eluted from the column (1 ml/min) using the isocratic buffer. Labeled PGE$_2$ metabolites were detected using a radiation flow-through monitor (RAMOND-D, IN/US, Fairfield, NJ) and identified by coelution with known standards.

RESULTS

The stability of PGE_2 in medium alone and in medium when exposed to metabolically active V79 cells was monitored using tritium-labeled PGE_2. Radiolabeled medium was incubated at 37°C for 0 or 24 hr in the presence of $1.0-10 \times 10^5$ cells. After each exposure (indicated in Figure 1), medium samples were analyzed for PGE_2 metabolites. All of the samples analyzed contained 10%-25% PGA_2 as well as PGE_2. Only when the PGE_2 was exposed to 1×10^6 cells or more for a minimum of 24 hr were metabolites of PGE_2 detectable. 15-Keto PGE_2 (peak II), 13,14-dihydro-15-keto PGE_2 (peak III), and PGA_2 (peak IV) were the major metabolites observed. No other PGE_2 metabolites were detected, nor was PGE_2 converted to $PGF_{2\alpha}$. PGE_2 was stable at 37°C in cell-free medium containing 10% fetal bovine serum for up to 1 week.

An acute exposure to PGE_2 (14 μM) did not confer radioprotection *in vitro* (Figure 2). Exponentially growing cells were incubated for 2 hr in medium with PGE_2 before irradiation, but the radiosensitivity of the cells was not modified. In addition, when the concentration and pretreatment time were increased to 30 μM and 8 hr, respectively, no modification in radiation resistance was observed (data not shown).

The intracellular content of glutathione and cAMP was measured after the acute exposure to PGE_2 to determine if an increase in intracellular sulfhydryl content might increase the cellular free radical-scavenging capability after radiation exposure. A small increase in GSH and cAMP content was observed after an acute PGE_2 exposure (Table 1), but no increase in the GSH:protein ratio was evident.

In order to mimic the environmental conditions that PGE_2-secreting tumor cells might create (13), cells were continuously subcultured in medium containing PGE_2 for a minimum of 3 weeks. Continuous exposure to PGE_2 marginally increased the doubling time of the cells from 10.5 to 11 hr but did not alter the plating efficiency of the cells. The tissue culture medium was replaced with fresh medium in order to remove metabolite by-products of PGE_2 that might interfere with any radioprotection that might be induced. The data in Figure 3 indicate that chronic exposure to PGE_2 does not confer any additional radioprotection to the cells.

The intracellular GSH and cAMP contents were also measured after the chronic exposure to PGE_2 (Table 2). Under these conditions, glutathione and cAMP contents increased 82% and 75%, respectively, with no appreciable increase in cellular protein.

DISCUSSION

Prostaglandin-induced radioprotection of hematopoietic and intestinal stem cells *in vivo* was expressed as an increase in the D_0 and shoulder of the radiation survival

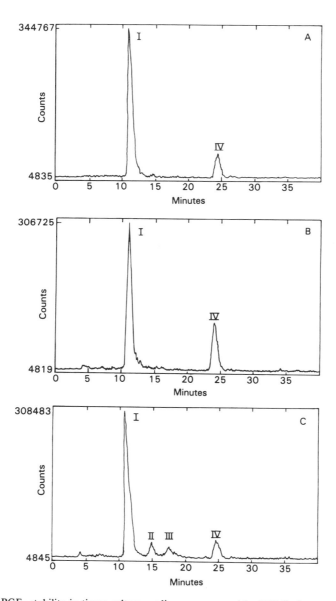

Fig. 1. PGE₂ stability in tissue culture medium as measured by HPLC chromatography. PGE₂ (5 μg/ml) was incubated at 37°C in presence of 0 cells (panel A), 1.0 x 10⁵ cells (panel B), or 1.0 x 10⁶ cells (panel C) for 24 hr. Four products eluted from chromatography columns were peak I: PGE₂ (11.5 min); peak II: 15-keto PGE₂ (13.8 min); peak III: 13,14-dihydro-15-keto PGE₂ (16.9 min); and peak IV: PGA₂ (24.1 min).

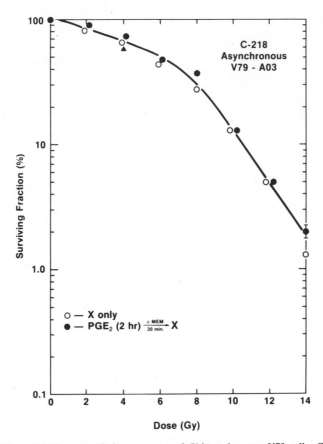

Fig. 2. Effect of PGE$_2$ on radiation response of Chinese hamster V79 cells. Cells were inoculated into flasks, incubated for 18 hr, and then exposed (2 hr) to either 0 μM (o) or 14 μM (●) PGE$_2$ before X irradiation. Each datum point represents mean survival of six replicate flasks.

Table 1. Effect of Acute PGE$_2$ Exposure on Intracellular cAMP and GSH Content[1]

	nmoles GSH/ 10^6 cells	μg protein/ 10^6 cells	nmoles GSH/ mg protein	pmoles cAMP/ 10^6 cells
Control	1.36 ± 0.07	164 ± 11	8.29 ± 0.31	2.0 ± 0.14
PGE$_2$	1.52 ± 0.10	178 ± 8	8.54 ± 0.55	2.5 ± 0.18
Relative Change	1.12	1.08	1.02	1.25

[1]Reported values represent mean of at least two separate measurements from each of three different experiments.

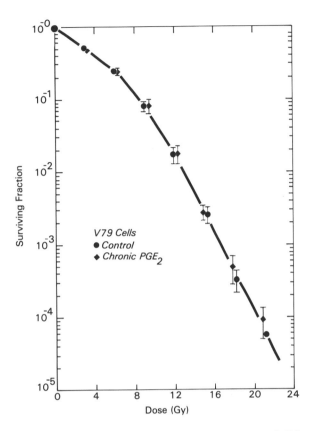

Fig. 3. Effect of chronic exposure to PGE₂ on radiation response of Chinese hamster cells. Exponentially growing cells were continuously subcultured for 3-5 wk in medium containing 5 μg/ml PGE₂. Two hr before irradiation, medium was exchanged with fresh medium containing PGE₂ (◆) and compared with survival of cells not exposed to PGE₂ (●). Datum points represent mean survival of four experiments, each with at least five replicate flasks per radiation exposure.

Table 2. Effect of Chronic PGE₂ Exposure on Intracellular cAMP and GSH Content[1]

	nmoles GSH 10^6 cells	μg protein 10^6 cells	nmoles GSH mg protein	pmoles cAMP 10^6 cells
Control	6.1 ± 0.53	163 ± 18	37.8 ± 2.7	2.0 ± 0.14
PGE₂	11.1 ± 1.33	169 ± 6.5	65.8 ± 7.1	3.5 ± 0.02
Relative Change	1.82	1.03	1.74	1.75

[1]Reported values represent mean of at least two different measurements from each of three different experiments.

curve (8,9). The *in vivo* radioprotection was postulated to result from a rapid increase in intracellular sulfhydryl compounds, rather than direct free radical scavenging by DiPGE$_2$, localized hypoxia, or a drug-induced cell cycle redistribution.

In this study we exposed Chinese hamster cells to concentrations of PGE$_2$ that were significantly greater than the serum concentrations of the C3H mice used in the *in vivo studies* (8,9), to determine if the radioprotective activity had a cellular or molecular mechanism of action. PGE$_2$ (C20:H32:05) does not contain any sulfhydryl moieties, and although it may interact with free radicals by attacking unsaturated bonds, it is unlikely that this proceeds at a significant rate. Exposure of cells to PGE$_2$, either acutely or chronically, failed to stimulate any increased radiation resistance (Figures 2 and 3). Under both conditions, the PGE$_2$ was not metabolically degraded (Figure 1), an increase was seen in the cellular cAMP content (Tables 1 and 2), and a significant increase in glutathione content was demonstrated after chronic PGE$_2$ exposure (Table 2).

The increase in cellular GSH may have been induced by chronic exposure to PGE$_2$, but it is also possible that the increased level of GSH may have been stimulated by a PGA$_2$ contaminant that could also have been introduced to the medium or the conversion of PGE$_2$ to PGA$_2$ in a slightly acidic medium environment. Millar and Jinks (14) demonstrated that exposure to 10 μg/ml PGA$_2$ for 46 hr could stimulate a 64% increase in the amount of GSH that could be isolated from V79-753B cells. However, this PGA$_2$-induced stimulation of GSH synthesis did not modify radiation sensitivity. Russo and Mitchell (15) also reported the failure of elevated GSH levels to alter radiosensitivity in a V79 cell line. Oxo-thiazolidine-4-carboxylate (10 mM) stimulated GSH levels to 200%-300% above controls. No decrease in radiosensitivity was observed, suggesting that GSH-mediated free radical scavenging and/or GSH peroxidase-mediated detoxification of hydrogen peroxide were operating at maximum velocity. In contrast, GSH exogenously supplied to a human lymphoid has been shown to confer some radioprotection (16). It appears that the radiation resistance of V79 cells may not be easily manipulated by simply varying cellular GSH content. While the increased radiation resistance observed for mouse stem cells *in vivo* might be the result of increased GSH content, PGE$_2$ exposure induced no additional radioresistance in V79 cells *in vitro*.

Another popular hypothesis regarding prostaglandin-induced radiation resistance involves the stimulation of cAMP content. Although the specific mechanism has not been deduced, variations in intracellular cAMP have been temporally correlated with changes in radiation resistance (6,7,17,18). This does not appear to be the case with our V79 cells. The chronic exposure to PGE$_2$ increased cAMP by 75% (Table 2), had no effect on radiation resistance (Figure 3), and increased the cell doubling time by approximately 30 min compared to the control cells. Although these data are not entirely inconsistent with other reports, it does appear that cAMP may be the important prostaglandin (E$_1$ and E$_2$) second messenger and not the only factor involved for radioprotection.

ACKNOWLEDGMENTS

The authors thank E. J. Golightly, J. J. Raymond, M. A. Behme, D. P. Dodgen, and C. L. Hollies for their excellent technical support and Dr. L. K. Steel for the serum PGE₂ measurements. Supported by the Armed Forces Radiobiology Research Institute, Defense Nuclear Agency, under Research Work Unit 4640-B5104. The views presented in this paper are those of the authors; no endorsement by the Defense Nuclear Agency has been given or should be inferred.

REFERENCES

1. Samuelsson, B., Granstrom, E., Green, K., Hamberg, M., and Hammarstrom, S. Prostaglandins. Ann. Rev. Biochem. 44: 669-695, 1975.
2. Samuelsson, B., Goldyne, M., Granstrom, E., Hamberg, M., Hammarstrom, S., and Malmsten, C. Prostaglandins and thromboxanes. Ann. Rev. Biochem. 47: 997-1029, 1978.
3. Makman, M. H. Conditions leading to enhanced response to glucagon, epinephrine, or prostaglandins by adenyl cyclase of normal and malignant cultured cells. Proc. Nat. Acad. Sci. USA 68: 2127-2130, 1971.
4. Maganiello, V., and Vaughan, M. Prostaglandin E₁ effects on adenosine 3':5'-cyclic monophosphate concentration and phosphodiesterase activity in fibroblasts. Proc. Nat. Acad. Sci. USA 69: 269-273, 1972.
5. Penit, J., Cantau, B., Huot, J., and Jard, S. Adenylate cyclase from synchronized neuroblastoma cells: Responsiveness to prostaglandin E₁, adenosine, and dopamine during the cell cycle. Proc. Nat. Acad. Sci. USA 74: 1575-1579, 1977.
6. Lehnert, S. Modification of postirradiation survival of mammalian cells by intracellular cyclic AMP. Radiat. Res. 62: 107-116, 1975.
7. Prasad, K. N. Radioprotective effect of prostaglandin and an inhibitor of cyclic nucleotide phosphodiesterase on mammalian cells in culture. Int. J. Radiat. Biol. 22: 187-189, 1972.
8. Hanson, W. R., and Thomas, C. 16,16-dimethyl prostaglandin E₂ increases survival of murine intestinal cells when given before photon irradiation. Radiat. Res. 96: 393-398, 1983.
9. Hanson, W. R., and Ainsworth, E. J. 16,16-dimethyl prostaglandin E₂ induces radioprotection in murine intestinal and hematopoietic stem cells. Radiat. Res. 103: 196-203, 1985.
10. Meister, A., and Anderson, M. E. Glutathione. Ann. Rev. Biochem. 53: 711-760, 1983.
11. Sinclair, W. K., and Morton, R. A. X-ray and ultraviolet sensitivity of synchronous chinese hamster cells at various stages of the cell cycle. Biophys. J. 5: 1-25, 1965.
12. Tietz, F. Enzymatic method for quantitative determination of nanogram amounts of total and oxidized glutathione: Applications to mammalian blood and other tissues. Anal. Biochem. 27: 502-522, 1969.

13. Seyberth, H. W., Segre, G. V., Morgan, J. L., Sweetman, J. R., Potts, J. T., and Oates, J. A. Prostaglandins as mediators of hypercalcemia associated with certain types of cancer. N. Engl. J. Med. 293: 1278-1283, 1975.

14. Millar, B. C., and Jinks, S. Do prostaglandins affect cellular radiosensitivity in vitro? Int. J. Radiat. Biol. 46: 367-373, 1984.

15. Russo, A., and Mitchell, J. B. Radiation response of Chinese hamster cells after elevation of intracellular glutathione levels. Int. J. Radiat. Oncol. Biol. Phys. 10: 1243-1247, 1984.

16. Jensen, G. L., and Meister, A. Radioprotection of human lymphoid cells by exogenously supplied glutathione in mediated γ-glutamyl transpeptidase. Proc. Nat. Acad. Sci. USA 80: 4714-4717, 1983.

17. Lehnert, S. Intracellular cyclic AMP levels and radiosensitivity in synchronized V-79 cells. Radiat. Res. 64: 394-398, 1975.

18. Chirkov, Y. Y., and Sobolev, A. S. Prostaglandin E_1: cAMP level and radiosensitivity of cultured B-82 cells. Strahlentherapie 160: 521-522, 1984.

A PARADOXICAL ROLE FOR EICOSANOIDS: RADIOPROTECTANTS AND RADIOSENSITIZERS

T. L. Walden, Jr.

Department of Radiation Biochemistry
Armed Forces Radiobiology Research Institute
Bethesda, Maryland 20814-5145

ABSTRACT

Understanding the radiobiology of eicosanoids is complicated by their ability to act as mediators of damage and recovery and as radioprotective agents. Changes in the tissue concentrations of eicosanoids following irradiation are dependent on several factors, including the type of eicosanoid, time postirradiation, radiation dose, and other contributing mediators and enzyme changes in the surrounding microenvironment. Many of these same prostaglandins and the leukotrienes have been shown to be radioprotective when given before irradiation.

INTRODUCTION

The eicosanoids are a group of biological mediators that have received attention in the field of radiobiology as mediators of radiation injury (1-3) and recently as radioprotective agents (4-6). They are metabolites of arachidonic acid (Figure 1), an essential 20-carbon fatty acid containing four unsaturated double bonds (7). Arachidonic acid is primarily esterified in the second position of the glycerol backbone of phospholipids in the cell membranes. Ultraviolet and ionizing radiation stimulate the release of free arachidonic acid through the action of phospholipases (8-10). Increased calcium concentrations also stimulate the phospholipase release of arachidonic acid. Following arachidonic acid release, one of three events occurs (outlined in Figure 1): (a) re-esterification of the free arachidonic acid back into the cell membrane, or (b) and (c) metabolism through the arachidonic acid cascade. Arachidonic acid is metabolized (Figure 1) through either the cyclooxygenase pathway (B), leading to the formation of prostaglandins, thromboxane, and prostacyclin, or the lipoxygenase pathway (C), leading to the formation of leukotrienes, lipoxins, and hydroxy fatty acids. These compounds have a number of important physiological roles in vasoregulation, smooth muscle regulation, electrolyte balance, and neuro-regulation, as well as pathological roles in inflammation, fever, pain, and shock (reviewed in reference 7).

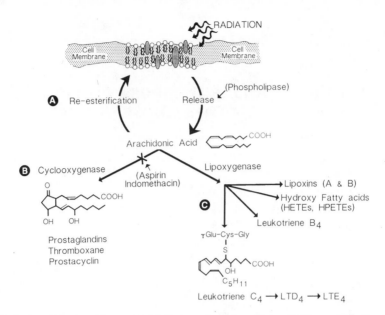

Fig. 1. Arachidonic acid cascade. Radiation stimulates release of free arachidonic acid from phospholipids of cell membrane. Once released, arachidonic acid may be (a) re-esterified back into cell membrane, (b) metabolized through cyclooxygenase pathway, or (c) metabolized through lipoxygenase pathway. Compounds that block either pathway may be exerting a beneficial effect by inhibiting production of a particular eicosanoid or by shunting more arachidonic acid into products of other pathway.

PROSTAGLANDINS IN RADIATION INJURY

The role of prostaglandins in radiation injury has been recently reviewed (11). Irradiation has been shown to alter tissue prostaglandin levels, producing both elevations (12-17) and decreases (1,17,18). The response following irradiation depends on dose received (18,19), tissue irradiated (1,13-19), and time of determination postirradiation (1). Some radiation-induced alterations in prostaglandin synthesis may persist for months after exposure to radiation or, conversely, may not become altered until several months postexposure (12). Prostaglandin changes following irradiation have been demonstrated *in vitro* in fibroblasts (11), endothelial cells (20), tissue slices, homogenates (21,22), and urine (23). Several effects of radiation on prostaglandin metabolism in the blood vessels are summarized in Figure 2. Alteration of tissue prostaglandin levels may result from either anabolic or catabolic processes, or from combinations of the two. Exposure of minipig skin to X radiation results in the increased production of prostaglandin E_2 (PGE$_2$) over the first 24 hours postirradiation (13). Early PGE$_2$ elevation in X-irradiated minipig skin is followed by a progressive decrease in PGE$_2$, which corresponds to an increase in prostaglandin F_2 (PGF$_2$) tissue levels. These latter processes are the result of an increase in 9-keto prostaglandin reductase, an enzyme that converts PGE$_2$ to PGF$_2$. There is an association between alterations in the spleen prostaglandin levels of irradiated mice and alterations in the activity of prostaglandin dehydrogenase (24), an enzyme responsible for prostaglandin inactivation. Increases in prostaglandin E_1 (PGE_1) in the spleen postirradiation are caused by decreased enzymatic catabolism, leading to accumulation.

A role for prostaglandins as mediators of radiation injury has been suggested in experiments using cyclooxygenase-inhibiting drugs, such as indomethacin (3) or aspirin (2,25), to reduce specific radiation-induced inflammatory responses. These experiments indicate that prostaglandins contribute to radiation-induced ocular tissue inflammation in rabbits (2), mucositosis in humans (14), esophagitis in opossums (3), and gastrointestinal syndrome in mice (26). Prostaglandins may mediate inflammation through increased extravasation, vasoregulation, and fever and as chemotactic factors for phagocytic white blood cells (reviewed recently in references 7,27). Radiation may also affect the ability of the receptor to bind the prostaglandin and induce a specific function. The specific binding of prostaglandin E_2 is decreased in the spleens and small intestines of irradiated mice (28). The effects of indomethacin on alteration of whole animal survival are contradictory, and have been shown to either enhance survival (29) or have no effect (30). The reasons for the different responses to nonsteroidal anti-inflammatory drugs are not

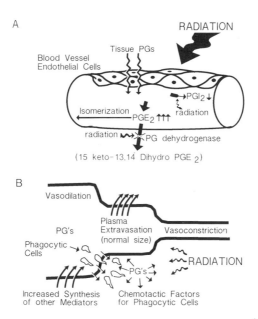

Fig. 2. A: Effects of radiation on prostaglandin metabolism by blood vessels. This figure presents a brief summary and is based on observations observed in human umbilical arteries (ref. 18) and in mouse spleen (ref. 24). B: Prostaglandin effects on blood vessels. Radiation induces an alteration of prostaglandin concentrations. These compounds may be either vasodilatory or vasoconstrictive, depending on class of prostaglandin released, particular vascular site, and prostaglandin concentration. As indicated, they may induce synthesis of other mediators or act as chemotactic agents.

known, but one may speculate that they are related to the dosage and schedule of administration.

PROSTAGLANDINS AND LEUKOTRIENES IN RADIOPROTECTION

Paradoxically, the prostaglandins have pathological roles in damage, but they also function as radioprotectants for cells in culture (6,31) and hematopoietic and intestinal stem cells (4) *in vivo,* and also enhance whole animal survival (5). Several studies on prostaglandin-induced modification of radiosensitivity in cell culture provide evidence that protection may be associated with elevations in cyclic AMP (6,31). These studies have centered on the use of PGE_1, a potent cAMP stimulus. For prostaglandin enhancement of whole animal survival from otherwise lethal exposure to ionizing radiation, the processes are more complicated, and the basic underlying mechanism(s) remains unknown. The most effective prostaglandin in terms of whole animal survival (31; Walden et al., submitted) is 16,16-dimethyl prostaglandin E_2 ($DiPGE_2$) (Figure 3), an analog of the naturally occurring prostaglandin E_2. Misoprostil, an analog of PGE_1, appears to be more effective for protection of the intestinal crypt cells (32). $DiPGE_2$ has a biological half-life in the tens of minutes (5), as illustrated and explained in Figure 3. The naturally occurring prostaglandin E_2 has a half-life of 2 minutes (7). Forty μg of $DiPGE_2$/mouse (1.6 mg/kg body weight) enhances the LD50/30 of mice exposed to cobalt-60 gamma radiation, providing a dose modification factor of 1.72 (5). Radioprotection by eicosanoids is both time and dose dependent, and must be

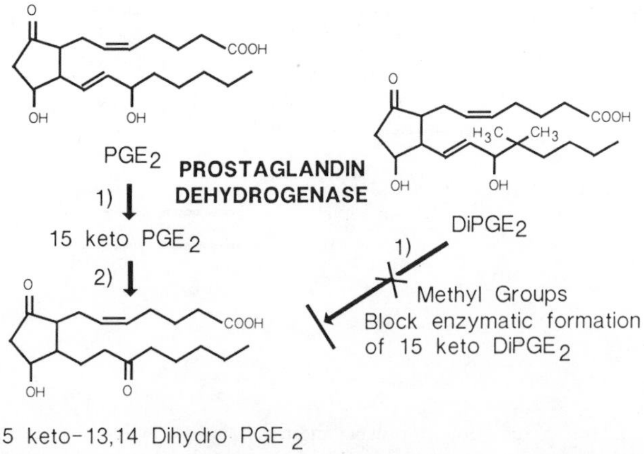

Fig. 3. Inactivation of prostaglandin E_2 but not 16,16-dimethyl prostaglandin E_2 ($DiPGE_2$). Prostaglandin E_2 in vasculature is normally inactivated by lung through action of prostaglandin dehydrogenase. Methyl groups in 16 position of $DiPGE_2$ prevent this inactivation step from occurring at 15 position, extending biological half-life. $DiPGE_2$ is primarily inactivated by liver (ref. 39).

administered just prior (approximately 15 min) to irradiation (5,33; Walden et al., submitted). There does not appear to be a common vascular end point induced by radioprotective eicosanoids, since some are vasodilatory and others are vasoconstrictive, and there is no general effect on the hematocrit (Walden et al., submitted). Pharmacological studies indicate that the $DiPGE_2$ protection results from the parent analog rather than a metabolite. However, the optimal period of radioprotection does not correlate with the optimal concentrations of $DiPGE_2$ in the tissue (5).

A decrease in locomotor activity (33) is produced by radioprotective concentrations of $DiPGE_2$. The optimal time for protection (5-60 min) is shorter than the duration of the locomotor behavior decrement (33). A marked depression of the locomotor activity occurred within 5 min postadministration of $10\,\mu g$ $DiPGE_2$/mouse or greater. The behavior of $DiPGE_2$-treated mice returns to control levels by 6 hours following a 10-μg $DiPGE_2$ dose and 30 hours following a 40-μg dose (33). Diarrhea is another undesirable effect of some but not all of the radioprotective eicosanoids (Walden et al., submitted). It may be possible in the future to separate the detrimental effects without a decrease in protective efficacy.

Most work on the radiobiology of arachidonic acid metabolites to date has centered on the prostaglandins, primarily because the standards and radioimmunoassay kits for analyses are commercially available. Arachidonic acid may also be metabolized by the lipoxygenase pathway (Figure 1) to form leukotrienes and hydroxy fatty acids. Ultraviolet-B irradiation of human skin produced a threefold elevation of 12-hydroxy eicosatetraenoic acid 24 hours postirradiation, although no changes in leukotriene B_4 levels were observed (34). Leukotriene C_4, a thiol ether of glutathione and the triene-containing 20-carbon backbone, is elevated in the plasma of mice receiving hematoporphyrin derivative-induced phototherapy (400- to 410-nm wavelength radiation) and may be related to mast cell degranulation (35). Leukotriene C_4 modifies the radiosensitivity of V79A03 Chinese hamster cells in culture (36). Two-hour pretreatment with $2.5\,\mu M$ leukotriene C_4 doubles the number of cells surviving a subsequent X irradiation (reproductive survival, based on colony formation). This protection is concentration dependent and appears to be associated with specific leukotriene C4-binding sites on the cell surface. Leukotriene C_4 also induces radioprotection and enhances the LD50/30 of mice (Walden, submitted).

Future research directions in eicosanoid radioprotection need to focus on the mechanism of action for this novel class of biological mediators/radioprotective agents. Effective radioprotective concentrations of arachidonic acid metabolites are in the μg/animal range (1.6 μg/kg body weight) (4,5), compared to mg/mouse (200-800 mg/kg body weight) required for the classical thiol radioprotectants. Several mechanisms may be involved in protection. The role of specific receptor activation provides the opportunity to enhance and control the protection. If a primary component of protection is receptor mediated, it may be possible to selectively protect all normal tissues that have receptors, but not those normal and tumor tissues that lack the receptor. Eicosanoid-induced events mediated on the cellular level by a receptor may enhance their effects on these cells and also other tissues through systemic responses to eicosanoids that might occur through the cardio-vascular system. Actions at these higher levels may play significant roles in the radioprotection. The success of the long-lived DiPGE$_2$, an analog of PGE$_2$, as a protective agent points to the need to develop other biologically stable eicosanoid analogs that retain radioprotective efficacy with minimal side effects.

Tumors secreting biological mediators with protective activity may modify the efficacy of the therapy. This appears to be the case for several of the prostaglandin-secreting tumors (37). The prostaglandin may conceivably protect the tumor at the cellular level by elevation of cyclic AMP (5,31) or glutathione (38), or at a systemic level by suppression of the immune system (1). An interesting set of experiments relating to this issue were conducted using the HSDM1C1 mouse fibrosarcoma cell line, which has been shown to secrete high levels of PGE$_2$ (38). Radiation clonogenic survival curves were performed in the presence or absence of flurbiprofen, a cyclooxygenase inhibitor. The two survival curves do not significantly differ, and may indicate that the PGE$_2$ produced by this tumor cell line does not feed back and modify its own cellular radiosensitivity. It is not known whether this cell line has receptors to PGE$_2$, and if the radioprotection is receptor mediated, then the lack of PGE$_2$ receptors may explain the inability of PGE$_2$ to act as a radioprotectant in this cell line.

ACKNOWLEDGMENTS

This research was supported by the Armed Forces Radiobiology Research Institute (AFRRI), Defense Nuclear Agency, under Research Work Unit 000153. The views presented in this paper are those of the authors; no endorsement by the Defense Nuclear Agency has been given or should be inferred. Any AFRRI research included in this review was performed according to the principles enunciated in the "Guide for the Care and Use of Laboratory Animals," prepared by the Institute of Laboratory Animal Resources, National Research Council.

REFERENCES

1. Trocha, P. J., and Catravas, G. N. Prostaglandins, lysosomes, and radiation injury. In: "Advances in Prostaglandin and Thromboxane Research," Volume 7. B. Sammuelsson, P. W. Ramwell, and R. Paoletti, eds. Raven Press, New York, 1980, pp. 851-856.
2. Bito, L. Z., and Klein, E. M. The role of arachidonic acid cascade in the species-specific X-ray-induced inflammation of the rabbit eye. Invest. Opthalmol. Vis. Sci. 22: 579-587, 1982.
3. Northway, M. G., Libshitz, H. I., Osborne, B. M., Feldman, M. S., Mamel, J. J., West, J. H., and Szwarc, I. A. Radiation esophagitis in the opossum: Radioprotection with indomethacin. Gastroenterology 78: 883-892, 1980.

4. Hanson, W. R., and Ainsworth, E. J. 16,16-Dimethyl prostaglandin E2 induces radioprotection in murine intestinal and hematopoietic stem cells. Radiat. Res. 103: 196-203, 1985.

5. Walden, T. L., Jr., Patchen, M. L., and Snyder, S. L. 16,16-Dimethyl prostaglandin E_2 increases survival in mice following irradiation. Radiat. Res. 109: 440-448, 1987.

6. Prasad, K. N. Radioprotective effect of prostaglandin and an inhibitor of cyclic nucleotide phosphodiesterase on mammalian cells in culture. Int. J. Radiat. Biol. 22: 187-189, 1972.

7. Ogburn, P. L., and Brenner, W. E."The Physiologic Actions and Effects of Prostaglandins." UpJohn, Kalamazoo, 1981.

8. Camp, R. D., Greaves, M. W., Hensby, C. N., Plummer, N. A., and Warin, A. P. Irradiation of human skin by short wavelength ultraviolet radiation (100-290 nm) (u.v.C): Increased concentrations of arachidonic acid and prostaglandins E2 and F2a. Br. J. Clin. Pharmac. 6: 145-148, 1978.

9. Black, A. K., Greaves, M. W., Hensby, C. N., Plummer, N. A., and Warin, A. P. The effects of indomethacin on arachidonic acid and prostaglandins E2 and F2a levels in human skin 24 hours after u.v.B. and u.v.C. irradiation. Br. J. Clin. Pharmac. 6: 261-266, 1978.

10. Snyder, D. S. Effect of topical indomethacin on UVR-induced redness and prostaglandin E levels in sunburned guinea pig skin. Prostaglandins 11: 631-643, 1976.

11. Polgar, P., Hahn, G., and Taylor, L. Radiation effects on eicosanoid formation. In: "Biochemistry of Arachidonic Acid Metabolism." W. E. M. Lands, ed. Kluwer Academic Publishers, Boston, 1985, pp. 161-173.

12. Ts'Ao, C. H., Ward, W. F., and Port, C. D. Radiation injury in rat lung. I. Prostacyclin (PGI2) production, arterial perfusion and ultrastructure. Radiat. Res. 96: 284-293, 1983.

13. Ziboh, V. A., Mallia, C., Morhart, E., and Taylor, J. R. Induced biosynthesis of cutaneous prostaglandins by ionizing radiation. Proc. Soc. Exp. Biol. Med. 169: 386-391, 1982.

14. Tanner, N. S., Stamford, I. F., and Bennett, A. Plasma prostaglandins in mucositis due to radiotherapy and chemotherapy for head and neck cancer. Br. J. Cancer 43: 767-771, 1981.

15. Eisen, V., and Walker, D. I. Effect of ionizing radiation on prostaglandin-like activity in tissues. Br. J. Pharmac. 57: 527-532, 1976.

16. Steel, L. K., Rafferty, M. A., Wolfe, W. W., Egan, J. E., Kennedy, D. A., Catravas, G. N., Jackson, W. E., III, and Dooley, M. A. Urinary excretion of cyclic nucleotides, creatinine, prostaglandin E2, and thromboxane B2 from mice exposed to whole-body irradiation from an enhanced neutron field. Int. J. Radiat. Biol. 50: 695-715, 1986.

17. Trocha, P. J., and Catravas, G. N. Effect of radioprotectant WR2721 on cyclic nucleotides, prostaglandins, and lysosomes. Radiat. Res. 94: 239-251, 1983.

18. Allen, J. B., Sagerman, R. H., and Stuart, M. J. Irradiation decreases vascular prostacyclin formation with no concomitant effect on platelet thromboxane production. Lancet 2: 1193-1195, 1981.

19. Pausescu, E., Chirvasie, R., Popescu, M.-V., and Tedodsiu, T. Dynamics of the changes in the cerebral amounts of cyclic AMP and some prostaglandins during cobalt-60 gamma-radiation-induced brain edema. Physiologie 14: 283-289, 1977.

20. Rubin, D. B., Drab, E. A., Ts'Ao, C. H., Gardner, D., and Ward, W. F. Prostacyclin synthesis in irradiated endothelial cells cultured from bovine aorta. J. Appl. Physiol. 58: 592-597, 1985.

21. Steel, L. K., Sweedler, I. K., and Catravas, G. N. Effects of ^{60}Co radiation on synthesis of prostaglandins F2a, E, and thromboxane B2 in lung airways of the guinea pigs. Radiat. Res. 94: 156-165, 1983.

22. Gal, D., Strickland, D. M., Lifshitz, S., Buchsbaum, H. J., and Mitchell, M. D. Effect of radiation on prostaglandin production by human bowel *in vitro*. Int. J. Radiat. Oncol. Biol. Phys. 10: 653-657, 1984.

23. Donlon, M., Steel, L. K., Helgeson, E. A., Shipp, A., and Catravas, G. N. Radiation induced alterations in prostaglandin excretion in the rat. Life Sci. 32: 2631-2639, 1983.

24. Walker, D. I., and Eisen, V. Effect of ionizing radiation on 15-hydroxy prostaglandin dehydrogenase (PGDH) activity in tissue. Int. J. Radiat. Biol. 36: 399-407, 1979.

25. Mennie, A. T., and Dalley, V. M. Aspirin in radiation-induced diarrhea. Lancet 1: 1131, 1973.

26. Borowska, A., Sierakowski, S., Mackowiak, J., and Wisiniewski, K. A prostaglandin-like activity in small intestine and postirradiation gastrointestinal syndrome. Experientia 35: 1368-1370, 1979.

27. Weissman, G. "Prostaglandins in Acute Inflammation." UpJohn, Kalamazoo, 1980.

28. Prianishikova, E. N., Zhulanova, Z. I., and Agaf'eva, V. S. Effect of different doses of ionizing radiation on the binding of prostaglandin E2 in mouse tissues. Radiobiologiia (USSR) 26: 652-655, 1986.

29. Popisil, M., Netikova, J., and Kozubik, A. Enhancement of hematopoietic recovery by indomethacin after sublethal whole-body gamma irradiation. Acta Radiol. Oncol. 25: 195-198, 1987.

30. Steel, L. K., and Ledney, G. D. Effects of neutron irradiation on PGE2 and TXB2 levels in biological fluids: Modification by WR-2721. In: "Prostaglandins and Lipid Metabolism in Radiation Injury." T. L. Walden and H. N. Hughes, eds. Plenum Press, New York, 1987 (this volume).

31. Lehnert, S. Modification of postirradiation survival of mammalian cells by intracellular cyclic AMP. Radiat. Res. 62: 107-116, 1975.

32. Hanson, W. R., Houseman, K. A., and Collins, P. W. *In vivo* radiation protection by prostaglandins and related compounds of the arachidonic acid cascade. Pharmacol. Ther., in press.

33. Landauer, M. R., Walden, T. L., Davis, H. D., and Dominitz, J. A. Alterations in locomotor activity induced by radioprotective doses of 16,16-dimethyl prostaglandin E2. In: "Prostaglandins and Lipid Metabolism in Radiation Injury." T. L. Walden and H. N. Hughes, eds. Plenum Press, New York, 1987 (this volume).

34. Black, A. K., Barr, R. M., Wong, E., Brain, S., Greaves, M. W., Dickinson, R., Shroot, B., and Hensby, C. N. Lipoxygenase products of arachidonic acid in human inflamed skin. Br. J. Clin. Pharmacol. 20: 185-190, 1985.

35. Kerdel, F. A., Soter, N. A., and Lim, H. W. *In vivo* mediator release and degranulation of mast cells in hematoporphyrin derivative-induced phototoxicity in mice. J. Invest. Dermatol. 88: 277-280, 1987.

36. Walden, T. L., Jr., Holahan, E. V., and Catravas, G. N. Development of a model system to study leukotriene-induced modification of radiation sensitivity in mammalian cells. Prog. Lipid Res. 25: 587-590, 1986.

37. Bennett, A. Cancer. In: "Biological Protection With Prostaglandins," Volume I. M. M. Cohen, ed. CRC Press, Boca Raton, Florida, 1986, pp. 73-80.

38. Millar, B. C., and Jinks, S. Do prostaglandins affect cellular radiosensitivity *in vitro?* Int. J. Radiat. Biol. 46: 367-373, 1984.

39. Weeks, J. R., Ducharme, D. W., Magee, W. E., and Miller, W. L. The biological activity of the (15S)-15-methyl analogs of prostaglandin E2 and F2. J. Pharmac. Exp. Ther. 186: 67-74, 1973.

CLINICAL IMPLICATION

EICOSANOIDS AND ELECTIVE IMMUNOSUPPRESSION

P. W. Ramwell, M. L. Foegh, and P. Kot

Georgetown University Medical Center
Department of Physiology and Biophysics
and
Department of Surgery Division of Transplantation
Washington, DC 20007

ABSTRACT

The eicosanoids are the largest class of biologically active lipid mediators. They are believed to constitute one of the oldest and most ubiquitous of all the physiologic. homeostatic mechanisms. An increasing number of plant products are being found to relate to eicosanoid synthesis and metabolism. Tissue dysfunction may readily occur following injury because the eicosanoids are not stored per se, but are rapidly released in large amounts for prolonged periods of time from the readily available long-chain polyenoic precursors. The pharmacological characterization of the eicosanoids permits the simplistic but useful division of the eicosanoids into "cytoprotective" and "pathogenic." This has led to a pharmaceutical strategy of synthesizing stable analogs of the former and inhibitors and receptor antagonists of the latter. Such drugs facilitate the determination in different types of injury of the degree of involvement of the pathogenic mediators and also the therapeutic potential of the cytoprotective drugs. In general, the cytoprotective analogs are vasodilators, promote increase in cyclic adenosine monophosphate (AMP), enhance immunosuppression, and prolong allograft survival. In contrast, the pathogenic eicosanoids are vasoconstrictors and are associated with increased calcium input and lymphocyte proliferation. Some pathogenic mediators (e.g., platelet-activating factor, bradykinin) may act in part by activating acylhydrolases and thus promoting eicosanoid synthesis. However, other mediators do not, and therefore calcium entry-blocking drugs may be more generally useful than eicosanoid synthase inhibitors and receptor antagonists.

INTRODUCTION

This introduction briefly outlines the accepted and well-recognized biological properties of the eicosanoids, describes their relationships, discusses procedures for amplifying or interdicting their actions, and finally draws attention to special features and situations that may be relevant to radiobiology, such as immunosuppression.

The eicosanoids constitute a subclass of lipids that demonstrate high potency and wide biological activity (1). This class of lipids has been termed lipid mediators and includes platelet-activating factor (PAF), acetal plasmalogens such as Darmstoff, diacyl glycerol, and the polyphosphoinositides. The eicosanoids are distinguished from these other lipid mediators in not containing phosphate or glycerol and, moreover, are formed by oxygenation of long-chain polyunsaturated fatty acids. The term icosanoid was coined originally by E. J. Corey, but eicosanoid is now the preferred spelling. The term is valuable because it covers the oxygenation products of not only the three well-known $C20\omega6$ carboxylic acids (homolinolenic, arachidonic, and eicosapentaenoic acids) but also docosahexanoic and other substrates for the different oxygenases.

The oxygenases discussed here are the cyclooxygenases and lipoxygenases. Regarding the significance of the cyclooxygenase, not only is it the means of synthesis of the prostaglandins and thromboxane but also it is easily blocked by a large number of clinically available nonprescription and prescription drugs. Further, it is an enzyme that is activated by peroxides, a prominent feature of radiation injury.

Most is known about the lipoxygenase that attacks the 5 position of arachidonic acid. An important feature of this enzyme is the requirement for Ca^{++} activation (2). Thus one might expect elevated Ca^{++} entry and also Ca^{++} mobilization following irradiation to increase leukotriene synthesis.

The relation of radiation to Ca^{++} is central to many of the ideas regarding radiation and eicosanoids since Ca^{++} is also a requirement for phospholipase A_2 activation and subsequent de-esterification of arachidonate. Radiation may also mobilize Ca^{++} through increased phosphoinositide turnover. Thus it is not unexpected that acute radiation injury is accompanied by eicosanoid release (3). Another reason for focusing on Ca^{++} is that it is a key determinant for the action of thromboxane on vascular tissue and many cell types. For this reason, the Ca^{++}-blocking drugs have been used to prevent eicosanoid synthesis and also inhibit thromboxane expression, especially in vascular tissue (4). Similarly, eicosanoids and drugs that elevate intracellular cyclic AMP and lower free intracellular Ca^{++} are frequently conceived to be protective. The term cytoprotective was coined originally by A. Robert (5) to describe this property of the prostaglandin E (PGE) compounds in preventing gastric ulcers. The term is now far more widely used, especially with respect to prostacyclin.

Thus, a case can be made for classifying the eicosanoids on the basis of their role on Ca^{++}. While those prostaglandins that increase cyclic AMP are protective, the leukotrienes and thromboxane appear to be pathophysiologic mediators since they promote lymphocyte proliferation, bronchoconstriction, edema, platelet aggregation, etc. This concept has proved useful as a working hypothesis in a number of different fields such as organ transplantation (Figure 1). The hypothesis has also been of considerable utility in designing experiments in other areas, including asthma, thrombosis, and cancer. However, the relationship between radiation-induced peroxidation, free radicals, eicosanoid synthesis, and P450 oxidation is

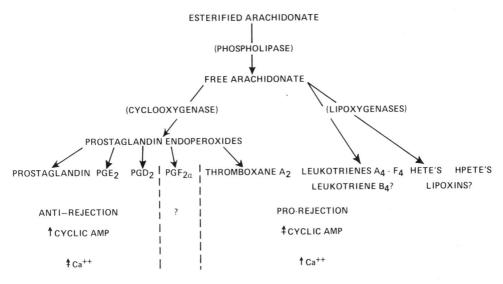

Fig. 1. Arachidonic acid cascade with metabolites arranged in anti-rejection (suppressors of immune response) and pro-rejection (promoters of immune response) eicosanoids

obviously too complex for such a simple hypothesis to apply to ionizing radiation injury. This is especially true when one considers the conflicting data arising from the use of cyclooxygenase inhibitors.

In studying the relationship of eicosanoids to radiation injury, the use of antioxidants and free radical scavengers is a primary means of investigation as well as use of lipoxygenase and cyclooxygenase inhibitors. However, one class of drugs that still awaits deployment in studying the relation of eicosanoids to radiation injury is the receptor antagonists. Specific antagonists are now available for thromboxane and some of the leukotrienes as well as for PGE and prostaglandin $F_{2\alpha}$ compounds. Such antagonists should prove a more definitive means of exploring the pathophysiologic role of individual eicosanoids and deserve to be considered.

Another point that should be addressed is based on the idea that eicosanoids are derived from a precursor by a cascade mechanism. This point is especially apparent in terms of the endoperoxides and their subsequent metabolism: inhibition of thromboxane synthase has been shown in a number of situations to divert the precursor endoperoxides to prostacyclin (6). Some workers believe that a similar situation pertains to cyclooxygenase inhibition in that arachidonate is thought to be diverted to the lipoxygenase. However, if separate pools of arachidonate are closely associated with the individual oxygenases, this divergence may not occur. Further, the 5-lipoxygenase, for example, may also need to be specifically activated.

Finally, the release of arachidonate and other polyunsaturated fatty acids can occur following stimulation by individual eicosanoids. For example, leukotrienes can release thromboxane from pulmonary tissue (7) and, in fact, some of the bronchoconstrictor activity of leukotrienes is attenuated by aspirin and indomethacin. This situation also occurs with other lipid mediators such as PAF. Some of the bronchoconstrictor effects of PAF can be shown to be due to platelet mobilization of arachidonate and release of thromboxane (8).

Elective immunosuppression as used in organ transplantation is useful to illustrate some of these points (9). This model has the advantage that it is directly relevant to the problem of radiation-induced immunosuppression. The drugs used for immunosuppression in patients receiving organ transplants are azathioprine, cyclosporine A, and corticosteroids. They exert their suppressive effect on the immune system by inhibiting DNA replication (azathioprine) or by inhibiting the formation of interleukin 1 (IL-1) (corticosteroids) and interleukin 2 (IL-2) (corticosteroids and cyclosporin A). Both IL-1 and IL-2 are essential for lymphocyte proliferation in response to foreign antigen (organ transplant). Figure 2 shows the cellular elements involved in cell-mediated transplant rejection and the attenuation of this event by corticosteroids as well as the differential effect of eicosanoids.

The initiation of the clonal expansion of T-lymphocytes requires two signals from the macrophage; one signal is IL-1 and the other is foreign antigen presenta-

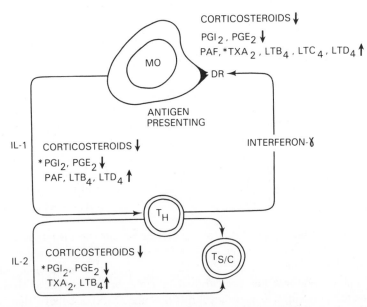

Fig. 2. T-lymphocyte proliferation in response to foreign antigen presentation by macrophage and modulation by corticosteroids, eicosanoids, and platelet-activating factor (PAF)

tion. The eicosanoids affect this interaction by modulating the formation and action of monokines and lymphokines as well as antigen expression (Figure 2). Prostaglandin E_2 (PGE_2) and corticosteroids inhibit IL-1 and IL-2 formation whereas leukotrienes have been shown both directly and indirectly to promote IL-1 and IL-2 formation. The effect of leukotriene B_4 (LTB_4) on lymphocyte proliferation has been explored by Goodwin (10) and by Rola-Pleszczynski (11), who found LTB_4 to promote lymphocyte proliferation. Initially, LTB_4 was found to inhibit both mitogen- and antigen-induced T_4 lymphocyte proliferation. However, further studies indicated that LTB_4 also facilitates the proliferation of T_4 (T-helper) lymphocytes when endogenous PGE_2 synthesis is blocked by indomethacin. Recent studies (12) introduced a new concept: T_4 lymphocytes undergo phenotypic conversion to T_8 (T-cytotoxic/suppressor) lymphocytes without de novo protein synthesis and, moreover, this conversion is enhanced by LTB_4. Another aspect of immune enhancement is the promotion of lymphocyte migration, which is also reported to be facilitated by LTB_4.

Thromboxane A_2 (TXA_2) synthase inhibitors have been shown to inhibit lymphocyte proliferation, which infers that TXA_2 promotes lymphocyte proliferation. However, this effect could also be explained by the shunting of endoperoxides into PGE_2 and PGI_2. More recently, a TXA_2 agonist was shown to promote mitogen-induced lymphocyte proliferation and to reverse the inhibition of lymphocyte proliferation obtained with a TXA_2 synthase inhibitor (13). In order for the lymphocyte to undergo clonal expansion in response to foreign antigen and IL-1, the lymphocyte needs to recognize DR antigen on the antigen-presenting macrophage. The DR antigen expression can be modulated by the cyclooxygenase products and also by gamma-interferon. Unanue and co-workers (14) showed in vitro that PGE_2, PGI_2, and corticosteroids decrease Ia or DR antigen expression on macrophages. Stable TXA_2 agonists may promote antigen expression, but this has not been described yet. Gamma-interferon enhances the immune response by increasing DR antigen expression by macrophages and endothelial cells. Leukotriene B_4, C_4, and D_4 all increase the release of gamma-interferon from lymphocytes (15). The in vivo findings in rat cardiac transplant model using corticosteroids, lipoxygenase inhibitors, cyclooxygenation, and TXA_2 synthase inhibitors or antagonists support the in vitro findings of an immunomodulatory role of the eicosanoids (16-18).

In addition to the effect of the dihydroxyeicosanoids such as LTB_4, further studies have demonstrated the synthesis of trihydroxy eicosanoids, the lipoxins, which are derived from the 15- as well as the 5-lipoxygenase pathway. These lipoxins are reported to inhibit the action of natural killer cells and, in lower doses, to promote natural killer cell activity. A biological role for the lipoxins is disputed since measurable amounts of lipoxins can only be formed by adding 15-hydroperoxyeicosatetraenoic acid to granulocytes. The eosinophils are the most likely cells to produce biologically important amounts of lipoxins due to the activity of its 15-lipoxygenase, but lipoxins have not yet been demonstrated to be released from eosinophils (19).

Recently PAF has been implicated as one of the mediators of allograft rejection (20). The mechanisms are unknown. PAF has direct effects on cells and tissues, but it also stimulates arachidonic acid release. Depending on the cell type, this latter effect of PAF leads to leukotriene, prostaglandin, or thromboxane synthesis. A stimulatory effect of PAF on lymphocyte proliferation would be expected. This is not the case with *in vitro* experiments, where inhibition has been described (21). This inhibition is attenuated *in vitro* with indomethacin, indicating that the inhibitory effect of PAF on lymphocyte proliferation may be indirect through PGE_2. The *in vivo* effect of PAF antagonists in a rat cardiac transplant model is a synergistic effect with cyclosporin A in prolonging graft survival, which supports the suggestion that PAF enhances the immune response.

In conclusion, the experimental models of transplantation and the *in vitro* data support the suggestions that prostaglandins that elevate cyclic AMP attenuate rejection by inhibiting the immune response. In contrast, TXA_2 and the leukotrienes act as pro-rejection compounds in promoting mediator release and initiating lymphocyte proliferation by increasing cytosolic calcium and preventing an increase in cyclic AMP. In addition, these products have vascular effects that are similar to rejection-related events, such as edema and decreased blood flow to the allograft.

ACKNOWLEDGMENT

This work was supported by grants from the National Institutes of Health HL 32319, HL 34974, and HL 31241.

REFERENCES

1. Foegh, M. L., and Ramwell, P. W. Physiological implications of products in the arachidonic acid cascade. In: "Prostaglandins and Related Substances." C. Pace-Asciak and E. Granstrom, eds. Elsevier, Amsterdam, 1983, pp. xiii-xvi.
2. Rouzer, C. A., and Samuelsson, B. 5-Lipoxygenase from human leukocytes associates with membrane in the presence of calcium. Adv. Prostaglandin Thromboxane Leukotriene Res. 17A: 60-63, 1987.
3. Schneidkraut, M. J., Kot, P. A., Ramwell, P. W., and Rose, J. C. Regional release of cyclooxygenase products after radiation exposure of the rat. J. Appl. Physiol. 61: 1264-1269, 1986.
4. Angerio, A. D., Fitzpatrick, T., Kot, P. A., Ramwell, P. W., and Rose, J. C. Effect of verapamil on the pulmonary vasoconstrictor action of prostaglandin $F_{2\alpha}$ and a synthetic PGH_2 analogue. Br. J. Pharmacol. 73: 101-103, 1981.
5. Robert, A. Prostaglandins and the digestive system. In: "The Prostaglandins," Vol. 3. P. W. Ramwell, ed. Plenum Press, New York, 1977, pp. 225-266.
6. Foegh, M. L., Maddox, Y. T., Winchester, J., Rakowski T., Schreiner, G., and Ramwell, P. W. Adv. Prostaglandin Thromboxane Leukotriene Res. 12: 45-49, 1983.

7. Iacopino, V. J., Kot, P. A., and Ramwell, P. W. Systemic and pulmonary vascular effects of leukotrienes. In: "Prostaglandins and Cardiovascular Diseases." T. Ozawa, K. Yamada, and S. Yamamoto, eds. Taylor & Francis Ltd, Philadelphia, 1986, pp. 203-211.

8. Braquet, P., Shen, T. Y., Touqui, L., and Vargaftig, B. B. Perspectives in platelet activity factor research. Pharmacol Rev., in press.

9. Foegh, M. L., Alijani, M. R., Helfrich, G. B., Khirabadi, B. S., Lim, K., and Ramwell, P. W. Elective immunosuppression. In: "Lipid Mediators in Immunology of Burn and Sepsis." M. Paubert-Braquet and P. Braquet, eds. Plenum Press, London, in press.

10. Goodwin, J. S. Role of leukotriene B_4 in T cell activation. Transplant. Proc. 18 (Suppl 4): 49-51, 1986.

11. Rola-Pleszczynski, M., and Gagnon, L. Natural killer cell function modulated by leukotriene B_4: Mechanisms of action. Transplant. Proc. 18 (Suppl 4): 44-48, 1986.

12. Atluru, D., Lianos, E. A., and Goodwin, J. S. Control of polyclonal immunoglobulin production from human lymphocytes by an OKT8(+), radiosensitive suppressor cell from resting human OKT8(-) T cells. J. Clin. Invest. 74: 1444-1450, 1984.

13. Ceuppens, J. L., Vertessen, S., Deckmyn, H., and Vermylen, J. Effect of thromboxane A_2 on lymphocyte proliferation. Cell Immunol. 90: 458-462, 1985.

14. Snyder, D. S., Beller, D. I., and Unanue, E. R. Prostaglandins modulate macrophage Ia expression. Nature 299: 163-165, 1982.

15. Johnson, H. M., and Torres, B. A. Leukotrienes: Positive signals for regulation of gamma-interferon production. J. Immunol. 132: 413-415, 1984.

16. Foegh, M. L., Khirabadi, B. S., and Ramwell, P. W. Prolongation of rat cardiac allograft survival with thromboxane related drugs. Transplantation 40: 124-125, 1985.

17. Foegh, M. L., Khirabadi, B. S., Rowles J. R., Braquet, P., and Ramwell, P. W. Inhibition of PAF and leukotriene expression in acute cardiac allograft rejection in rats. Pharmacol. Res. Commun. 18 (Suppl): 127-132, 1986.

18. Foegh, M. L., Khirabadi, B. S., and Ramwell, P. W. Improved rat cardiac allograft survival with non-steroidal pharmacologic agents. Transplant. Proc. 19: 1297-1300, 1987.

19. Rokash, J., and Fitzsimmons, B. Lipoxins—Do they have a biological role? Tranplant. Proc. 18 (Suppl 4): 7-9, 1986.

20. Foegh, M. L., Khirabadi, B. S., Rowles, J. R., Braquet, P., and Ramwell, P. W. Prolongation of rat cardiac allograft survival with BN 52021, a specific platelet activating factor. Transplantation 42: 86-88, 1986.

21. Foegh, M. L., Hartmann, D.-P., Rowles, J. R., Khirabadi, B. S., Alijani, M. R., Helfrich, G. B., and Ramwell, P. W. Leukotrienes, thromboxane and platelet activating factor in organ transplantation. Adv. Prostaglandin Thromboxane Leukotriene Res. 17A: 140-146, 1987.

PROSTAGLANDINS AND HEMOPOIETIC CELL RECOVERY FOLLOWING IRRADIATION

R. L. DeGowin

The University of Iowa Cancer Center
and
Radiation Research Laboratory and
Division of Hematology-Oncology
The University of Iowa College of Medicine
Iowa City, Iowa

INTRODUCTION

— Suppressed hemopoietic recovery in patients with cancer
— Tumor bearing and inflammation induce microenvironmental changes
— PGE_2 as modulator of hemopoiesis in mice with inflammation

1. Effects of tumor bearing and chronic inflammation on hemopoiesis and survival after irradiation

   a. Anemia, leukocytosis, with tumor bearing in mice
   b. Hypocellularity of the bone marrow and hyperplasia of the spleen
   c. Impaired survival from sublethal total-body irradiation
   d. Similar changes with inflammation

2. Effects of endotoxin on hemopoiesis and postirradiation survival

   a. Anemia, leukocytosis, medullary hypocellularity, and splenic hyperplasia
   b. Impaired survival and suppressed hemopoietic recovery
   c. Effects similar to tumor bearing and inflammation

3. Role of stromal cells in the hemopoietic microenvironment

   a. Repopulation of irradiated bone marrow
   b. Support of hemopoiesis in culture
   c. Biphasic response of erythropoiesis to increasing MSC and PGE_2

4. Differential synthesis of PGE_2 by stromal cells of the hemopoietic microenvironment

   a. Opposite responses to endotoxin of hemopoiesis in marrow and spleen
   b. Differential synthesis of PGE$_2$ by stromal cells of the marrow and spleen

SUMMARY

— A variety of lesions that activate the inflammatory response in mice suppress hemopoietic recovery and impair survival after irradiation.

— Adherent stromal cells of the marrow's microenvironment, required for hemopoiesis *in vivo* and *in vitro*, synthesize PGE$_2$ in response to endotoxin.

— Low concentrations of PGE$_2$ enhance but high concentrations suppress hemopoiesis *in vitro*.

— Stromal cells of the marrow synthesize more PGE$_2$ than those of the spleen in response to endotoxin, providing an explanation for the differential modulation of hemopoiesis.

INTRODUCTION

One of the major impediments to the administration of optimal regimens of radiation (1) or chemotherapy to patients with cancer is myelosuppression and persistent pancytopenia. Diminished doses or delayed cycles of treatment to permit recovery of peripheral blood cells frequently result in failure to control tumor growth. The cause of the pancytopenia is usually obscure. Although initial cytotoxic therapy may transiently diminish the hemopoietic stem cell pool in a dose-related fashion, the anemia of chronic disorders (2) accompanying a recalcitrant tumor or inflammation in some patients suggests that marginal bone marrow reserves have been depleted by another mechanism.

Mice bearing either a tumor (3) or a chronic inflammatory lesion (4) exhibit impaired survival from suppressed hemopoietic recovery following a dose of ordinarily sublethal total-body radiation. Although these lesions cause hypocellularity of the bone marrow, they are associated with a compensatory hyperplasia of hemopoietic tissues in the spleen (3,4). These opposite responses to the same stimulus (e.g., inflammation) suggest that a microenvironmental change in the marrow's ability to support hemopoiesis has occurred. The circumstances under which these changes in the bone marrow's microenvironment have been observed implicate components of the inflammatory response as mediators.

A serious candidate for such a mediator should play a major role in the inflammatory response, act as a short-range messenger, be synthesized by cells of the bone marrow's microenvironment, and affect hemopoietic cellular proliferation and differentiation. Because prostaglandin E$_2$ (PGE$_2$) serves as an important mediator of the inflammatory response (5), is synthesized by bone marrow stromal cells (6), and profoundly affects hemopoiesis (5-14), we have concentrated on studies of tumor bearing, inflammation, and PGE$_2$ in relation to recovery from exposure to myelosuppressive radiation. In the following, I review some of the work that points to a role of PGE$_2$ in the modulation of hemopoietic recovery in irradiated mice that exhibit signs of activation of the inflammatory response.

Effects of Tumor Bearing and Chronic Inflammation on Hemopoiesis and Survival After Irradiation

Studies initiated to elucidate the mechanisms of anemia in tumor-bearing mice were undertaken with Ehrlich's ascites carcinoma (15). In the ascitic form, death occurred in about 2 weeks from an intraperitoneal inoculation of 1×10^6 tumor cells. When 4×10^6 tumor cells were injected intramuscularly in the hind leg to produce a solid tumor, tumor burden could be estimated by measuring leg diameter, and death was delayed for 4 to 6 weeks (15,16).

Anemia and neutrophilia became progressive after the first week of bearing a solid tumor (15-17). The anemia resembled the anemia of chronic disorders (2) in that erythrocyte survival was slightly shortened, and the anemia was hypoproliferative and refractory to the administration of vitamin B$_{12}$, folate, and iron (15). The erythropoietic response to hypoxia or exogenous erythropoietin was undoubtedly due to splenic erythropoiesis, for splenectomy made the anemia much more severe (16,17). Leukocytosis approximated levels of $20,000/\mu l$ and resulted from a neutrophilia (15-17). After 2-4 weeks, tumor-bearing mice all experienced progressive hypocellularity of the femoral bone marrow, diminishing by one-fourth to one-third of normal, with profound erythroblastopenia and lymphopenia. Compensatory hemopoietic hyperplasia of the spleen followed, with increases in granulocytes, megakaryocytes, lymphocytes, and erythroblasts, i.e., a panhyperplasia (16).

Finally, we determined that these remarkable changes in hemopoiesis occurred not only in mice bearing Ehrlich's ascites carcinoma but also in mice bearing a sarcoma (S-180) or a leukemia (L-1210) (18). Other investigators have reported that a variety of transplanted tumors induce anemia and leukocytosis in rodents (19,20). Certainly, anemia and leukemoid reactions are commonly seen in nontreated patients with bulky tumors (21).

Since our other studies had shown reduced numbers of hemopoietic stem cells (CFU-S) in bone marrow and increased numbers in the spleen of mice bearing S-180 (3), we asked whether these changes in hemopoiesis were important. We reasoned that if the total body pool of hemopoietic stem cells were reduced in tumor-bearing mice, recovery of hemopoiesis and survival after ordinarily sublethal total-body irradiation would be impaired. Indeed, survival and hemopoietic recovery of mice bearing S-180 as a solid tumor for 1 week before receiving 600 cGy total-body radiation were markedly diminished (3). All tumor-bearing mice died with pancytopenia during the second and third weeks postirradiation when death from bone marrow failure was expected (22). Inoculation of tumor cells after total-body irradiation produced the same result (3). In contrast, 60%-70% of normal controls survived more than 48 days postirradiation (3). Nonirradiated mice did not begin to die of their tumors until after 3-4 weeks (3).

Coincident with the observations of the hematological consequences of tumor bearing in our laboratory, we were studying the postirradiation repopulation of murine femoral bone marrow by endogenous hemopoietic stem cells (23). By serendipity, we discovered that doses of 5,000-10,000 cGy to the leg induced a chronic inflammatory lesion in the soft tissues that produced similar hematologic changes in the peripheral blood and nonirradiated bone marrow as the tumor bearing did (23). The inflammatory lesion (characterized first by erythema and later by epilation, edema, and ulceration) developed during the second week postirradiation and induced a neutrophilic leukocytosis by 30 days postirradiation (23,24).

The same microenvironmental changes noted with tumor bearing occurred with hypocellularity of the bone marrow and compensatory hemopoietic hyperplasia of the spleen (23,24). Stem cells (CFU-S), erythroblasts, and lymphocytes were greatly depressed in the marrow but markedly increased in the spleen (7,23,24). Like tumor-bearing mice, mice with a chronic inflammatory lesion or a sterile abscess displayed impaired survival and suppressed hemopoietic recovery following ordinarily sublethal total-body irradiation (600 cGy). Death occurred from bone marrow failure during the second and third weeks (4). The transplantation of bone marrow cells to mice with chronic inflammatory lesions that had received 750 cGy total-body radiation restored survival to control values. This confirmed that the impaired survival in mice with inflammation was due to a depleted stem cell pool.

Effects of Endotoxin on Hemopoiesis and Postirradiation Survival

It appeared that the similarity of effects on hemopoiesis noted with tumor bearing or chronic inflammation might be explained by activation of the inflammatory response. If endotoxin injections could produce the same hematological and microenvironmental changes in the intact animal that tumor bearing and the chronic inflammatory lesion did, then the effects of a relatively pure substance (lipopolysaccharide, LPS) might be examined *in vitro* as well as *in vivo* to elucidate the mechanisms involved.

Although single injections of endotoxin given before total-body irradiation were shown by Smith and colleagues to enhance survival (25), transient depletion of marrow hematopoietic stem cells had been observed (26). Therefore, we studied the effects of eight daily intraperitoneal injections of 100 μg LPS from *E. coli* (27). Twenty-four hours after completing the endotoxin injections, anemia, leukocytosis, medullary hypocellularity, and compensatory splenic hemopoietic hyperplasia were similar to those noted with tumor bearing or the chronic inflammatory lesion (27). Reissmann and colleagues (28) and others (29) have reported hyperplasia of erythroblastic precursors in the spleen of mice injected with endotoxin from *S. typhosa*.

Because survival after total-body irradiation improved if tumors regressed or chronic inflammatory lesions healed, we decided to administer the eight daily injections of endotoxin after 600 cGy of total-body irradiation to assess the effects on survival. The idea was to see if we could suppress postirradiation hemopoietic recovery. Although the results were not as striking as the effects of bearing a growing tumor or a chronic inflammatory lesion, only 20% of mice injected with endotoxin recovered whereas 60% of saline-injected controls survived (27). Four daily injections of 200 μg LPS per mouse produced similar results, but 10% of nonirradiated control mice injected with such high doses died. Therefore, the lower dose, from which nonirradiated mice seemed to recover their activity after about 5 days of injections, was used for the studies. The administration of 100 μg LPS per mouse for eight daily doses before 600 cGy total-body irradiation, without continuing the injections during the period of postirradiation hemopoietic recovery, did not impair survival over that of controls injected with isotonic saline (27).

We concluded that serial injections of endotoxin produced hematological changes in the peripheral blood and bone marrow like those observed in mice bearing tumor or a chronic inflammatory lesion. Moreover, the effects were important in that the endotoxin suppressed hemopoietic recovery and impaired survival from ordinarily sublethal irradiation (27).

Role of Stromal Cells in the Hemopoietic Microenvironment

The importance of marrow stromal cells in the support of hemopoiesis has been amply demonstrated in the intact animal (23,24,30), in ectopic sites of hemopoiesis (31), and in tissue culture (32-37). Fibroblastoid cells from the bone marrow (MSC) adhere to plastic culture dishes and grow in colonies, permitting quantitation and *in vitro* studies of function (7,16,23,24,32,33). The early studies of Werts et al. in my laboratory revealed that doses of 1,000 cGy acutely depleted femoral bone marrow of marrow stromal cells (MSC) (23,24). The partial repopulation of irradiated hemopoietic bone marrow was preceded by circulating MSC, and the number of erythroblasts, lymphocytes, and CFU-S closely approximated the number of MSC (23,24,30). A noncirculating or matrical component of the microenvironment was required to support MSC capable of circulating. The matrical component was more resistant to radiation than circulating MSC, and doses of 5,000 cGy or greater inhibited the repopulation of femoral marrow by MSC and hemopoietic cell precursors (23,24,38,39).

Coculture experiments showed that increasing the number of MSC in the underlayer of split-phase cultures resulted in a biphasic response of suspended bone marrow cells to erythropoietin (6,7). Increasing concentrations of MSC first enhanced erythroblastic precursor colony (BFU-E, CFU-E) formation and later suppressed it (6,7). However, increasing numbers of MSC progressively suppressed granulocyte-macrophage colony (GM-CFU) formation (6,7). Concentrations of PGE₂, in the supernatants of culture dishes containing increasing numbers of MSC, progressively increased as the number of cells increased (6,18). Supernatant media from culture dishes containing indomethacin (5 μg/ml) did not contain PGE₂ and did not suppress erythropoiesis (6,18), confirming that MSC synthesized PGE_2 which, in turn, suppressed colony formation of erythroid precursors.

When PGE_2 was added to cultures (containing suspended bone marrow cells and erythropoietin) in the same concentrations as those that had been measured in the supernatants over MSC, the same biphasic response of erythropoietic precursors was observed (6,18). Levels of 0.5-1.0 ng PGE_2/ml enhanced, while those exceeding 1.0 ng PGE_2/ml suppressed (18,40). Other investigators (11,12) have demonstrated the suppressive effect of increasing concentrations of PGE_2 on GM-CFU. We concluded that increasing numbers of stromal cells from the bone marrow's microenvironment suppressed erythropoiesis and granulocytopoiesis by synthesizing increasing amounts of PGE_2. The effect could be abrogated by inhibiting PGE_2 synthesis with indomethacin, and it could be reproduced in the absence of MSC by incubating bone marrow cells with incremental concentrations of PGE_2.

Differential Synthesis of PGE_2 by Stromal Cells of the Hemopoietic Microenvironment

We reflected on the similar hemopoietic effects of tumor bearing, chronic inflammation, and serial endotoxin injections. In each of these circumstances, hypocellularity of the bone marrow was accompanied by compensatory hemopoietic hyperplasia of the spleen. Since all of the lesions were remote from the marrow and the spleen, presumably the same message reached the spleen as reached the marrow. However, the responses of the two sites of hemopoietic tissues were completely opposite. If stromal cells from the marrow's microenvironment responded to endotoxin by synthesizing PGE_2, which in turn suppressed hemopoiesis, then stromal cells from the splenic microenvironment should elaborate less PGE_2 to permit the compensatory hemopoietic hyperplasia that we observed.

After permitting bone marrow cells (1 x 10^6 cells/ml MEM-α) to adhere to plastic culture dishes for 3 hr, medium was replaced with fresh MEM-α with or without 1 μg LPS/ml and incubated at 37.0°C for another 24 hr. Thereafter, the concentration of PGE_2 in the supernatant was determined with a sensitive ^{125}I-radioimmunoassay. Unstimulated cells elaborated 0.3-0.4 ng/ml PGE_2, approximately 8-10 times baseline concentrations observed in medium without cells (0.04 ng/ml) (27). Cultures to which LPS was added contained concentrations of PGE_2 in excess of 50-100 times baseline (2.0-4.32 ng/ml) (27). A cyclooxygenase inhibitor (indomethacin, 5 μg/ml) completely inhibited the production of PGE_2 in stimulated and nonstimulated cultures (27). Thus, adherent stromal cells from the hemopoietic microenvironment of the bone marrow were capable of vigorous synthesis of PGE_2 in response to endotoxin.

To compare the response of adherent cells from two different sites of hemopoiesis, 1 x 10^6 cells/ml medium from the spleen and the marrow of the same mice were separately exposed to 1 μg LPS/ml for 24 hr following adherence time of 3 hr. Nonstimulated baseline levels of PGE_2 in supernatants of cells from the spleen and marrow were similar in magnitude, and they exceeded those in dishes devoid of cells by 8-10 times (27). However, the concentrations in LPS-stimulated splenic stromal cells never exceeded 0.5 ng/ml, while PGE$_2$ levels in the supernatants of marrow stromal cells surpassed 2 ng/ml in several replicate experiments (27). As noted in the earlier studies referred to before (18,40), low concentrations of PGE$_2$ enhanced erythropoiesis in culture while concentrations greater than 1 ng/ml suppressed it. These findings represent a striking difference in the responses of adherent cells (in fresh primary cultures) from the hemopoietic microenvironments of the marrow and spleen. To permit more time for cells to settle out and adhere from the spleen-cell suspension, adherence times were extended from 3 to 24 and 48 hr, without changing the differential response of marrow and splenic stromal cells (27).

SUMMARY

We have studied a variety of lesions that activate the inflammatory response in mice. To a greater or lesser degree, tumor bearing, chronic inflammation, sterile abscess, and endotoxin injections impaired survival by suppressing hemopoietic recovery following ordinarily sublethal total-body irradiation. These lesions produced anemia, leukocytosis, hypocellularity of the bone marrow, and hemopoietic hyperplasia of the spleen. The diminished cellularity of the bone marrow and the compensatory hemopoietic hyperplasia in the spleen revealed the remarkably different modulations of hemopoiesis by the two different microenvironments in response to systemic stimuli of the inflammatory reaction.

Adherent stromal cells are an integral part of the bone marrow's microenvironment, and they are required to support hemopoiesis in ectopic sites and *in vitro* as well as *in situ*. They modulate hemopoiesis in cocultures and synthesize prostaglandin E$_2$ (PGE$_2$) in response to endotoxin. Prostaglandin E$_2$, an important mediator of the inflammatory response, enhances erythropoiesis in low concentrations and suppresses hemopoiesis in high concentrations.

The fact that adherent stromal cells from the bone marrow vigorously synthesized PGE$_2$ in response to endotoxin when splenic stromal cells failed to respond suggests that suppressive concentrations of PGE$_2$ were generated in the marrow but lower concentrations in the spleen may have permitted the compensatory hemopoietic hyperplasia there.

With such a complex process as inflammation, there are undoubtedly many mediators, but the studies I have discussed support a major role for prostaglandin in the modulation of postirradiation hemopoietic recovery and survival.

ACKNOWLEDGMENTS

I wish to acknowledge and thank the following persons for their assistance in these studies: D. An, P. G. Fisher, D. P. Gibson, F. M. Grund, M. J. Johnson,

S. A. Knapp, L. M. Wathen, and E. D. Werts. Supported in part by U.S. Public Health Service Grants CA 11472, HL 31882, and CA 28848.

REFERENCES

1. Parker, R. G. Principles of radiation oncology. In: "Cancer Treatment," 2nd ed. C. M. Haskell, ed. W. B. Saunders Co., Philadelphia, 1985, pp.14-20.

2. Cartwright, G. E., and Lee, G. R. The anaemia of chronic disorders. Brit. J. Haematol. 21: 147-152, 1971.

3. DeGowin, R. L., Gibson, D. P., and Knapp, S. A. Tumor bearing impairs hemopoietic recovery and survival after irradiation. Exp. Hemat. 11: 305-314, 1983.

4. DeGowin, R. L., and Lass, S. L. Chronic inflammation impairs hemopoiesis and survival after irradiation. J. Lab. Clin. Med. 105: 299-304, 1985.

5. Gerrard, J. M., ed. The role of prostaglandins in host defense against bacteria, viruses, and tumors and in inflammation. In: "Prostaglandins and Leukotrienes." Marcel Dekker, Inc., New York, 1985, pp. 239-256 and 265-273.

6. DeGowin, R. L., and Gibson, D. P. Prostaglandin-mediated enhancement of erythroid colonies by marrow stromal cells (MSC). Exp. Hemat. 9: 274-280, 1981.

7. Werts, E. D., DeGowin, R. L., Knapp, S. A., and Gibson, D. P. Characterization of marrow stromal (fibroblastoid) cells and their association with erythropoiesis. Exp. Hemat. 8: 423-433, 1980.

8. Schooley, J. E., and Mahlmann, L. J. Stimulation of erythropoiesis in plethoric mice by prostaglandins and its inhibition by antierythropoietin. Proc. Soc. Exp. Biol. Med. 138: 523-524, 1971.

9. Dukes, P. P., Shore, N. A., Hammond, G. D., and Ortega, J. A. Prostaglandins and erythropoietin action in erythropoiesis. In: "Erythropoiesis." K. Nakao, J. W. Fisher, and F. Takaku, eds. University Park Press, Baltimore, 1975, pp. 3-14.

10. Feher, I., and Gidali, J. Prostaglandin E_2 as stimulator of haemopoietic stem cell proliferation. Nature 247: 550-551, 1974.

11. Kurland, J. I., and Moore, M. A. S. Modulation of hemopoiesis by prostaglandins. Exp. Hematol. 5: 357-373, 1977.

12. Kurland, J. I., Broxmeyer, H. E., Pelus, L. M., Bockman, R. S., and Moore, M. A. S. Role for monocyte-macrophage-derived colony-stimulating factor and prostaglandin E in the positive and negative feedback control of myeloid stem cell proliferation. Blood 52: 388-407, 1978.

13. Rossi, G. B., Migliaccio, M. G., Lettieri, F., DeRosa, M., Mastroberardino, G., and Peschle, C. In vitro interactions of PGE and cAMP with murine and human erythroid precursors. Blood 56: 74-79, 1980.

14. Chan, H. S. L., Saunders, E. F., and Freedman, M. H. Modulation of human hematopoiesis by prostaglandins and lithium. J. Lab. Clin. Med. 95: 125-132, 1980.

15. DeGowin, R. L., Grund, F. M., and Gibson, D. P. Erythropoietic insufficiency in mice with extramedullary tumor. Blood 51: 33-43, 1978.

16. DeGowin, R. L., and Gibson, D. P. Suppressive effects of an extramedullary tumor on bone marrow erythropoiesis and stroma. Exp. Hemat. 6: 568-575, 1978.

17. DeGowin, R. L., and Gibson, D. P. Erythropoietin and the anemia of mice bearing extramedullary tumor. J. Lab. Clin. Med. 94: 303-311, 1979.

18. DeGowin, R. L., Gibson, D. P., and Knapp, S. A. Prostaglandin E and the erythropoietic and stromal insufficiency induced by extramedullary tumor. J. Lab. Clin.

19. Delmonte, L., Liebelt, A. G., and Liebelt, R. A. Granulopoiesis and thrombopoiesis in mice bearing transplanted mammary cancer. Cancer Res. 26: 149-159, 1966.

20. Boggs, D. R., Mallory, E., Boggs, S. S., Chervenick, P. A., and Lee, R. E. Kinetic studies of a tumor-induced leukemoid reaction in mice. J. Lab. Clin. Med. 89: 80-92, 1977.

21. Leukemoid blood pictures. In: "Clinical Hematology," 8th ed. M. Wintrobe et al., eds. Lea & Febiger, Philadelphia, 1981, p. 1312.

22. Hall, E. J. The hematopoietic syndrome. In: "Radiobiology for the Radiologist," 2nd ed. Harper and Row, Hagerstown, MD, 1978, pp. 213-214.

23. Werts, E. D., Johnson, M. J., and DeGowin, R. L. Postirradiation hemopoietic repopulation and stromal cell viability. Radiat. Res. 71: 214-244, 1977.

24. Werts, E. D., Johnson, M. J., and DeGowin, R. L. Abscopal suppression of bone marrow erythropoiesis. Radiat. Res. 76: 206-218, 1978.

25. Smith, W. W., Alderman, I. M., and Gillespie, R. E. Increased survival in irradiated animals treated with bacterial endotoxins. Am. J. Physiol. 191: 124-130, 1957.

26. Smith, W. W., Brecher, G., Fred, S., and Budd, R. A. Effect of endotoxin on the kinetics of hemopoietic colony-forming cells in irradiated mice. Radiat. Res. 27: 710-717, 1966.

27. DeGowin, R. L., and Fisher, P. G. Differential elaboration of prostaglandin-E₂ (PGE₂) by cells of the hemopoietic microenvironment. Exp. Hematol. 14: 499, 1986.

28. Reissmann, K. R., Udupa, K. B., and Labedzki, L. Induction of erythroid colony forming cells (CFU-E) in murine spleen by endotoxin. Proc. Soc. Exp. Biol. Med. 153: 98-101, 1976.

29. Molendijk, W. J., Ploemacher, R. E., and Erkens-Versluis, M. E. Mediatory role of stem cell derived cells in LPS-induced splenic CFUs accumulation. Exp. Hematol. 10: 499-504, 1982.

30. Werts, E. D., Gibson, D. P., Knapp, S. A., and DeGowin, R. L. Stromal cell migration precedes hemopoietic repopulation of the bone marrow after irradiation. Radiat. Res. 81: 20-30, 1980.

31. Friedenstein, A. J., et al. Stromal cells responsible for transferring the microenvironment of the hemopoietic tissues. Transplantation 17: 331-340, 1974.

32. Friedenstein, A. J., et al. Precursors for fibroblasts in different populations of hematopoietic cells as detected by the *in vitro* colony assay method. Exp. Hemat. 2: 83-92, 1974.

33. Wilson, F. D., et al. The formation of bone marrow derived fibroblastic plaques *in vitro*: Preliminary results contrasting these populations to CFU-C. Exp. Hemat. 2: 343-354, 1974.

34. Bentley, S. A., Knutsen, T., and Whang-Peng, J. The origin of the hematopoietic microenvironment in continuous bone marrow culture. Exp. Hemat. 10: 367-372, 1982.

35. Castro-Malaspina, H., et al. Characteristics of bone marrow fibroblast colony-forming cells (CFU-F) and their progeny in patients with myeloproliferative disorders. Blood 59: 1046-1054, 1982.

36. Dexter, T. M., Allen, T. D., and Lajtha, L. D. Conditions controlling hemopoietic stem cells *in vitro*. J. Cell Physiol. 91: 335-344, 1977.

37. Greenberger, J. S. Sensitivity of corticosteroid-dependent, insulin-resistant lipogenesis in marrow preadipocytes of mutation diabetic-obese mice. Nature 275: 752-754, 1978.

38. Wathen, L. M., Knapp, S. A., and DeGowin, R. L. Suppression of marrow stromal cells and microenvironmental damage following sequential radiation and cyclophosphamide. Int. J. Radiat. Oncol. Biol. Physics 7: 935-941, 1981.

39. Wathen, L. M., DeGowin, R. L., Gibson, D. P., and Knapp, S. A. Residual injury to the hemopoietic microenvironment following sequential radiation and busulfan. Int. J. Radiat. Oncol. Biol. Physics 8: 1315-1322, 1982.

40. Gibson, D. P., DeGowin, R. L., and Knapp, S. A. Effect of X-irradiation on release of prostaglandin E from marrow stromal cells in culture. Radiat. Res. 89: 537-545, 1982.

IONIZING RADIATION, PROSTAGLANDINS, AND GASTRIC FUNCTION IN MONKEYS

A. Dubois[1], E. D. Dorval[1], N. Fiala[1], and L. K. Steel[2]

[1]Digestive Diseases Division
Department of Medicine
Uniformed Services University of the Health Sciences
Bethesda, Maryland 20814

[2]Radiation Biochemistry Department
Armed Forces Radiobiology Research Institute
Bethesda, Maryland 20814

ABSTRACT

The prodromal syndrome of radiation sickness is characterized by nausea, vomiting, and suppression of gastric function, but the mediators responsible for this entity remain unknown. We investigated the possible involvement of prostaglandins in this syndrome by determining the concurrent effects of ionizing radiation on gastric function and on prostaglandins. We measured gastric electrical control activity (ECA, waves/min), fractional emptying rate (FER, %/min), acid output (AO, μEq/min), and plasma and gastric juice levels of prostaglandin E_2 (PGE_2) and 6-keto-$PGF_{2\alpha}$, the stable metabolite of PGI_2. Nine conscious, chair-adapted rhesus monkeys were studied once before, once immediately after, and once 2 days after a single 8-Gy total-body irradiation (800 rads) of cobalt-60. In addition to causing vomiting, total-body irradiation transiently suppressed gastric electrical control activity, gastric emptying, and gastric secretion. Levels of PGE_2 and 6-keto-$PGF_{2\alpha}$ were increased in the gastric juice but not in the plasma. Thus, PG's may be responsible for the gastric symptoms of the prodromal syndrome.

INTRODUCTION

Emesis occurs immediately after total-body irradiation, and it is the most obvious and best-documented prodromal symptom of radiation sickness (1,2). Time of onset, duration, and intensity of this vomiting depend on the species (3) as well as the type, dose rate, and total dose of radiation (2). These symptoms clearly differ from those observed during the gastrointestinal syndrome, which occurs 7-15 days after

293

irradiation and is characterized by diarrhea, often accompanied by intestinal bleeding (4).

We recently studied emesis produced in dogs by total-body gamma radiation (5). We observed that gastric emptying was suppressed during the prodromal syndrome of radiation sickness and that prevention of radiation-induced vomiting with domperidone did not improve the suppression of gastric emptying.

The present report summarizes subsequently published observations (6,7) concerning the relation between radiation-induced vomiting and stomach function in an animal model that appears to be closer to man in terms of brain organization and gastric function. We produced vomiting in rhesus monkeys with a single dose of total-body irradiation, and we measured gastric acid secretion, gastric emptying of liquids, and gastric electrical control activity (ECA) before, during, and after the acute prodromal syndrome of radiation sickness. In addition, we determined plasma and gastric juice levels of prostaglandin (PG) E_2 and 6-keto-PGF$_{1\alpha}$.

MATERIALS AND METHODS

Nine conscious, chair-adapted rhesus monkeys were studied on three separate days after an overnight fast: on control preirradiation day, irradiation day, and 2 days after irradiation. Studies were performed in the morning and started 20 min after either sham irradiation on control day or after irradiation on irradiation day. On control days, the animals were brought to the exposure room and the doors were closed for 3 min. On irradiation day, each monkey was exposed to 8 Gy (800 rads) total-body radiation delivered at 5 Gy/min by two 10^5-Ci cobalt-60 irradiators placed anteriorly and posteriorly. Phantom studies demonstrated that the midline abdomen received 8 Gy and that the head received 6 Gy.

Each monkey was visually monitored for 3 hours on control days and 6 hours on irradiation days. Bipolar electrical potentials were recorded from two abdominal disposable skin electrodes on a multichannel recorder (Beckman R612, Beckman Instr., Schiller Park, IL) (6). Mean fasting and postload frequencies obtained in each animal on each study day were used to compute the grand mean ($\pm$ SE) for each type of study in each group of animals.

Vomiting was defined as a succession of strong and brief contractions of thoracic and abdominal muscles leading to the expulsion of gastric contents through the mouth; retching was defined as nonproductive vomiting (6,8). During both events, recordings displayed a succession of brief bursts of high potential spikes, clearly different from the movement artifacts that were sometimes superimposed.

A previously described and validated marker dilution technique (9,10) was used concurrently to determine gastric secretion and gastric emptying during a 40-min fasting period and for 60 min after the injection of an 80-ml water load (postload period). In the present studies, this technique was modified slightly in that ^{99m}Tc-

DTPA (diethylene triamine pentaacetic acid) was used instead of phenol red as the marker. This intubation method requires only the sequential sampling of the gastric contents and permits concurrent measurement of intragastric volume, gastric emptying, and gastric secretion. A 12 French double lumen nasogastric tube was placed in the stomach, and its position was verified by the water recovery test (11). Starting 45 min later, samples of the mixed gastric contents were aspirated just before and immediately after intragastric administration of 5-20 ml of a ^{99m}Tc-DTPA test solution (30 μCi/100 ml H_2O, pH 7.4, 37°C) and were centrifuged. The clear supernatant of each sample was assayed for ^{99m}Tc concentrations using an Ultrogamma autogamma counter (LKB Instr, Turku, Finland) and for titratable acidity using electrometric titration to pH 7.4 (Radiometer, Copenhagen, Denmark). These determinations were repeated every 10 min during the basal period and after intragastric instillation of an 80-ml water load containing ^{99m}Tc-DTPA (3 μCi/100 ml, pH 7.4, 37°C).

Intragastric volumes of fluid (V_1, V_2 . . .) and amounts of ^{99m}Tc (Tc_1, Tc_2 . . .) were determined at the time of each sampling using the dilution principle (9,10,12,13). Fractional emptying rate (g) was then determined for each 10-min interval (t) between two dilutions, assuming that emptying was a first-order process (exponential) during a given 10-min interval and using the equation

$$g = - [\log_e (Tc_2/Tc_1)]/t$$

Since g is allowed to vary from interval to interval, no general assumption has to be made regarding emptying over the total duration of the experiment. Net fluid output (R_v) in ml/min was then determined for the corresponding interval, assuming that it remained constant over the given interval and using the equation

$$R_v = [V_2 - V_1 \cdot \exp(-gt)] \cdot g / [1 - \exp(-gt)]$$

Intragastric volumes of fluid and masses of ^{99m}Tc were then recalculated, taking into account these first estimates of fractional emptying and fluid output, which were in turn recalculated. This iterative process was repeated until the improvement of the solution became less than 1% per iteration. Having previously determined intragastric concentrations of acid (A_1, A_2 . . .), we then calculated net acid output (R_A) using the equation

$$R_A = [V_2 \cdot A_2 - V_1 \cdot A_1 \cdot \exp(-gt)] \cdot g / [1 - \exp(-gt)]$$

These calculations were performed using a locally developed program and a PDP-10 computer (Division of Computer Research and Technology, National Institutes of Health, Bethesda, MD). The assumptions involved have been described and discussed elsewhere (9,10), and they are based on the original contributions by Hildes and Dunlop (12) and George (13). However, in contrast to their method, the present technique allows correction for emptying and secretion occurring during the 1-min dye dilution interval and can be applied during fasting. On irradiation day, intervals with occurrence of vomiting were not taken into account for calculation of g, R_V or R_A.

Before each of the studies, 2 ml of blood was drawn and placed in tubes containing 2% EDTA (80 μg/ml blood) and 1.6% solution of indomethacin in $NaHCO_3$ buffer (5 μg/ml blood) for determination of levels of 6-keto-$PGF_{1\alpha}$ (a stable metabolite of PGI_2) and PGE_2 in the plasma. One-ml samples of gastric juice were placed in tubes containing 1.6% solution of indomethacin in $NaHCO_3$ buffer (5 μg/ml). The samples were immediately centrifuged and the supernatants frozen at -80°C for subsequent analysis. At the time of analysis, the samples were thawed, acidified to pH 3.2 with 2.0 M citric acid, and centrifuged for 10 min at 100 x g and 4°C. The clear supernatants were then applied to individual C18 solid-phase extraction columns (Bond-Elut C18, 200 mg sorbent mass, Analytichem International, Harbor City, CA) previously washed with methanol and equilibrated with H_2O adjusted to pH 3.2 with citric acid. When the supernatants had passed through the columns, each was sequentially washed with 2 ml of distilled water at pH 3.2, 2 ml 12.5% methanol, and 2 ml benzene. PG's were eluted off the C18 column with four 0.5-ml aliquots of ethyl acetate. Slight vacuum pressure was applied to capture the remaining solvent from the columns. Ethyl acetate extracts were dried under N_2 and reconstituted in 1.0 ml of phosphate-buffered saline at pH 6.8 containing 1% bovine gamma globulin and 0.05% sodium azide. PGE_2 and 6-keto-$PGF_{1\alpha}$ levels were determined by radioimmunoassay using ^{125}I-radioimmunoassay kits (New England Nuclear, Boston, MA) (14).

Prior to processing, extraction efficiency was determined by adding to representative samples of plasma and gastric juice 10,000 CPM ^{3}H-PGE_2 or ^{3}H-6-keto-$PGF_{1\alpha}$. Recovery of all four PG tracers was greater than 95% in plasma samples. Similarly, samples were analyzed for nonspecific interfering materials. Samples of plasma and gastric juice were depleted of PG's by charcoal adsorption (Norit A, Fisher Chemical Co., Fairlawn, NJ) (14). One-ml plasma samples and 3-ml gastric juice samples were incubated for 2 hr at 4°C with 0.15 or 0.45 ml, respectively, of a 100 gm/ml charcoal-saline solution. Tracer amounts of labeled PG's added to the samples revealed this procedure to completely adsorb the radioactivity. PG-depleted plasmas and gastric juices were processed on extraction columns as described. Evaporated ethyl acetate eluates, reconstituted in assay buffer and analyzed by radioimmunoassay, revealed less than 5% decrease in binding of all three radiolabeled ligands to their respective antibodies. A number of plasma and gastric juice PG extracts were analyzed at two dilutions. The results, when corrected for the dilution factor, yielded sample values within 10% of one another.

The statistical significance of differences observed for each measurement of gastric function (fractional emptying rate, acid output, etc.) was evaluated using a three-factor (treatment, time, and monkey) analysis of variance with repeated measures on the last two factors (9,10), the program LDU-040 (K. L. Dorn), and an IBM 370 computer (Division of Computer Research and Technology, National Institutes of Health, Bethesda, MD).

RESULTS

No vomiting or retching was observed on control days, i.e., after sham irradiation. On irradiation day, episodes of retching and/or vomiting were observed and recorded in 6 of 9 monkeys. These episodes started at 30 ± 4 min (mean $\pm$ SE) and ended 72 ± 8 min after irradiation. By 2 days after irradiation, vomiting occurred in only one monkey; this was at the end of the study and after most of the stomach contents had been aspirated.

Fasting gastric electrical control activity (ECA) was $3.31 \pm 0.10/$min on control days, was significantly decreased after irradiation ($2.86 \pm 0.19/$min), and returned to control level 2 days after irradiation ($3.11 \pm 0.09/$min). ECA tended to be lower after the water load, but the difference was not statistically significant.

Values for the fractional emptying rate of liquids (FER) are shown in Table 1. On control days, FER was stable during the fasting period but more than doubled after the water load, compared to fasting. On irradiation day, fasting FER was suppressed as soon as the study started (20 ± 2 min after exposure) and was not significantly stimulated by the water load. As a result, there was almost no emptying of the water load on irradiation day in the absence of vomiting.

Acid output was significantly stimulated after the load, compared to fasting on control days. After irradiation, acid output was suppressed in all the monkeys that had secreted on control day; the suppression started 39 ± 4 min after irradiation and persisted after load stimulation (Table 1).

Table 1. Effect of Domperidone and Irradiation on Fractional Emptying Rate, Acid Output, and Fluid Output

		Control Day	Irradiation Day	2 Days After Irradiation
FER:	Fasting (1)	4.1 ± 1.3	0.4 ± 0.3[1]	3.4 ± 1.5
	Fasting (2)	5.5 ± 1.2	0.3 ± 0.2[1]	3.3 ± 1.2
	Postload	11.8 ± 1.8	0.8 ± 0.2[1]	12.9 ± 2.2
AO:	Fasting (1)	8.5 ± 3.0	5.4 ± 2.2	15.2 ± 6.5
	Fasting (2)	9.7 ± 3.8	0.0 ± 0.0[1]	14.6 ± 5.0
	Postload	24.4 ± 7.0	1.6 ± 1.6[1]	27.8 ± 8.1
FO:	Fasting (1)	0.21 ± 0.06	0.40 ± 0.12	0.24 ± 0.06
	Fasting (2)	0.20 ± 0.04	0.23 ± 0.06	0.24 ± 0.06
	Postload	0.30 ± 0.04	0.12 ± 0.01	0.23 ± 0.05

FER, fractional emptying rate, %/min; AO, acid output, μEq/min; FO, fluid output, ml/min. Values are mean ± SE.

[1]$p \pm 0.05$ compared to corresponding value on control day

Fasting (1) corresponds to period 0-20 min after start of study. On day of irradiation, this period was 20-40 min after exposure.
Fasting (2) corresponds to period 20-40 min after start of study. On day of irradiation, this period was 40-60 min after exposure.

Table 2. Effects of Irradiation on Prostaglandin Levels In Plasma and Gastric
Juice

	Control Day	Irradiation Day (2 Hr After Irradiation)	2 Days Postradiation
Plasma:			
PGE$_2$	425 ± 29 N = 9	460 ± 38 N = 8	397 ± 15 N = 8
6-Keto-PGF$_{1\alpha}$	415 ± 59 N = 9	469 ± 44 N = 9	380 ± 27 N = 9
Gastric Juice:			
PGE$_2$	318 ± 80 N = 9	523 ± 94[1] N = 8	305 ± 65 N = 8
6-Keto-PGF$_{1\alpha}$	230 ± 36 N = 9	346 ± 57[1] N = 9	320 ± 40 N = 9

[1]Values are pg/ml, mean ± SE.

As previously reported (7), plasma PGE$_2$ and 6-keto-PGF$_{1\alpha}$ were not significantly different 2 hours after irradiation compared to those on control day or at 2 days postirradiation (Table 2). In contrast, a significant increase ($p < 0.05$) in the concentrations of PGE$_2$ and 6-keto-PGF$_{1\alpha}$ was observed in gastric juice at 30 min after irradiation; both had returned to basal levels 2 days later (Table 2).

There was no significant linear correlation between the levels of PGE$_2$ or 6-keto-PGF$_{1\alpha}$ in the gastric juice and either acid output (R = 0.10 and 0.23, respectively; not significant) or FER (R = 0.10 and 0.32, respectively; NS).

DISCUSSION

In the present studies, we report the precise and objective measurements of the immediate occurrence of retching and vomiting in rhesus monkeys exposed to total-body radiation as well as the relation between these events, gastric function, and PG levels in the plasma and gastric juice.

The visual distinction between retching and vomiting may be difficult in fasting monkeys, although it is easy in the presence of a large vomitus of food. Animals can either store a small vomitus into their cheek pouches and then swallow it, or they can emit foamy saliva after nonproductive retching. In the present studies, recording of skin potential helped in differentiating between these two types of events, demonstrating retching and vomiting in 6 of the 9 monkeys and retching alone in only 2 monkeys. This dose of 800 cGy has been selected because it is twice the ED50 for vomiting as previously determined by others (1,8) for monkeys.

Retching and vomiting started after a delay of about 30 min and disappeared after 70 min, which agrees with the observations of Middleton and Young for similar doses of exposure, but markedly differs from a delay of almost 1 hour with doses 4-5.5 Gy (1) and a delay of less than 5 min following a dose of 12 Gy (8). Thus, the interval between irradiation and vomiting appears to be inversely proportional to the dose received.

The incidence of vomiting after irradiation appears to increase if monkeys are fed solid food 1 to 2 hours before irradiation (8). However, in our study, a 16-hour fast before irradiation did not reduce the incidence of this "radioemesis." Moreover, intragastric administration of a water load after irradiation was associated with only 6 vomiting episodes versus 18 during fasting. This suggests that the incidence of "radioemesis" is actually decreased by gastric distension with noncaloric liquids when radiation is delivered at a high dose rate (5 Gy/min).

Gastric ECA may be recorded via skin electrodes as first described in 1922 (quoted in reference 15). The frequency of this cutaneous electrogastrogram (EGG) is correlated with the slow wave frequency recorded from gastric serosal electrodes (6,15-18). In the present studies, ECA frequency was significantly decreased on the day of irradiation and returned to basal values 2 days later. This finding was confirmed by preliminary results obtained in monkeys with implanted gastric serosal electrodes (17). The radiation-induced decrease of ECA frequency accompanied a concurrent decrease of gastric fractional emptying rate both during fasting and after the load. This latter finding is similar to that observed in dogs (5,19) and rats (20).

Our observations demonstrated that ionizing radiation had a different effect on gastric acid output and on nonparietal secretions. Immediately after irradiation, acid output was suppressed both during fasting and after a water load (Table 1). This suppression of acid could be due to ultrastructural changes of parietal cells similar to those observed in the mouse within 30 min of exposure (21), but it clearly differed from the hypochlorhydria due to gastric atrophy that appears several weeks after irradiation (22-24). In contrast, fluid output was suppressed only after the water load whereas it remained unchanged or even tended to increase during fasting (Table 1). Thus, fasting nonparietal fluid secretions appeared to increase immediately after irradiation, thereby masking the concurrent suppression of the parietal component of fluid output. As nonparietal secretions are not stimulated after a water load (9), no change of fluid output is expected during the postload period if acid output is suppressed. In addition, we found previously (25,26) that mucus output is increased immediately after irradiation. Two days later, both parietal and nonparietal secretions return to basal values, while mucus output remains suppressed (25,26).

The relation between radiation-induced emesis and gastric inhibition is unclear. Since acid suppression started at about the same time as vomiting and persisted after its disappearance, these two symptoms may be closely related. This possibility is also supported by our anecdotal finding that acid output was unchanged after irradiation in the only monkey that did not retch or vomit. In contrast, radiation-induced suppression of gastric emptying and of ECA was observed even in the three monkeys that did not vomit; furthermore, in the animals that vomited, suppression of gastric emptying started before emesis and persisted after its disappearance. Thus, radiation-induced slowing of gastric emptying may appear independently from vomiting.

Concurrently with these changes of gastric function, we observed that the concentrations of PGE_2 and 6-keto-$PGF_{1\alpha}$ were increased in gastric juice. Since radiation was previously found to not modify fluid output (6), these increased PG concentrations appear to reflect an increase of the gastric output of PG. This observation is consistent with the previous findings that PG-like material is elevated in the rat small intestine following irradiation (27).

Therefore, it is tempting to propose that the local release of PG's is related to the gastric effects of radiation. This hypothesis is supported by the observation that the exogenous administration of sufficient doses of PG's produces vomiting (28-30), inhibits gastric acid secretion (28,31-34), and stimulates gastric mucus output (35). In addition, PGE_2 increases (32) and PGI_2 decreases (33) gastric emptying, although the doses of exogenously administered PG that are necessary to cause these effects are likely to produce higher concentration of PG's in the plasma and gastric juice. However, no information is available regarding this point, and PG's generated within gastrointestinal tissue may be effective at much lower concentrations than when given exogenously. Finally, the absence of a significant correlation between the concentration of PG in the gastric juice and the acid output suggests that PG's are not the only factors responsible for the gastric effects of radiation.

In contrast to the effect of radiation on the concentration of PG in gastric juice, the plasma PGE_2 and PGI_2 remained unchanged before or after vomiting and suppression of gastric acid output. This observation agrees with the absence of changes in plasma PG reported (36) in human radiation-induced enteritis. It could mean that these *circulating* PG's are not involved in radiation-induced emesis or in the suppression of secretion and emptying. However, the possibility remains that the effect of radiation is mediated by a *local release* of gastrointestinal PG's that would not be reflected by circulating PG's. In addition, a transient increase in plasma PG's could have occurred during or after irradiation but would not be detected with this sampling regimen.

In conclusion, radiation-induced emesis was accompanied by a suppression of gastric emptying and acid secretion. The concurrent slowing of gastric ECA suggests that an alteration of the motility of the stomach was responsible for the suppression of gastric emptying. The time of onset of each symptom after irradiation and the transiency of the acute prodromal syndrome to radiation sickness suggest the involvement of neural (37) and/or neurohormonal mechanisms (38) or a receptor inactivation (39). The local release of PGE_2 and PGI_2 demonstrated by the present studies could mediate these effects.

ACKNOWLEDGMENTS

We thank M. E. Flynn, J. Stewart, J. L. Warrenfeltz, and N. L. Fleming for their valuable support in animal handling and radiopharmaceutical preparation and administration, and S. Bailey for her expert editorial assistance. The opinions and assertions contained herein are the private ones of the authors and are not to be construed as official policy or as reflecting the views of the Department of Defense. Supported by the Armed Forces Radiobiology Research Institute, Defense Nuclear Agency, under work unit B2152. Research was conducted according to the principles enunciated in the "Guide for the Care and Use of Laboratory Animals" prepared by the Institute of Laboratory Animal Resources, National Research Council.

REFERENCES

1. Middleton, G. R., and Young, R. W. Emesis in monkeys following exposure to ionizing radiation. Aviat. Space Environ. Med. 46: 170-172, 1975.
2. Conard, R. A. Some effects of ionizing radiation on the physiology of the gastrointestinal tract: A review. Radiat. Res. 5: 167-188, 1956.
3. Cordts, R. E. Animal-model studies of radiation-induced emesis and its control. Technical Report SAM-TR-82-26. USAF School of Aerospace Medicine, San Antonio, TX, 1982.
4. Bond, V. P., Swift, M. N., Allen, A. C., and Fishler, M. C. Sensitivity of abdomen of rat to X-irradiation. Am. J. Physiol. 161: 323-330, 1950.
5. Dubois, A., Jacobus, J. P., Grissom, M. P., Eng, R., and Conklin, J. J. Altered gastric emptying and prevention of radiation induced vomiting in dogs. Gastroenterology 86: 444-448, 1984.
6. Dorval, E. D., Mueller, G. P., Eng. R. R., Durakovic, A., Conklin, J. J., and Dubois, A. Effect of ionizing radiation on gastric secretion and gastric motility in monkeys. Gastroenterology 89: 374-380, 1985.
7. Dubois, A., Dorval, E. D., Steel, L., Fiala, N. P., and Conklin, J. J. Effect of ionizing radiation on prostaglandins and gastric secretion in rhesus monkeys. Radiat. Res. 110: 289-293, 1987.
8. Mattson, J. L., and Yochmowitz, M. G. Radiation-induced emesis in monkeys. Radiat. Res. 82: 191-199, 1980.
9. Dubois, A., Natelson, G. H., Van Eerdewegh, P., and Gardner, J. D. Gastric emptying and secretion in the rhesus monkey. Am. J. Physiol. 232: 186-192, 1977.
10. Dubois, A., Van Eerdewegh, P., and Gardner, J. D. Gastric emptying and secretion in Zollinger-Ellison syndrome. J. Clin. Invest. 59: 255-263, 1977.
11. Findlay, J. M., Prescott, R. J., and Sircus, W. Comparative evaluation of water recovery test and fluoroscopic screening in positioning a nasogastric tube during gastric secretory studies. Br. Med. J. 4: 458-461, 1972.
12. Hildes, J. A., and Dunlop, D. L. A method for estimating the rates of gastric secretion and emptying. Can. J. Med. Sci. 29: 83-89, 1951.

13. George, J. D. New clinical method for measuring the rate of gastric emptying: The double sampling test meal. Gut 9: 237-242, 1968.

14. Steel, L., and Kaliner, M. Prostaglandin-generating factor of anaphylaxis. J. Biol. Chem. 256: 12692-12698, 1981.

15. Smout, A. J. P. M., Van Der Schee, E. J., and Grashuis, J. L. What is measured in electrogastrography? Dig. Dis. Sci. 25: 179-187, 1980.

16. Smout, A. J. P. M. Myoelectric activity of the stomach: Gastro-electromyography and electrogastrography. Thesis, Delft University Press, Delft, The Netherlands, 1980.

17. Laporte, J. L., O'Connell, L., Durakovic, A., Sjogren, R., Conklin, J. J., and Dubois, A. Cross correlation analysis of gastric motility following exposure to ionizing radiation in primates. Dig. Dis. Sci. 29: 565A, 1984.

18. Kim, M. S., Lee, Y. L., Park, H. J., and Choi, H. Role of PGs on experimental tachygastria in cats. Gastroenterology 84: 1208A, 1983.

19. Conard, R. A. Effect of gamma radiation on gastric emptying time in the dog. J. Appl. Physiol. 9: 234-236, 1956.

20. Hulse, E. V. Gastric emptying in rats after part-body irradiation. Int. J. Radiat. Biol. 10: 521-532, 1966.

21. Helander, H. F. Early effects of X-irradiation on the ultrastructure of gastric fundus glands. Radiat. Res. 26: 244-262, 1965.

22. Regaud, C., Nogier, T., and Lacassagne, A. Sur les effets redoutable des irradiations etendues de l'abdomen et sur les lesions du tube digestif determinees par les rayons de rontgen. Arch. Elect. Med. 21: 321-334, 1912.

23. Ivy, A. C., Orndoff, B. H., Jacoby, A., and Whitlow, J. E. Studies of the effect of X-rays on glandular activity. III. The effect of X-rays on gastric secretion. Radiology 1: 39-46, 1923.

24. Palmer, W. L., and Templeton, F. The effect of radiation therapy on gastric secretion. JAMA 112: 1429-1434, 1939.

25. Dubois, A., Dorval, E. D., Wood, L. R., Rogers, J. E., O'Connell, L., Durakovic, A., and Conklin, J. J. Effect of gamma irradiation on the healing of gastric biopsy sites in monkeys. An experimental model for peptic ulcer disease and gastric protection. Gastroenterology 88: 375-381, 1985.

26. Shea-Donohue, T., Dorval, E. D., Montcalm, E., El-Bayar, H., Durakovic, A., Conklin, J. J., and Dubois, A. Alterations in gastric mucus secretion in rhesus monkeys following exposure to ionizing radiation. Gastroenterology 88: 685-690, 1985.

27. Borowska, A., Sierakowski, S., Mackowiak, J., and Wisniewski, K. A PG-like activity in small intestine and postirradiation gastrointestinal syndrome. Experientia 35: 1368-1370, 1979.

28. Robert, A., and Yankee, E. W. Gastric antisecretory effect of 15(R)-15-methyl PGE_2, methyl ester and of 15(S)-15-methyl PGE_2, methyl ester (38707). Proc. Soc. Exper. Biol. Med. 148: 1155-1158, 1975.

29. Ganeson, P. A., and Karim, S. A. A. Acute toxicity of prostaglandins E_2, $F_{2\alpha}$, and 15(S), 15-methyl PGE_2 methyl ester in the baboon. Prostaglandins 7: 215-221, 1974.

30. Smith, E. R., and Mason, A. A. Toxicology of the prostaglandins. Prostaglandins 7: 247-268, 1974.

31. Wilson, D. E. Prostaglandins: Their actions on the gastrointestinal tract. Arch. Intern. Med. 133: 112-118, 1974.

32. Nompleggi, D., Myers, L., Castell, D. O., and Dubois, A. D. Effect of a PGE$_2$ analog on gastric emptying and secretion in rhesus monkeys. J. Pharmacol. Exp. Ther. 212(3): 491-495, 1980.

33. Shea-Donohue, P. T., Myers, L., Castell, D. O., and Dubois, A. Effect of prostacyclin on gastric emptying and secretion in rhesus monkeys. Gastroenterology 78: 1476-1479, 1980.

34. Shea-Donohue, P. T., Nompleggi, D., Myers, L., and Dubois, A. A comparison of the effects of prostacyclin and the 15(S), 15-methyl analogs of PGE$_2$ and PGF$_{2\alpha}$ on gastric parietal and nonparietal secretion. Dig. Dis. Sci. 27: 17-22, 1982.

35. Mahoney, J. M., and Waterbury, L. D. The effect of orally administered PGs on gastric mucus secretion in the rat. Prostaglandins Med. 7: 101-107, 1981.

36. Lifschitz, S., Savage, J. E., Taylor, K. A., Tewfik, H. H., and Van Orden, D. E. Plasma prostaglandin levels in radiation-induced enteritis. Int. J. Radiat. Oncol. Biol. Phys. 8: 275-277, 1982.

37. Dubois, A., and Natelson, B. H. Habituation of gastric function suppression in monkeys after repeated free-operant avoidance sessions. Physiol. Psychology 6: 524-528, 1978.

38. Shea-Donohue, P. T., Adams, N., Arnold, J., and Dubois, A. Effects of met-enkephalin and naloxone on gastric emptying and secretion in rhesus monkeys. Am. J. Physiol. 245: G196-G200, 1983.

39. Garvey, T. Z., Kempner, E. S., Steer, C. J., et al. Molecular sizes of receptors for vasoactive intestinal peptide (VIP) and secretin determined by radiation inactivation. Gastroenterology 82: 1064A, 1982.

CHANGES IN RENAL BLOOD FLOW, GLOMERULAR FILTRATION, AND VASOACTIVE HORMONES IN BONE-MARROW-TRANSPLANT RECIPIENTS AFTER TOTAL-BODY IRRADIATION

S. C. Textor[1], S. J. Forman[2], R. D. Zipser[3], and J. E. Carlson[1]

[1]Department of Consultative Medicine/Nephrology
[2]Department of Bone Marrow Transplantation
City of Hope National Medical Center
Duarte, California 91010

[3]Department of Medicine
Harbor-UCLA Medical Center
Torrance, California

ABSTRACT

Although the potential for radiation to produce renal damage at high doses has been well established, there are a few data regarding early functional and hormonal changes in the normal human kidney exposed to "tolerable dose" levels. We measured renal blood flow, glomerular filtration rate (GFR), plasma renin activity, and 6-keto-PGF$_{1\alpha}$ in patients undergoing fractionated total-body irradiation (1320 rads, 13.2 Gy) prior to allogeneic bone marrow transplantation. Both renal plasma flow and GFR rose (23.8% and 28.4%, respectively), while renal vascular resistance fell. GFR rose to levels well above normal, associated with a fall in serum creatinine. These changes were accompanied by a rise in plasma renin activity and urinary 6-keto-PGF$_{1\alpha}$. We propose that irradiation at these doses induces mild renal vascular injury, allowing net vasodilation and hyperfiltration to occur. Such changes represent a hitherto-unrecognized response to whole-body irradiation.

INTRODUCTION

It has long been recognized that the kidney is sensitive to ionizing radiation and that avoidance of high exposures is essential in order to limit radiation damage. Early reports (1-3) indicated that doses above 2000 rads (20 Gy) to the kidney regularly induced vascular and tubular injury that led to loss of functioning tissue, accelerated hypertension, and occasionally renal death.

Histologic studies of kidneys exposed to lower doses rarely showed signs of irreversible injury, although subtle changes in vascular endothelial cells could be detected—hence the designation "tolerable dose." Most such studies represent data after fractionated radiotherapy delivered by cobalt-60, orthovoltage, megavoltage radiation, or high-energy particle beam therapy (Table 1). Recent studies (4) have suggested that under certain conditions (e.g., in the prenatal or immediate postnatal period), even low-dose exposure to the kidney may produce widespread parenchymal changes later evident as renal failure. The mechanisms underlying such changes have not been fully established.

It is of interest that normal tissues often demonstrate capillary dilation and enhanced blood flow following modest irradiation (5). It is uncertain whether these changes are related directly to vascular effects of radiation or partly reflect the action of local vasoactive hormones. The kidney represents a rich source of vasoactive substances, including prostacyclin and other vasodilator prostaglandins. Since renal functions, in particular glomerular filtration and solute reabsorption, are closely regulated by vasomotor tone within the organ, it should not be surprising that these may be affected by a variety of noxious stimuli affecting renal hemodynamics. The kidney is also the source of renin, which leads to both systemic and local generation of the potent vasoconstrictor angiotensin II. Many forms of renal injury result in the release of renin and the activation of prostaglandins, whose effects tend to buffer each other (6-8).

Table 1. Histopathologic Changes Following Ionizing Irradiation

Tissue	Early Effects	Late Effects
Lung	Capillary swelling; septal enlargement; alveolar exudation	Pulmonary fibrosis; vascular thickening
Kidney	<2000 rads "barely detectable"; microvascular swelling; decrements in flow; tubular cell atypism; rare: malignant hypertension	Random distribution of damaged nephrons. Severity of vascular damage may appear later. Vessels postirradiation are more sensitive to hypertension.
Liver	Single exposure of 600 rads; direct hepatic endothelial injury; 3000 rads plus: major hepatitis; veno-occlusive disease: central veins	Little known; occasionally cirrhosis

After White, 1976 (reference 3)

There are few data available regarding renal hormonal and vascular changes associated with ionizing radiation at "tolerable doses" in man. Total-body irradiation (TBI) of 750-1440 rads (7.5-14.4 Gy) has become a standard component of the preparative regimen before bone marrow transplantation.

The present studies resulted from the observation that measured glomerular filtration rates (GFR) were unexpectedly high in patients after preparative TBI. We undertook to evaluate prospectively changes in renal hemodynamics, function, vasoactive hormones, and tubular lysosomal enzyme release associated with this period. The results demonstrated substantial renal vascular effects characterized by increases in renal perfusion and filtration above normal levels. These were associated with rises of plasma renin activity and urinary 6-keto-PGF$_{1\alpha}$ but without evidence of direct tubular injury.

METHODS

Sixteen patients admitted for allogeneic bone marrow transplantation (BMT) at the City of Hope National Medical Center were studied. Ages ranged between 18 and 43 years (mean 27 years). There were 10 males and 6 females. Karnofsky performance status in these patients was 100% prior to transplantation. Blood pressures and renal function as determined by serum creatinine were normal. The studies described had been approved by the Institutional Review Board, and written informed consent was obtained.

Preparatory Regimen

The marrow-conditioning regimen used in this institution has been described in detail elsewhere (9). In brief, day 0 represents the day of actual marrow transplantation. Before then, fractionated total-body irradiation (FTBI) totalling 1320 rads was administered between days -7 and -4. This was delivered in 11 fractions with boosts to the testes and chest. Nonabsorbable antibiotics were given during that interval. This protocol is illustrated in Figure 1. Initial renal hemodynamic studies were performed on day -11 relative to BMT. Renal studies were then repeated on day -4 after completion of radiotherapy.

Renal Hemodynamic Studies

Clearance studies were performed as follows: After the administration of a water load (20 ml/kg) over 1 hour, loading doses of 125-I-iothalamate (0.5 μCi/kg) and 131-I-iodohippuran (0.6 μCi/kg) were administered intravenously. A maintenance infusion containing 45 μCi of iothalamate and 80 μCi of iodohippuran in normal saline was started at a rate of 1 ml/min. Following an equilibration period of 45 minutes, three precisely timed, spontaneously voided samples were obtained. Plasma samples were obtained at the beginning and end of each period for determination of blood levels. The average value of beginning and end samples

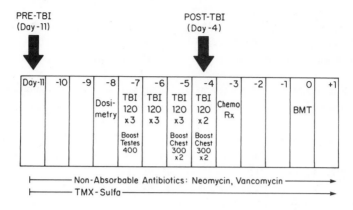

Fig. 1. Preparative regimen for allogeneic bone marrow transplantation at City of Hope National Medical Center. Total-body irradiation (FTBI) was administered in fractionated doses over 4 days to a total of 1320 rads (13.2 Gy). Renal function was determined before (day -11) and immediately after FTBI (day -4).

was used for calculations. Interperiod variability for studies of this kind is less than 8%. Variability for repeat studies on the same patients (interstudy variability) in 19 subjects was 7%.

Other Studies

Twenty-four-hour urine collections were obtained on days -11 and -4 for urinary sodium excretion, urinary 6-keto-$PGF_{1\alpha}$, n-acetyl-beta-glucosaminidase, creatinine clearance, and protein excretion. Samples of blood were obtained from a right atrial catheter in the fasting, supine state for determination of hematocrit, albumin, plasma renin activity, and electrolytes.

Laboratory Determinations

Serum electrolytes, CBC, and albumin were determined in the clinical laboratory. Urinary electrolytes were determined by a flame photometer. Plasma renin activity was determined by radioimmunoassay of the rate of generation of angiotensin I generation (10). Urinary n-acetyl-glucosaminidase was measured by a kinetic fluorimetric assay and corrected for simultaneously measured urinary creatinine concentration (11). Concentrations of 125-I-iothalamate and 131-I-iodohippuran were determined by gamma scintillation counting with corrrection for spillover of 131-I into 125-I. Clearance was calculated as [U][V]/[P], where [V] is the urinary flow rate in ml/min and [U] and [P] are urinary and plasma concentrations of the clearance markers, respectively.

Urinary excretion of 6-keto-$PGF_{1\alpha}$ was determined by radioimmunoassay after extraction with ethylacetate and purification by high-performance liquid chromatography (12).

STATISTICAL METHODS

Results are presented as mean ± SEM. Values obtained before and after irradiation were compared using repeated measures analysis of variance. Specific comparisons between individual time points were made by analysis of variance using the Bonferroni correction. Comparison between groups was made using the nonpaired t test or Mann-Whitney test for parametric or nonparametric analysis, as appropriate (13).

RESULTS

Body weight, albumin, hematocrit, blood pressure, and urinary sodium excretion before and after FTBI are listed in Table 2. There were modest decrements in these indices consistent with catabolic effects of radiation. Urinary sodium excretion during 24-hour collections did not change, nor did the fractional excretion measured during formal renal hemodynamics. Patients received a normal diet during this period, although gastrointestinal symptoms were common, including loss of appetite and occasional nausea. Five of 16 patients were given total parenteral nutrition during this interval.

There were no changes in urinary urea excretion during this period (709 ± 152 versus 732 ± 121 mg/day, not significant). Serum creatinine fell from 0.94 ± .05 to 0.75 ± .05 mg/dl (p <.01) while urinary creatinine did not change. Hence it

Table 2. Changes in Weight, Blood Pressure, Albumin, and Sodium Excretion Before and After Total-Body Irradiation

	Pre TBI (Day -11)	Post TBI (Day -4)	
Body weight (kg)	70.0 ± 3.6	69.1 ± 3.5	p < 0.05
Serum albumin (g/dl)	4.0 ± 0.2	3.6 ± 0.1	p < 0.01
Hematocrit (%)	39 ± 1	35 ± 1	p < 0.01
Blood pressure (mm Hg)	115 ± 3	111 ± 2	p < 0.05
	74 ± 2	68 ± 2	p < 0.01
Urinary sodium excretion (mEq/24°)	164 ± 29	110 ± 15	NS
Fractional excretion of sodium (%)	1.13 ± 0.17	0.82 ± 0.15	NS
Urea nitrogen (mg/24°)	710 ± 152	732 ± 121	NS
Creatinine (mg/24°)	1480 ± 233	1650 ± 166	NS

NS, not significant

appeared that no major protein breakdown developed during this interval. Effective renal plasma flow, glomerular filtration rate, and renal vascular resistance before and after TBI are shown in Figure 2. Pretreatment values of GFR, corrected for body surface area, were normal, compared to 19 normal subjects (120 ± 8 versus 126 ± 6 ml/min/1.73 m², not significant). After FTBI, both effective renal plasma flow (ERPF) and GFR increased considerably (23.8% and 28.4%, respectively). Sustained changes in GFR ranged from +5 ml/min (3.5%) to +90 ml/min (64.3%) and bore no evident relationship to initial values. Filtration fraction was unchanged. Urinary flow rates were not different (8.64 ± 1.23 versus 7.38 ± 1.13 ml/min, not significant). Since arterial pressures fell slightly during this period, these changes reflect a considerable reduction in calculated renal vascular resistance (Figure 2) and therefore net vasodilation. In those patients who underwent repeat studies at a later date, ERPF and GFR were reduced back to baseline or below. Serum creatinine values for this group of patients plus three additional patients not having formal renal hemodynamic studies before and after TBI and BMT are illustrated

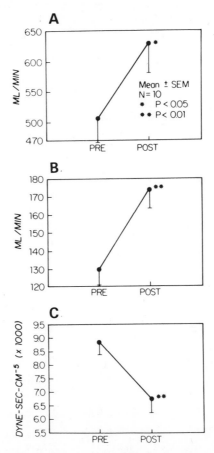

Fig. 2. Renal hemodynamics and glomerular filtration rate (GFR) before and after total-body irradiation. Effective renal plasma flow (A) and GFR (B) rose during this interval, while calculated renal vascular resistance (C) fell. No changes in urinary flow rate or fractional excretion of sodium were observed during these studies.

in Figure 3. There was a fall in creatinine consistent with the changes in measured GFR, which persisted through day +6 after BMT. Thereafter levels rose to baseline or above.

Changes in plasma renin activity, urinary 6-keto-PGF$_{1\alpha}$, and NAGA (n-acetyl-glucosaminidase) are shown in Figures 4-6. Both renin and the prostacyclin metabolite rose after FTBI, while NAGA excretion remained within the normal range. NAGA levels rose promptly after the intravenous administration of chemotherapeutic agents.

DISCUSSION

The results of this study indicate that patients with initially normal renal function undergoing fractionated total-body irradiation of 1320 rads (13.2 Gy) developed substantial renal vasodilation and glomerular hyperfiltration. These changes have been hitherto unrecognized and were associated with increases in urinary 6-keto-PGF$_{1\alpha}$ and plasma renin activity. We propose that these changes represent vascular responses within the kidney to ionizing radiation below doses known to induce major toxicity. Furthermore, these studies indicated that the functional and vascular responses observed were dissociated from tubular injury, as reflected by sodium reabsorption and tubular lysosomal enzyme release (NAGA).

High-dose irradiation of normal kidney tissue has been long appreciated to produce both vascular and tubular changes (1-5,14,15). Such changes were rarely seen at exposure levels less than 2000-2500 rads (20-25 Gy). Histopathologic studies revealed few changes at lower levels of radiation, and it appeared that recovery from any such changes was relatively certain.

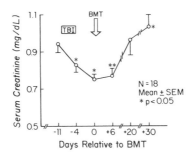

Fig. 3. Serum creatinine values before and after preparation and bone marrow transplantation (BMT) in the 16 patients studied and 3 others. Although basal levels were within normal range, a fall in serum creatinine regimen persisted until at least day +6. It should be noted that after BMT, marked immunosuppression, infection, and exposure to nephrotoxic antibiotics were universal and may have affected renal function.

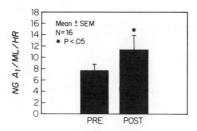

Fig. 4. Plasma renin activity (PRA) before and after FTBI. Rise in PRA occurred despite no change in urinary sodium excretion.

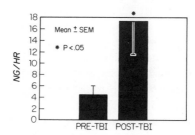

Fig. 5. Urinary 6-keto-PGF$_{1\alpha}$ measured by radioimmunoassay before and after FTBI in bone-marrow-transplant recipient candidates

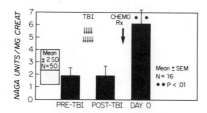

Fig. 6. Urinary excretion of N-acetyl-glucosaminidase (NAGA) corrected for urine creatinine. Shaded zone (left) represents normal range (mean ± 2 SD) of 50 normal subjects. No changes in NAGA were directly related to FTBI, although levels rose immediately after subsequent chemotherapy with cyclophosphamide or etoposide.

Nonetheless, occasional reports from postmortem studies in persons who had received 1500 rads (15 Gy) to the kidneys suggested that vascular dilation and hyperemia may occur (3). Similar indications of transient rises in blood flow after irradiation have been encountered in other organs. It appears likely that vasodilation may represent part of a spectrum of vascular response to radiation, which may proceed to endothelial cell damage if further injury occurs. Such injury may be manifest as cell swelling and obliteration of the vascular lumen, eventually producing overt tissue ischemia.

The mechanisms underlying these changes have not been fully elucidated. It is of interest that the vascular release of prostaglandins after irradiation has been consistently observed. Urinary PGE rises dramatically following whole-body irradiation (16). Vascular endothelial beds are recognized to be a rich source of prostaglandin production under many circumstances (17). Most often they are released in response to some pathologic stimulus. Within the kidney, many exogenous nephrotoxic insults induce a wide array of hormonal changes, often leading to the release of both vasoconstrictor (e.g., angiotensin II) and vasodilator substances (e.g., prostacyclin). Under most circumstances, the result is net vasoconstriction or, at best, an even balance without reduction in blood flow. Under such circumstances, the inhibition of prostaglandin production by administration of cyclooxygenase inhibitors has produced major decrements in blood flow and reduction of renal function. Hence it appears that local prostaglandin production serves a protective role under some conditions in which either hemodynamic or toxic insults to the kidney would otherwise produce renal failure (6-8,18,19).

Our results during total-body irradiation in man may be exceptional in that net vasodilation appears to predominate. We propose that radiation injury with the production and release of vasodilator prostaglandins may be an instance wherein vasodilation overwhelms vasoconstrictor influences, i.e., activation of the renin-angiotensin system. Additional vasodilator mechanisms cannot be excluded because no studies were performed under conditions of cyclooxygenase inhibition. Moreover, we cannot exclude the possibility that the rise in plasma renin activity observed under these circumstances may not lead to a corresponding release of angiotensin II. Measurements of angiotensin II levels might clarify this issue.

Few stimuli are known that produce sustained increments in blood flow and filtration to this degree in normal kidneys (Table 3). In recent years, attention has been focused on the potential for dietary or intravenous protein loading to induce a rise in GFR comparable to the levels achieved in the current study. The relative increase in GFR after a standardized meal has been proposed to reflect the maximum functional potential of the kidney and therefore to be a measure of "renal reserve" (20,21). The authors emphasized that baseline GFR is widely variable and that measurement under such a stimulus provides useful information about the maximal filtration capacity. Another example of this effect has been described in patients with widespread burns, whose glomerular filtration rates and measured kidney length expand during hospitalization to levels comparable to those reported here. Such persons were markedly catabolic and received substantial intravascular protein support (22). Although our patients may have been mildly catabolic, as judged by weight loss and decrement in serum albumin, it is unlikely that this was sufficient to explain the renal changes observed. Urinary excretion of urea nitrogen did not change appreciably, nor did daily creatinine excretion. Only a few patients were deemed anorectic enough to warrant initiation of intravenous nutrition during this phase of the protocol, and changes in serum protein were relatively modest (Table 2).

Table 3. Clinical Examples of Increased Glomerular Filtration
in a Previously Normal Kidney

Pregnancy

Compensatory hypertrophy following nephrectomy

Compensatory hypertrophy associated with renovascular disease

Protein loading

    Burn patients
    ? Amino acid load

Diabetes mellitus

Total-body irradiation

The other situations in clinical medicine in which overt hyperfiltration has been observed without a reduction in nephron number have been in patients with early diabetic nephropathy and during normal pregnancy. The mechanisms underlying these observations have not been fully established, although glucose control and growth hormone secretion have been suggested as major control factors in diabetics (23,24). No abnormalities in glucose homeostasis were seen in our subjects. Pregnancy is associated with widespread hormonal alterations, including increased synthesis of prostaglandins. The administration of indomethacin in pregnant animal models has prevented the rise in blood flow and GFR (25-27). It appears that whole-body radiation exposure at the doses used constitutes an additional instance of renal hyperfiltration in normal kidneys.

Measurements of serum creatinine and creatinine clearance in our patients agreed with the magnitude of changes in glomerular filtration measured by iothalamate clearance. Such changes may easily go unnoticed in subjects with normal renal function, because they represent relatively minor changes within the "normal range." Careful tracking of creatinine measurements in this group indicates that increased glomerular filtration was sustained for at least 1 and perhaps 2 weeks after irradiation (Figure 3). Thereafter, group averages tended to rise back toward pretreatment levels and beyond. This period was associated with chemotherapy, immunosuppressive therapy including cyclosporine A, clinical infection, and exposure to nephrotoxic antibiotics. Thus the interpretation of a later rise in creatinine must be qualified. Nonetheless, the fall in creatinine observed during the preparatory regimen persisted for 1-2 weeks.

The clinical consequences of these results are not fully known. At the very least, the recognition of a change in renal function facilitates the dosage selection and anticipated clearance during administration of nephrotoxic agents, e.g., aminoglycoside antibiotics. Clinicians must recognize that rises of serum creatinine to "normal" ranges under these conditions may represent a major loss of renal function. No adverse effects attributable to hyperfiltration were evident during these studies.

Of greatest potential concern is the evidence these studies provide of underlying vascular changes, probably representing mild injury to vascular endothelium in the kidney. Early studies (28,29) in radiation nephropathy implicated an exaggerated sensitivity of the exposed kidney to hypertensive nephrosclerosis as a factor in the development of malignant hypertension and renal failure. Whether renal vasodilation in bone-marrow-transplant recipients predisposes, or even protects, the kidney from subsequent nephrotoxic injury merits further study (30).

Taken together, these data provide evidence that fractionated total-body irradiation at doses below those thought to produce long-term damage does produce substantial vascular and hormonal changes in the normal kidney. Recognition of these changes may provide the basis for improved understanding and further study of the complex alterations induced by the regimen of bone marrow transplantation and other conditions associated with renal irradiation.

REFERENCES

1. Luxton, R. W., and Kunkler, P. B. Radiation nephritis. Acta Radiol. Ther. Phys. Biol. 2: 169-177, 1964.
2. Rosen, S., Swerdlow, M. A., Muehrcke, R. C., and Pirani, C. Radiation nephritis: Light and electron microscopic observations. Am. J. Clin. Pathol. 41: 487-502, 1964.
3. White, D. C. The histopathologic basis for functional decrements in late radiation injury in diverse organs. Cancer 37: 1126-1143, 1976.
4. Jaenke, R. S., Phemister, R. D., and Norrdin, R. W. Progressive glomerulosclerosis and renal failure following perinatal gamma radiation in the beagle. Lab. Invest. 42: 643-655, 1980.
5. Blake, D. B. Radiobiologic aspects of kidney. Radiol. Clin. North Am. 3: 75-87, 1965.
6. Terragno, N. A., Terragno, A., and McGiff, J. A. Contribution of prostaglandins to the renal circulation in conscious, anesthetized and laparotomized dogs. Circ. Res. 40: 590-595, 1977.
7. Henrich, W. L., Anderson, R. J., and Berenes, A. S. The role of renal nerves and prostaglandins in control of renal hemodynamics and plasma renin activity during hypotensive hemorrhage in the dog. J. Clin. Invest. 61: 744-750, 1978.
8. Edwards, R. M. Effects of prostaglandins on vasoconstrictor action in isolated renal arterioles. Am. J. Physiol. 248: F779-784, 1985.
9. Forman, S. J., Spruce, W. E., Farbstein, M. J., et al. Bone marrow ablation followed by allogeneic marrow grafting during first complete remission of acute nonlymphocytic leukemia. Blood 61: 439-442, 1983.
10. Bravo, E. L., Tarazi, R. C., Fouad, F. M., Textor, S. C., Gifford, R. W., and Vidt, D. G. Blood pressure regulation in pheochromocytoma. Hypertension (Supp II) 4(3): II193-II199, 1982.
11. Powell, S. C., Scaro, J., Wilson, E., and Shihabi, Z. K. Assay of urinary n-acetyl-beta-glucosaminidase in a centrifugal analyzer. Clin. Chem. 29: 1717-1719, 1983.

12. Zipser, R. D., Morrison, A., Laffi, J., and Duke, R. Assay methods for 6-keto-prostaglandin-F-1-alpha in human urine: Comparison of chromatographic techniques with radioimmunoassay and gas chromatography-negative-ion-space-chemical-ionization-mass-spectrometry. J. Chromatogr. 339: 1-9, 1985.

13. Miller, R. G. "Simultaneous Statistical Inference." 2nd ed. Springer-Verlag, New York, 1981.

14. Keane, W. F., Crosson, J. T., Staley, N. A., Anderson, W. R., and Shapiro, F. L. Radiation-induced renal disease. Am. J. Med. 60: 127-137, 1976.

15. Scanlon, G. T. Vascular alteration in the irradiated rabbit kidney. Radiology 94: 401-406, 1970.

16. Donlon, M., Steel, L., Helgeson E. A., Shipp, A., and Catravas, G. N. Radiation-induced alterations in prostaglandin excretion in the rat. Life Sci. 32: 2631-2639, 1983.

17. Dunn, M. J. Renal prostaglandins. In: "Contemporary Nephrology." S. Klahr and S. G. Massry, eds. Plenum Press, New York, 1983, pp. 145-191.

18. Belch, J. J., McLaren, M., Anderson, J., Lowe, G. D., Sturrock, R. D., Capell, H. A., Forbes, C. D., Mikhailidis, D. P., Jeremy, J. Y., and Dandona, P. Increased prostacyclin metabolites and decreased red cell deformability in patients with systemic sclerosis and Raynaud's syndrome. Prostaglandins Leukotrienes Med. 18: 401-402, 1985.

19. Lifschitz, M. D., and Basrnes, J. L. Prostaglandin I2 attenuates ischemic acute renal failure in the rat. Am. J. Physiol. 247: F714-717, 1984.

20. Davies, D. J., Brewer, D. B., and Hardwicke J. Urinary proteins and glomerular morphometry in protein overload proteinuria. Lab. Invest. 38: 232-243, 1978.

21. Bosh, J. P., Saccaggi, A., Lauer, A., Ronco, C., Belledonne, M., and Glabman, S. Renal functional reserve in humans: Effect of protein intake on glomerular filtration rate. Am. J. Med. 75: 943-950, 1983.

22. Goodwin, C. W., Aulick, L. H., Becker, R. A., and Wilmore, D. W. Increased renal perfusion and kidney size in convalescent burn patients. JAMA 244: 1588-1590, 1980.

23. Mogensen, C. E., and Andersen, M. J. F. Increased kidney size and glomerular filtration rate in early juvenile diabetes. Diabetes 22: 706, 1973.

24. Hostetter, T. H. "Diabetic Nephropathy in the Kidney." B. M. Brenner and F. C. Rector, eds. W. B. Saunders, Philadelphia, 1986, pp. 1377-1402.

25. Lindheimer, M. D., and Katz, A. I. The kidney in pregnancy. In: "The Kidney." B. M. Brenner and F. C. Rector, eds. W. B. Saunders, Philadelphia, 1986, pp. 1253-1295.

26. Baylis, C. The mechanism of the increase in glomerular filtration rate in the twelve-day pregnant rat. J. Physiol. 305: 405, 1980.

27. Conrad, K. P. Renal hemodynamics during pregnancy in chronically catheterized conscious rats. Kidney Int. 26: 24, 1984.

28. Fisher, E. R., and Hellstrom, H. R. Pathogenesis of hypertension and pathologic changes in experimental renal irradiation. Lab. Invest. 19: 530-538, 1968.

29. Ljungqvist, A., Unge, G., Lagergren, C., and Notter, G. The intrarenal vascular alterations in radiation nephritis and their relationship to the development of hypertension. Acta Path. Microbiol. Scand. Section A 79: 629-638, 1971.

30. Textor, S. C., Forman, S. J., Borer, W. Z., and Carlson, J. Sequential blood pressure, hormonal and renal changes during cyclosporine administration in bone marrow transplant recipients with normal renal function. Clin. Res. 34: 487A, 1986.

EICOSANOIDS AND RADIOTHERAPY IN HUMAN AND ANIMAL TUMORS

I. F. Stamford, J. D. Gaffen, P. B. Melhuish, and A. Bennett

Department of Surgery
King's College School of Medicine and Dentistry
The Rayne Institute
London, SE5 9NU, England

ABSTRACT

Since eicosanoids may influence tumor growth and spread and radiotherapy can increase tissue prostaglandins, it is important to investigate relationships between prostaglandins and radiotherapy in cancer. Our studies in mice with NC carcinoma show that local radiotherapy reduced tumor size and tumors were even smaller when the prostaglandin synthesis inhibitor flurbiprofen was also given. However, we do not know if this reduction was merely due to removal of radiotherapy-induced inflammation. It is of interest to note that prostaglandin synthesis inhibitors increased the response to chemotherapy and the accumulation of methotrexate in malignant cells.

Since prostaglandins can protect certain tissues such as the gastrointestinal mucosa from damage, we determined whether indomethacin affected the viability of normal and malignant cells exposed to 3 Gy X rays. Indomethacin (0.1 or 1 μg/ml) did not alter the viability of epithelial cells from human normal embryonic intestine or their injury by radiotherapy. Indomethacin alone reduced the viability of rat hepatic tumor cells, and this added to the effect of radiotherapy. In neither case did indomethacin alter the response to radiotherapy, and these results argue against "cytoprotection" by prostaglandins.

Studies in breast cancer patients showed primary tumor prostagandin-like material (PG-LM) to correlate with early death. Yields of tumor PG-LM from patients with head and neck cancer (whose tumors were irradiated prior to excision) correlate with the amount of necrosis, inflammation, and fibrosis but not with survival. Plasma PG-LM was measured before and after synchronous radiotherapy and chemotherapy, and both the amount of mucositis and the treatment correlated with the plasma measurements, but this was despite the intake of nonsteroidal anti-inflammatory drugs. The clinical implications of the tumor prostaglandin measurements are not clear. We need to know the influence of prostaglandin synthesis on the response to radiotherapy, and the ability of indomethacin to increase the response to methotrexate is potentially of great importance.

INTRODUCTION

This chapter outlines all the aspects of our cancer/prostaglandin work and reviews specific areas. The study of prostaglandins in cancer has expanded rapidly over the last 10 years. We became interested in this subject because of our early work on prostaglandins and the gastrointestinal tract (1) and our demonstration that the release of prostaglandin-like material (PG-LM) may explain bone resorption by human dental cysts (2).

We initially hypothesized that prostaglandins (PG's) released by tumor cells might contribute to the formation of bone metastases. Therefore, we studied human breast cancer, which often metastasizes to bone. The results of these experiments, started in 1973, have been published previously (3-6).

Studies parallel to those on breast tumors were also started in 1973 on human gastrointestinal cancer (7), including characterization of PG's extracted from normal and malignant gastrointestinal tissues (8,9). The encouraging work, which initially indicated a relationship of breast tumor PG-LM to bone metastasis, was followed by research into bone destruction by the VX2 tumor in rabbits and its reduction by indomethacin (10,11). Further investigation with PG synthesis inhibitors was conducted in the mouse NC tumor model. This murine adenocarcinoma partially mimics human breast cancer, since it metastasizes to the lungs and mediastinum (although not to bone). The early studies involved chemotherapy, radiotherapy (12,13), and the PG synthesis inhibitor flurbiprofen (14,15). Other drugs investigated for their effect on the NC tumor include 16,16-dimethyl PGE_2 (16), prednisolone (17), mepacrine (17), a thromboxane synthase inhibitor (18), and tamoxifen (19). The human breast cancer results and the beneficial effects of the PG synthesis inhibitors in the mouse studies led us into investigations of other human cancers that can destroy bone. In 1978 we began studies of previously untreated lung cancer, and of head and neck cancer patients given radiotherapy and chemotherapy (20-22). This work involved the quantitation of prostanoids in malignant and normal lung tissues as well as blood (23,24).

Our cell culture investigations (25), started in 1982, paralleled some of the previous *in vivo* studies. In addition, we have now studied chromosome damage and the effect of indomethacin on methotrexate uptake by NC cells (26). This work has now progressed into studies of human malignant and nonmalignant cell lines.

The work selected for review in somewhat more detail below includes our *in vivo* and *in vitro* cancer research involving PG's and radiotherapy either alone or in combination with PG synthesis inhibitors or cytotoxic drugs, as well as our recent work on indomethacin and methotrexate.

MOUSE CANCER *IN VIVO* (NC TUMOR)

The NC tumor arose spontaneously in the mammary region of a WHT/Ht mouse and has been passaged only in this strain (27). The transplanted tumor metastasizes by day 5 and is palpable by day 10.

Following local excision of this carcinoma, there is a high incidence of local lymphatic spread, recurrence in the excision scar, and metastasis mainly to the lungs and mediastinum. A 100% take is expected with 10^6 transplanted cells, consistent with the claim that the tumor is non-immunogenic. This reproducible animal tumor system is therefore suitable for studying the effects of drugs and radiotherapy on tumor growth and spread.

Effect of Chemotherapy and Radiotherapy on NC Tumors

Mice transplanted subcutaneously with 10^6 NC carcinoma cells were given drugs orally in raspberry syrup. The treatment regimen consisted of flurbiprofen 5 mg/kg daily from day 25 after tumor transplantation, plus chemotherapy with pulses of methotrexate and melphalan. Radiotherapy (2 x 5 Gy) was given locally to the tumor, using a restraining jig. Tumors removed from the mice on Day 42 showed, at most, a tendency to be lighter in the flurbiprofen-treated group, but were clearly lighter when radiotherapy ± chemotherapy was also given. However, we do not know whether flurbiprofen potentiated the effect of the radiotherapy, or whether the reduction in tumor weight was due to the inhibition of radiotherapy-induced edema.

CELL CULTURE

These *in vitro* studies were done because prostaglandins can protect the gastrointestinal mucosa from damage, and we wished to know how the viability of normal and malignant cells exposed to radiotherapy would be affected.

Cell Survival After X Irradiation

Cell lines from a mouse hepatic tumor and epithelium-like cells from normal human embryonic intestine were studied using clonogenic assays. After exposure to 0-8 Gy X rays, viability was determined as the percentage that formed colonies of >50 cells after incubation for 9 days. Normal cells were slightly more sensitive to X rays (LD50, 2.5 Gy) than were mouse hepatic tumor cells (LD50, 3.3 Gy).

Indomethacin (0.1-1 μg/ml) had little or no effect on the viability of normal human intestinal cells, but reduced that of the rat hepatic tumor cells (IC50, 0.45 μg/ml). Higher concentrations of indomethacin reduced the viability of both cell lines (Figure 1).

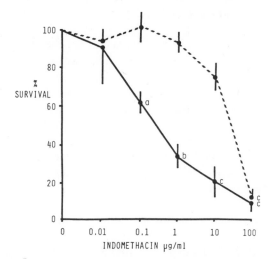

Fig. 1. Effect of 0.1-1 μg/ml indomethacin on viability of cell lines. Viability of normal intestinal cells (dotted line) was not affected, whereas viability of rat hepatic tumor cells was reduced. Each point is a mean ± SEM of 10-32 cultures in 3-4 experiments. a, P < 0.1; b, P < 0.01; c, P < 0.001.

In normal cells exposed to 3 Gy X rays, indomethacin (0.1 or 1 μg/ml) had little or no effect on viability judged by clonogenicity. In the hepatic cell line, the reduction of viability by indomethacin plus X rays seems to be mainly additive (Figure 2). Since there was no evidence of an altered response to radiotherapy, these results may argue against prostaglandin-induced cell "protection." The term "cytoprotection" is sometimes used for the prostaglandin-induced protection of the gastric mucosa. It obviously does not apply to the human intestinal cells, and since there is little evidence that in the gastric mucosa the protection occurs at the level of the cell being damaged, the term "cytoprotection" is best avoided.

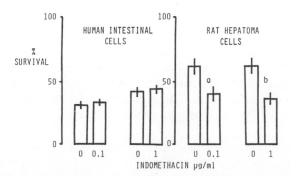

Fig. 2. Cell cultures exposed to 3 Gy X rays. Controls, 0, without indomethacin; test cultures 0.1 and 1 μg/ml indomethacin. Survival in irradiated control and indomethacin-treated normal intestinal cells was similar, but was smaller in the indomethacin-treated rat tumor cells. Results are mean ± SEM of 16-18 cultures frcm 3-4 experiments.

Table 1. Effect of X Irradiation and Indomethacin on Incidence of Chromosome Abnormalities[1]

Abnormality	3 Gy X Irradiation			
	Vehicle	Indo 0.1	Indo 1	Indo 5 $\mu g/ml$
Polyploids	1	0	1	2
Massive rearrangement	7	13	9	12
Gaps	9	10	10	8
Breaks and fragments	31	35	22	44
% abnormal spreads	48	58	42	66

[1]Epithelial cells from human normal embryonic intestine

CHROMOSOME ABNORMALITIES AFTER EXPOSURE TO X RAYS AND INDOMETHACIN

Three experiments, each with 50-140 metaphase spreads, were carried out to investigate the incidence of radiation-induced chromosome damage in the normal human embryonic intestinal cells exposed to 3 Gy X rays. Abnormalities, including polyploids, massive structural rearrangement, gaps, breaks, and fragments, were assessed to give the total percentage of abnormal spreads (Table 1). There was no evidence that indomethacin altered the effect of X rays on chromosomal damage.

HEAD AND NECK CANCER

This study was prompted by the observation that squamous carcinomas of the head and neck that invade bone are accompanied by pronounced osteoclastic activation. Patients with large T3/4 tumors, predominantly in the larynx, tongue, mouth, and oropharynx, received radiotherapy or synchronized radiotherapy and chemotherapy (vincristine, bleomycin, methotrexate, and 5-fluorouracil). The tumors and irradiated macroscopically normal mucosa and uninvolved lymph nodes were then excised, and PG's were extracted. The amounts of PG-LM from irradiated nodal metastases or uninvolved lymph nodes were similar, but were considerably lower than from the primary squamous carcinomas. Primary tumors removed within 3 months of treatment tended to yield more PG-LM than did those removed after longer intervals, and there was an inverse correlation between tumor PG-LM and an arbitrary morphological score that graded fibrosis, inflammation, and necrosis.

Blood draining from the tumor into the internal jugular vein (i.e., before metabolism in the pulmonary circulation) showed no consistent differences in the amounts of PG-LM compared to peripheral blood (22). However, any material released from the tumor would be greatly diluted by blood draining from other regions of the head and neck, and may have masked the release of tumor PG-LM.

PLASMA PROSTAGLANDINS IN RADIATION-INDUCED MUCOSITIS

Synchronous radiotherapy and chemotherapy in the treatment of squamous carcinoma of the head and neck can produce intense inflammation of the treated area (28). We investigated previously untreated patients with histologically proven primary squamous carcinoma of the head and neck (mainly in the larynx, oral cavity, or oropharynx). Pulses of vincristine, bleomycin, methotrexate, and leucovorin were integrated with 60-66 Gy radiotherapy over 6-7 weeks in daily fractions of 2 Gy. Blood was drawn from the antecubital vein of patients, before and during synchronous chemotherapy and radiotherapy. The blood was transferred to lithium heparin tubes containing 10 μg/ml indomethacin and extracted for PG's (29), which were quantitated by bioassay. Patients with severe mucosal discomfort often were prescribed aspirin mucilage or acetaminophen. The severity of mucositis (assessed by visual inspection) and the concentration of blood PG-LM correlated with the amount of treatment received, regardless of whether or not the patients received aspirin or acetaminophen.

Eisen and Walker (30,31) demonstrated that radiation increases the amount of PG-LM extracted from some mouse tissues, by inhibiting PG breakdown, and Levine (32) showed that chemotherapeutic drugs can release PG's from cells. Therefore, it would be expected that treatment given to patients with head and neck cancer would increase the amount of PG's in tissues and contribute to the signs and symptoms of inflammation. However, the mechanism for increased amounts of plasma PG-LM is more difficult to explain because many PG's (particularly PGE$_2$, to which the rat stomach bioassay is particularly sensitive) are inactivated during passage through the pulmonary circulation. However, perhaps radiotherapy reduced the activity of PG-inactivating enzymes (31) or increased PG formation and release from blood cells in response to the trauma of blood sampling. The elevated amounts of PG's might be directly attributable to the effect of the chemotherapy and/or to the irradiation of blood elements as they flow through the region during radiotherapy. However, irradiation of blood *in vitro* did not affect the amount of PG-LM released (33).

Although the rat fundus strip preparation is most sensitive to PGE$_2$, other PG's or fatty acids may contribute to the biological activity of these extracted blood samples, and it would be of value to use other methods to identify and quantitate the blood PG's before and after radiotherapy. Since some patients administered aspirin or acetaminophen also had elevated PG-LM as treatment progressed, higher doses of aspirin or different cyclo-oxygenase inhibitors might have given more relief from the mucositis.

BREAST CANCER

This aspect was the first of our PG cancer studies, and was chosen because breast carcinomas commonly metastasize to bone. We found that these tumors produce more PG-LM than do benign or normal breast tissue, and there is a higher amount of PG-LM in extracts of carcinomas from patients with evidence of bone metastases (3,4). Furthermore, studies in rats (34) and rabbits (10,11,35) indicated that PG's may be important in tumor-induced bone destruction. However, after more patients were studied, we revised our conclusions. Although tumors from the bone-scan-negative group yielded a lower median amount of PG-LM, this was because no positive scans occurred in patients whose tumors yielded very low amounts of PG-LM. Above 16 ng PGE_2 equivalents/g tumor, the groups with negative and positive scans overlapped. Furthermore, with regard to the site of tumor recurrence (local, bone, or visceral), the amount of PG-LM from the previously excised primary tumor tended to be highest in patients with local recurrence (regardless of local postoperative irradiation); skeletal recurrence occupied the intermediate position. The patients in this study are still being followed, and at the latest analysis the minimum followup time is more than 6 years. During the 11-year period of the study, 82 patients have now died: 69 of cancer and 13 of other causes. Figure 3 shows the striking inverse correlation between tumor PG-LM and patient deaths up to the 3-year postoperative period. This inverse correlation is not easy to explain, and does not occur in the period more than 3 years postoperatively. A simple relationship between PG's and cancer seems unlikely, since there are many different types of PG's, cancers, and processes involved in malignancy. Furthermore, various other factors are involved. One of these is menopausal status; the inverse relationship between PG's and death within 3 years postoperatively holds for the postmenopausal patients, but in the premenopausal group there is a tendency for a direct correlation.

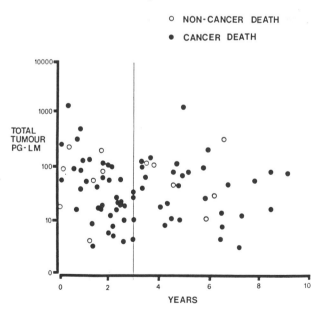

Fig. 3. Relationship between PG's and death. Each dot represents a patient (•, cancer death, n = 69; o, noncancer death, n = 13). Time to death is inversely related to tumor total PG-LM up to 3 years postoperatively (P < 0.001, n = 46) but not subsequently.

INDOMETHACIN AND CYTOTOXIC DRUGS

Survival of Mice With NC Carcinoma

Mice whose NC carcinomas were resected after metastasis had occurred survived longer when indomethacin was given with methotrexate and melphalan, compared to groups given only cytotoxic drugs (15). We did not know if there was an interaction with just one of the cytotoxic drugs, although some *in vitro* evidence suggested that this might be so (see later). More recent experiments have demonstrated that mice given only methotrexate with indomethacin survive longer than those treated with the highest tolerated dose of methotrexate alone (Figure 4). This argues against a pharmacokinetic explanation (e.g., increased bioavailability of methotrexate due to displacement from plasma protein-binding sites), unless there is simultaneous protection against the toxic effects of methotrexate.

NC carcinoma cells, taken into culture from a WHT/Ht mouse with peritoneal metastases, were killed by methotrexate in a concentration-dependent fashion judged by cloning assays. Indomethacin (1 μg/ml) alone had little or no effect, but it increased the cell killing by methotrexate.

Uptake of [³H] by NC Cells Incubated With [³H]-Methotrexate

Indomethacin increased the accumulation of tritium when NC cells were incubated with [³H]-methotrexate. This enhancement may be mediated by an energy-dependent folate uptake system and/or by decreased efflux of methotrexate metabolites from the cells, but the extent to which it involves inhibition of cyclo-oxygenase is not clear. Indomethacin does not seem to act via inhibition of cAMP phosphodiesterase, since theophylline at a concentration equipotent to indomethacin as an inhibitor of cAMP phosphodiesterase in toad bladder (36) did not affect the cell killing by methotrexate.

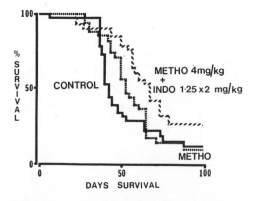

Fig. 4. Survival of mice with NC cancers given methotrexate (METHO) 4 mg/kg with indomethacin (INDO) 1.25 mg/kg twice daily. Survival was longer than in mice given either methotrexate alone (P = 0.04) or drug vehicle (P = 0.003).

In contrast, normal human embryonic intestinal cells behaved differently from mouse NC cells. Indomethacin did not affect the killing of these cells by methotrexate, or their ability to take up tritium when incubated with [³H]-methotrexate. These effects of indomethacin on malignant cells have obvious therapeutic potential, particularly since they did not occur with normal epithelial cells. An obvious possibility is to determine if indomethacin increases the sensitivity of human gastrointestinal cancers to methotrexate since chemotherapy currently is of little value in treating these tumors (37). In addition, we have preliminary data that indomethacin increases the cytotoxicity of methotrexate to human breast cancer cells in culture. Therefore, it may be of value to use this drug combination in breast cancer, after studying safety aspects.

SUMMARY

These studies evolved mainly from work showing that PG's can resorb bone, thus raising the possibility that the growth of tumor cells in bone involves PG production. We chose breast cancer for our first study because these patients often develop skeletal metastases. (The experiments on human gut cancer were started around the same time because of our long interest in this organ, and lung cancers were studied later because they also commonly spread to bone.)

Our early demonstration of a high production of PG-LM by breast cancers, and an apparent association between tumor PG-LM and bone metastasis, led to studies in rabbits indicating that bone destruction by VX2 tumors involves PG's. At this stage, it appeared likely that PG's were concerned in tumor growth and spread, and that PG synthesis inhibitors would be beneficial. We tested this in mice using the transplantable metastasizing NC tumor, and found that flurbiprofen or indomethacin reduced tumor growth and prolonged survival, particularly when given chemotherapy. The increased survival with indomethacin seemed likely to be due to inhibition of PG synthesis, since it was prevented by giving 16,16-dimethyl-PGE$_2$.

Studies of several other drugs, many of which affect PG metabolism, were then studied in mice bearing the NC carcinoma. At the same time, experiments were begun on malignant and normal cells in culture. The most exciting finding is the ability of indomethacin to increase the killing of some malignant cells by methotrexate, probably by increasing the uptake and/or retention of the cytotoxic drug, whereas this effect did not occur in the normal intestinal cells.

Studies in progress include a fuller analysis of breast tumor PG-LM in relation to steroid receptors, disease recurrence, and patient survival. Future human work will include the effect of anti-cancer drugs on the tumor formation of PG's, leukotrienes, and related substances, and a report of a trial with flurbiprofen in breast cancer. The mouse work will include the interaction of other cytotoxic drugs and X rays by drugs that alter prostaglandin and/or leukotriene synthesis. Cell cultures will be used to study the mechanism by which indomethacin increases the killing of malignant cells by methotrexate, and whether the response to other related and unrelated cytotoxic drugs and X rays are affected similarly. Various human malignant and normal cell lines will be examined with methotrexate and indomethacin, and if the results are consistent with those already obtained, this may lead to clinical trials of the drug combination.

REFERENCES

1. Bennett, A., Friedman, C. A., and Vane, J. R. Release of PGE$_1$ from rat stomach. Nature (Lond.) 216: 873-876, 1967.
2. Harris, M., Jenkins, M. V., Bennett, A., and Wills, M. R. Prostaglandin production and bone resorption by dental cysts. Nature 245: 213-215, 1973.
3. Bennett, A., McDonald, A. M., Simpson, J. S., and Stamford, I. F. Breast cancer, prostaglandins and bone metastases. Lancet 1: 1218, 1975.
4. Bennett, A., Charlier, E. M., McDonald, A. M., Simpson, J. S., Stamford, I. F., and Zebro, T. Prostaglandins and breast cancer. Lancet 2: 624-626, 1977.
5. Bennett, A., Berstock, D. A., Raja, B., and Stamford, I. F. Survival time after surgery is inversely related to the amounts of prostaglandins extracted from human breast tumours. Br. J. Pharmacol. 66: 451, 1979.
6. Stamford, I. F., Carroll, M. A., Civier, A., Hensby, C. N., and Bennett, A. Identification of arachidonate metabolites in normal, benign, and malignant human breast. J. Pharm. Pharmacol. 35: 48-49, 1983.
7. Bennett, A., del Tacca, M., Stamford, I. F., and Zebro, T. Prostaglandins from tumours of large bowel. Br. J. Cancer 35: 881-884, 1977.
8. Bennett, A., Stamford, I. F., and Stockley, H. L. Estimation and characterisation of prostaglandins in the human gastrointestinal tract. Br. J. Pharmacol. 61: 579-586, 1977.
9. Bennett, A., Hensby, C. N., Sanger, G. J., and Stamford, I. F. Metabolites of arachidonic acid formed by gastrointestinal tissues and their actions on the muscle layers. Br. J. Pharmacol. 74: 435-444, 1981.
10. Galasko, C. S. B., and Bennett, A. Relationship of bone destruction and skeletal metastases to osteoclast activation and prostaglandin. Nature 263: 508-510, 1976.
11. Galasko, C. S. B., Rawlins, R., and Bennett, A. Timing of indomethacin in the control of prostaglandins, osteoclasts, and bone destruction produced by VX2 carcinoma in rabbits. Br. J. Cancer 40: 360-364, 1979.
12. Leaper, D. J., Bennett, A., Charlier, E. M., and Stamford, I. F. The enhancement of radiotherapy and chemotherapy of a murine tumour using the prostaglandin synthetase inhibitor flurbiprofen. Br. J. Surg. 65: 369, 1978.
13. Berstock, D. A., Houghton, J., and Bennett, A. Improved anti-cancer effect by combining cytotoxic drugs with an inhibitor of prostaglandin synthesis. Cancer Treat. Rev. (Suppl.) 6: 69-71, 1979.
14. Bennett, A., Houghton, J., Leaper, D. J., and Stamford, I. F. Cancer growth, response to treatment and survival time in mice: Beneficial effects of the prostaglandin synthesis inhibitor flurbiprofen. Prostaglandins 17: 179-191, 1979.

15. Bennett, A., Berstock, D. A., and Carroll, M. A. Increased survival of cancer-bearing mice treated with inhibitors of prostaglandin synthesis alone or with chemotherapy. Br. J. Cancer 45: 762-768, 1982.

16. Bennett, A., Carroll, M. A., Melhuish, P. B., and Stamford, I. F. Treatment of mouse carcinoma *in vivo* with a prostaglandin E_2 analogue and indomethacin. Br. J. Cancer 52: 245-249, 1985.

17. Bennett, A., Melhuish, P. B., Patel, S., Randles, R., and Stamford, I. F. Cancer in mice: Effects of prednisolone or meparine alone with cytotoxic drugs. Br. J. Cancer 55: 385-388, 1987.

18. Stamford, I. F., Melhuish, P. B., Carroll, M. A., Corrigan, C., Patel, S., and Bennett, A. Survival of mice with NC carcinoma is unchanged by drugs that are thought to inhibit thromboxane synthesis or increase prostaglandin formation. Br. J. Cancer 54: 257-263, 1986.

19. Stamford, I. F., Tavares, I. A., Melhuish, P. B., Mascolo, N., Brown, J. A., Morgan, S., and Bennett, A. Tamoxifen has little effect on cyclo-oxygenase but can inhibit 5-lipoxygenase. In: "Prostaglandins and Cancer Research." E. Garaci, R. Paoletti, and M. G. Santoro, eds. Springer Verlag, Berlin, 1987, pp. 269-272.

20. Bennett, A., Carroll, M. A., Stamford, I. F., Whimster, W. F., Williams, F., and Wright, J. E. Prostaglandins and human lung cancer. Br. J. Pharmacol. 74: 207-208, 1981.

21. Bennett, A., Carroll, M. A., Stamford, I. F., Whimster, W. F., and Williams, F. Prostaglandins and human lung carcinomas. Br. J. Cancer 46: 888-893, 1982.

22. Bennett, A., Carter, R. L., Stamford, I. F., and Tanner, N. S. B. Prostaglandin-like material extracted from carcinomas of the head and neck. Br. J. Cancer 41: 204-208, 1980.

23. Hensby, C. N., Carroll, M. A., Stamford, I. F., Civier, A., and Bennett, A. Identification of arachidonate metabolites in normal and malignant human lung. J. Pharm. Pharmacol. 34: 811-813, 1982.

24. Tanner, N. S. B., Stamford, I. F., and Bennett, A. Plasma prostaglandins in mucositis due to radiotherapy and chemotherapy for head and neck cancer. Br. J. Cancer 43: 467-471, 1981.

25. Gaffen, J. D., Bennett, A., and Barer, R. A new method for studying cell growth in suspension, and its use to show that indomethacin enhances killing by methotrexate. J. Pharm. Pharmacol. 37: 261-263, 1985.

26. Gaffen, J. D., Tsang, R., and Bennett, A. Increased killing of malignant cells by giving indomethacin with methotrexate. Prog. Lipid Res. 25: 543-545, 1986.

27. Hewitt, H. B., Blake, E. R., and Walder, A. S. A critique of the evidence for active host defence against cancer, based on personal studies of 27 murine tumours of spontaneous origin. Br. J. Cancer 33: 241-259, 1976.

28. O'Connor, A. D., Clifford, P., Durden-Smith, D. J., Edwards, W., Hollis, B. A., and Dalley, V. M. Synchronous VBM and radiotherapy in the treatment of squamous carcinoma of head and neck. Clin. Otolaryngol. 1: 347-357, 1977.

29. Unger, W. G., Stamford, I. F., and Bennett, A. Extraction of prostaglandins from human blood. Nature 233: 336-337, 1971.

30. Eisen, V., and Walker, D. I. Effect of ionising radiation on prostaglandin-like activity in tissues. Br. J. Pharmacol. 5: 527-532, 1976.

31. Eisen, V., and Walker, D. I. Effect of ionizing radiation on prostaglandin 15-OH-dehydrogenase (PGDH). Br. J. Pharmacol. 62: 461, 1978.
32. Levine, L. Chemical carcinogens stimulate canine kidney (MDCK) cells to produce prostaglandins. Nature 268: 447-448, 1977.
33. Leaper, D. J. The relationship of prostaglandins to tumour growth. Doctor of Medicine degree, University of Leeds, England, 1978.
34. Powles, T. J., Clark, S. A., Easty, G. C., and Neville, A. M. The inhibition of aspirin and indomethacin of osteolytic tumour deposits and hypercalcaemia in rats with Walker tumour and its possible application to human breast cancer. Br. J. Cancer 28: 316-321, 1973.
35. Voelkel, E. F., Tashjian, A. H., Franklin, R., Wasserman, E., and Levine, L. Hypercalcemia and tumor-prostaglandins: The VX2 carcinoma model in the rabbit. Metabolism 24: 973-986, 1975.
36. Flores, A. G., and Sharp, G. W. Endogenous prostaglandins and osmotic water flow in the toad bladder. Am. J. Physiol. 233: 1392-1397, 1972.
37. Wrigley, P. F. M., and Slevin, M. L. Management of malignant tumour-chemotherapy. In: "Colorectal Disease: An Introduction for Surgeons and Physicians." J. P. S. Thompson, R. J. Nicholls, and C. B. Williams, eds. Heineman, London, 1981, pp. 284-286.

MODIFICATION OF RADIATION INJURY OF MURINE INTESTINAL CLONOGENIC CELLS AND B-16 MELANOMA BY PGI₂ OR FLURBIPROFEN

W. R. Hanson[1,2], W. Jarnagin[1], K. DeLaurentiis[1], and F. D. Malkinson[2]

Departments of [1]Therapeutic Radiology and [2]Dermatology
Rush-Presbyterian St. Luke's Medical Center
Chicago, Illinois 60612

ABSTRACT

Prostaglandins (PG's) protect several types of cells *in vivo* from ionizing radiation. Many human tumors secrete excessive PG's, which may result in some degree of tumor protection from radiation treatment. In contrast, PG-blocking agents might sensitize tumors to radiation. To assess a possible role of PG's and PG-blocking agents in modifying normal tissue or tumor radiation sensitivity, either PGI₂ (the most protective natural PG found to date) or Flurbiprofen (a cyclooxygenase inhibitor) was given to mice before irradiation with cesium-137. The effects of these agents on the radiosensitivity of intestinal clonogenic cells or a B-16 melanoma were measured. PGI₂ increased the D_o of intestinal clonogenic cell survival from about 1.15 ± 0.13 Gy in controls to 1.45 ± 0.15 Gy. Flurbiprofen had no effect on the intestinal clonogenic cell survival curve. PGI₂ also protected the B-16 melanoma, but Flurbiprofen protected the melanoma to an even greater extent, mainly through an increase in the shoulder. Flurbiprofen may switch the arachidonic acid cascade to other pathways such as the 5-lipoxygenase pathway, and leukotrienes have also been shown to protect from radiation. Tumor radiosensitization may occur only through blocking all pathways of the cascade or by blocking tumor receptor sites.

INTRODUCTION

Exogenously administered 16,16-dimethyl prostaglandin E₂ (dm PGE₂) has been shown to protect murine cells *in vivo* from radiation injury. The shoulder of the intestinal clonogenic cell survival curve was increased along with the slope (D_o) (1). A qualitatively similar increase in the shoulder and the D_o of the murine bone marrow stem cell (CFU-S) survival curve was seen in dm PGE₂-treated mice compared to controls (2). More recently, several PG's and leukotrienes (LT's) have

Table 1. Prostaglandin Secretion by Human Tumors

Tumor	Prostaglandin	Reference[1]
Breast	PGE_2, $PGF_{2\alpha}$	Bennett et al., Lancet 1: 1218, 1979
	15-Keto-13,14-dihydro $PGE_{2\alpha}$	Powles et al., Lancet 2: 138, 1977
Cervix	PGE_2, $PGF_{2\alpha}$	Nagasaki et al., PG's and Cancer, 1st Intl. Conf., 567, 1982
Kaposi's sarcoma	PGE_2, $PGF_{2\alpha}$	Bland et al., Cancer 27: 233, 1971
Lung	PGE_2, PGI_2, TxB_2	Stamford et al., PG's and Cancer, 1st Intl. Conf., 691, 1982
Thyroid	PGE_2, $PGF_{2\alpha}$,	Williams et al., Lancet 1: 22, 1968
Kidney	PGA_2, PGE_2, $PGF_{2\alpha}$	Beckman et al., PG's and Cancer, 1st Intl. Conf., 719, 1982

[1]From a list of over 100 references

been shown to protect murine intestinal stem cells from photon radiation (reference 3; see W. R. Hanson, this volume). With the exception of PGE_1, all natural PG's investigated were protective. PGI_2 was the most protective natural PG, increasing cell survival to about 300% of control values. Of the PG analogues tested, 15-deoxy,16-methyl,16-hydroxy PGE_1 methyl ester (misoprostol) was the most protective, increasing survival to 600% of controls (4).

Many human tumors secrete excessive PG's (5-10). Table 1 is a representative list. The possible consequences of increased tumor PG concentrations are many and varied, including changes in tumor metastatic potential (11), cell proliferation and growth (12), and immune function (13). The results showing that products of the arachidonic acid cascade protect tissue from radiation suggest that tumors with high concentrations of PG's may be protected, to some degree, from radiation injury. Conversely, then, cyclooxygenase-blocking agents should sensitize tumors to radiation.

This paper reports studies of the influence of PGI_2 and the nonsteroidal anti-inflammatory agent Flurbiprofen on the radiosensitivity of a murine B-16 melanoma tumor and intestinal clonogenic stem cells in non-tumor-bearing animals.

MATERIALS AND METHODS

Male B6D2F1 mice purchased from Jackson Laboratories (Bar Harbor, ME) were used for all studies. Mice were kept on a 12-hr light-dark cycle, and were fed Agway RMH-3000 chow and acidified tap water (pH 2.7) to prevent the infection and spread of *Pseudomonas*. The care and use of these mice were overseen by the Institutional Animal Care and Use Committee of Rush University to ensure that animals were kept in accordance with the recommendations as stipulated in DHEW Publication No. NIH 85-23.

B-16 Melanoma Studies

A B-16 melanoma cell line compatible with the B6D2F1 strain of mouse (kindly provided by Dr. A. Schaffner, Illinois Institute of Technology) was cultured through three passages in RPMI 1640 medium with 10% fetal calf serum. At a near confluent state of the third passage, cells were removed with trypsin and EDTA and then vortexed to prepare a single cell suspension. A 0.25-ml volume containing 10^5 melanoma cells was injected into the lateral aspects of the flanks of animals, three mice for each treatment group for each dose of radiation of the experimental design shown in Figure 1. After about 3 weeks, when the tumors were approximately 1 cm in diameter, the tumor-bearing mice were divided into three groups: (a) controls (phosphate-buffered saline), (b) PGI₂ (25 μg/mouse) given subcutaneously (s.c.) in the dorsal neck region 1 hr before irradiation (the preirradiation time shown to produce maximum radiation protection) (3), and (c) 2 mg/kg Flurbiprofen (a gift from Upjohn, Kalamazoo, MI) given intraperitoneally (i.p.) every 8 hr starting 24 hr before irradiation.

Graded doses of cesium-137 gamma radiation between 0 and 20 Gy at a dose rate of 1.25 Gy/min were delivered using a dual head cesium-137 small-animal irradiator (Atomic Energy of Canada, Ltd.). Dosimetry was done using thermal luminescence of irradiated lithium fluoride rods. The lung colony assay (14), as modified by Grdina et al. (15), was used to assay B-16 melanoma cell survival.

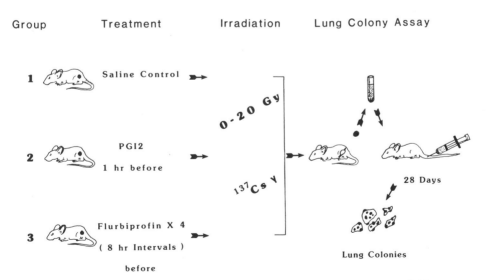

Fig. 1. Experimental design to investigate effect of PGI₂ or Flurbiprofen on B-16 melanoma tumor

Immediately after irradiation, the tumors were removed and the three separate tumors of each group at each radiation dose were pooled and minced, and single cell suspensions were made.

Various dilutions of B-16 melanoma cells were made, depending on the dose of radiation, so that tumors receiving the highest dose of radiation were diluted the least. These different dilutions of B-16 cells from the experimental animals were mixed with cultured B-16 cells irradiated with 100 Gy so that the total cell count (experimental plus heavily irradiated) was the same per volume. The various dilutions of cells in 0.2 cc were injected into the tail veins of recipient male B6D2F1 mice that had been exposed to 6 Gy radiation the previous day with a single hind limb shielded. These measures enhance the lung colony counts by filling the reticuloendothelial system and by inhibiting the immune response of the recipient mice. A group of control animals was given 10^6 heavily irradiated cells to ensure that none of these cells survived. Twenty-eight days later when the micrometastases were countable lung colonies, the lungs of the recipient mice were fixed in Bouin's solution. The number of colonies was counted, and the number of surviving cells in each dilution was plotted versus dose of cesium-137 radiation.

Intestinal Studies

To measure the effect of PGI_2 or Flurbiprofen on intestinal clonogenic cells, B6D2F1 male mice were divided into three groups: controls that received phosphate buffer, PGI_2 (25 μg/mouse s.c. 1 hr before irradiation), and Flurbiprofen (2 mg/ kg i.p. every 8 hr starting 24 hr before irradiation). Graded doses of 12-18 Gy cesium-137 were delivered to groups of five animals of each treatment. The intestinal microcolony assay of Withers and Elkind (16) was used to assay clonogenic cell survival. Briefly, 4 days after irradiation, the mice were killed by cervical fracture and the upper jejunum was fixed, embedded, and sectioned to give 18 independent cross sections of tissue. The number of regenerative foci of epithelial cells (microcolonies) was counted, averaged, and plotted versus dose of radiation. The D_o of the terminal portion of the survival curves was determined by least squares linear regression.

RESULTS

The number of B-16 melanoma cells that survived cesium-137 irradiation was increased at each radiation dose level by 25 μg PGI_2 (Figure 2). Both the shoulder and the D_o were increased. In contrast to our original hypothesis that a cyclooxygenase-blocking agent might increase tumor radiosensitivity, Flurbiprofen increased B-16 cell survival more than PGI_2 (Figure 2). The shoulder was markedly increased; however, the D_o appeared to be similar to that of the control curve.

Twenty-five micrograms of PGI_2 increased both the shoulder and the D_o of the intestinal clonogenic cell survival curve (Figure 3). The D_o was increased from

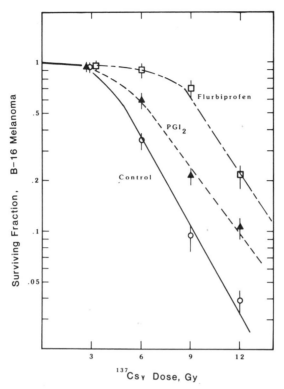

Fig. 2. Surviving fraction of cells (lung colony assay) in B-16 melanoma tumors grown in flanks of mice versus dose of cesium-137 in controls, in tumor-bearing mice treated with 25 μg PGI$_2$, or in tumor-bearing mice treated with Flurbiprofen. Symbols are mean values from 15 animals $\pm$ 1 SEM.

a control value of 1.15 $\pm$ 0.13 Gy to 1.45 $\pm$ 0.15 Gy in PGI$_2$-treated mice. These values were qualitatively similar to those found for the dm PGE$_2$-induced radiation protection reported earlier (1). In contrast, Flurbiprofen had no effect on the survival curves of this normal tissue (Figure 3). The shoulder and the D$_0$ were identical to control values.

DISCUSSION

The results reported here show that PGI$_2$ protected both the murine intestine and the B-16 melanoma from cesium-137 injury. Protection of the intestinal clonogenic population was similar to that seen previously (1). Flurbiprofen had no effect on this system, which agrees with the results of MacDougall et al. (17), who reported that Flurbiprofen did not modify the radiation response of either the intestine or the CFU-S (stem cells) of the murine bone marrow.

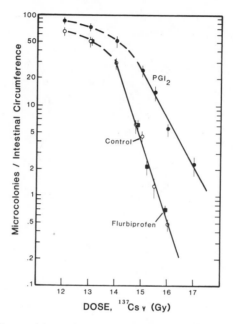

Fig. 3. Microcolonies per jejunal circumference in control mice or in mice treated with 25 μg PGI$_2$ or Flurbiprofen. Symbols represent mean values from five animals $\pm$ 1 SEM.

The hypothesis tested by the experiments reported here was that a nonsteroidal anti-inflammatory agent (Flurbiprofen) might sensitize tumors to radiation and reduce cell survival at any given radiation dose compared to tumors irradiated only. Not only was there no radiation sensitization, but there was marked protection of the B-16 melanoma by Flurbiprofen. A possible explanation for these results is that the arachidonic acid cascade was shifted to the 5-lipoxygenase pathway. LTC$_4$ and LTD$_4$ have been shown to protect cells *in vivo* (reference 3; see W. R. Hanson, this volume) and *in vitro* (18). The uniformity of the presence and the quantities of 5-lipoxygenase in tissues or tumors is unknown; however, if tumors contain enzymes of the lipoxygenase pathway, it is doubtful that cyclooxygenase-blocking agents would sensitize the tumor. In this case, blocking both major pathways may achieve the goal of tumor sensitization (assuming there are no other pathways that could produce radioprotective compounds). Another possible approach may be to block membrane receptor sites to prevent the action of PG's or LT's. Of course, the success of this approach depends on the ability to selectively block tumor receptor sites.

The data reported here must be confirmed in other tumors, especially since they appear to conflict with several other reports. Hellman and Pym (19) found that Flurbiprofen sensitized the S180 sarcoma to X rays although they saw no effect of Flurbiprofen on the Lewis lung carcinoma. Likewise, Weppelmann and Monkemeier (20) reported an increase in the 5- and 10-year survivals of women treated with oxyphenbutazone before radiation therapy of cervical cancer, thus demonstrating tumor radiosensitization by a nonsteroidal anti-inflammatory agent. It is possible that these tumors cannot metabolize arachidonic acid through the 5-lipoxygenase pathway. In apparent contrast to these results, Northway et al. (21) reported that indomethacin protected opossums from radiation esophagitis and that dm PGE_2 exacerbated the condition rather than protected the esophagus from injury. In subsequent studies, Northway et al. (22) reported that several nonsteroidal anti-inflammatory agents had little effect on a mouse mammary tumor. These results are difficult to compare directly with those reported here since the end points for radiation injury are different.

It is clear that products of the arachidonic acid cascade can modify the radiation response of both normal and tumor tissues. Further details of human tumor PG and LT levels and the mechanisms for radiation protection by these products are essential in order to begin to successfully manipulate this system to improve the therapeutic ratio of radiation treatment.

ACKNOWLEDGMENTS

This work was supported in part by the Departments of Dermatology and Therapeutic Radiology, RPSLMC, and by Contract No. DNA001-86-0038 from the Defense Nuclear Agency, DoD.

REFERENCES

1. Hanson, W. R., and Thomas, C. 16-16 dimethyl prostaglandin E2 increases survival of murine intestinal stem cells when given before photon radiation. Radiat. Res. 96: 393-398, 1983.
2. Hanson, W. R., and Ainsworth, E. J. 16-16 dimethyl prostaglandin E_2 induces radioprotection in murine intestinal and hematopoietic stem cells. Radiat. Res. 103: 196-203, 1985.
3. Hanson, W. R., this volume
4. Hanson, W. R., and DeLaurentiis, K. Comparison of *in vivo* murine intestinal radiation protection by the E-series prostaglandins E_1, E_2, 16-16 dm E_2, and Misoprostol. In preparation.
5. Bockman, R. S. Prostaglandins in cancer: A review. Cancer Invest. 1(6): 485-493, 1983.
6. Karmali, R. A. Prostaglandins and cancer. Ca: Cancer J. Clin. 33(6): 322-332, 1983.
7. Cummings, K. B., and Robertson, R. P. Prostaglandin: Increased production by renal cell carcinoma. J. Urol. 118: 720-723, 1977.
8. Powles, T. J., Coombes, R. C., Neville, A. M., Ford, H. T., Gazet, J. C., and Levine, L. 5-Keto-13,14-dihydroprostaglandin E_2 concentrations in serum of patients with breast cancer. Lancet 2: 138-142, 1977.
9. Bennett, A., Charlier, E. M., McDonald, A. M., Simpson, J. S., Stamford, I. F., and Zebro, T. Prostaglandins and breast cancer. Lancet 2: 624-626, 1977.

10. Bennett, A., Carroll, M. A., Stamford, I. F., Whimster, W. F., and Williams, F. Prostaglandins and human lung carcinomas. Br. J. Cancer 46: 888-893, 1982.

11. Bhuyan, B. K., Adams, E. G., Badiner, G. J., Li, L. H., and Barden, K. Cell cycle effects of prostaglandins A_1, A_2, and D_2 in human and murine melanoma cells in culture. Cancer Res. 46: 1688-1693, 1986.

12. Honn, K. V. Prostacyclin/thromboxane ratios in tumor growth and metastasis. In: "Prostaglandins and Related Lipids," Vol. 2. T. J. Powles, R. S. Bockman, K. V. Honn, and P. Ramwell, eds. Alan R. Liss, Inc., New York, 1982, pp. 733-752.

13. Goodwin, J. S. Prostaglandin E and cancer growth potential for immunotherapy with prostaglandin synthetase inhibitors. In: "Progress in Cancer Research and Therapy; Augmenting Agents in Cancer Therapy," Vol. 16. E. M. Hersh, M. A. Chirigos, and M. J. Mastrangelo, eds. Raven Press, New York, 1981, pp. 393-415.

14. Hill, R. P., and Bush, R. S. A lung colony assay to determine the radiosensitivity of the cells of a solid tumor. Int. J. Radiat. Biol. 15: 435-444, 1969.

15. Grdina, D. J., Basic, I., Guzzino, S., and Mason, K. A. Radiation response of cell populations irradiated *in situ* and separated from a fibrosarcoma. Radiat. Res. 66: 634-643, 1976.

16. Withers, H. R., and Elkind, M. M. Microcolony survival assay for cells of mouse intestinal mucosa exposed to radiation. Int. J. Radiat. Biol. 17: 261-267, 1970.

17. MacDougall, R. H., Duncan, W., and Clark, B. M. The effect of flurbiprofen on stem cell survival following X irradiation. In: Reference 12, pp. 819-823.

18. Walden, T. L., Jr., Holahan, E. V., Jr., and Catravas, G. N. Development of a model system to study leukotriene-induced modification of radiation sensitivity in mammaliam cells. Prog. Lipid Res. 25: 587-590, 1986.

19. Hellman, K., and Pym, B. A. Antitumor activity of flurbiprofen *in vivo* and *in vitro*. In: Reference 12, pp. 767-774.

20. Weppelmann, B., and Monkemeier, D. The influence of prostaglandin antagonists on radiation therapy of carcinoma of the cervix. Gynecol. Oncol. 17: 196-199, 1984.

21. Northway, M. G., Libshitz, H. I., Osborne, B. M., Feldman, M. S., Mamel, J. J., West, J. H., and Szwarc, I. A. Radiation esophagitis in the opossum: Radio-protection with indomethacin. Gastroenterology 78: 833-837, 1980.

22. Northway, M. G., Bennett, A., Carroll, M. A., Eastwood, G. L., Feldman, M. S., Libshitz, H. I., Mamel, J. J., and Szwarc, I. A. Effects of anti-inflammatory agents and radiotherapy on esophageal mucosa and tumors in animals. In: Reference 12, pp. 799-802.

INFLUENCE OF UV LIGHT (250 nm) ON PLATELET ACTIVATION

G. H. R. Rao[1], C. A. Cox[2], C. J. Witkop[3], and J. G. White[1]

[1]Department of Laboratory Medicine and Pathology
[3]Department of Oral Pathology
University of Minnesota Medical School
Minneapolis, Minnesota 55455

[2]Department of Biochemistry
University of Oklahoma
Oklahoma City, Oklahoma 73190

ABSTRACT

Previous investigations have shown that ultraviolet light (UVL) can cause irreversible aggregation in samples of stirred platelet-rich plasma (PRP). The precise mechanisms involved in UVL-induced platelet activation and the aspects of stimulus-activation-contraction secretion coupling involved were not determined. In the present study we have evaluated the response of normal and abnormal platelets to UVL and the influence of various inhibitors on the platelet reaction. UVL (250 nm) stimulated irreversible aggregation and secretion in stirred samples of PRP. Aspirin and Ibuprofen, which block cyclooxygenase and secretion, did not prevent UVL-induced aggregation. Agents that elevate levels of cAMP and free radical scavengers were also ineffective. Only ascorbic acid (5 mM) and EDTA effectively blocked UVL-stimulated platelet clumping. Thrombosthenic platelets lacking glycoproteins IIb-IIIa were not aggregated by UVL, but afibrinogenemic platelets responded like normal cells. UVL caused the formation of macroaggregates in washed platelet cytoskeletons that did not enter PAGE gels. Electron microscopy revealed no physical changes in unstirred platelets exposed to UVL. Stirring in the presence of UVL caused the development of stickiness, resulting in the aggregation of discoid platelets. Shape change and internal transformation occurred only after aggregates were well established. Results of the studies indicated that extracellular calcium and GPIIb-IIIa were required for UVL-induced aggregation, but thromboxane A_2, fibrinogen, and secretion were unnecessary. Inhibition by ascorbic acid suggests that oxidant-induced changes in the membrane leading to changes in the membrane cytoskeleton may underlie platelet activation by UVL.

INTRODUCTION

Earlier studies have shown that ultraviolet rays (UVL, 250 nm) induce the activation of mammalian blood platelets (1,2). UVL-induced platelet activation is unique, as it seems to occur independently of mediation by thromboxane on endogenously released ADP. The precise mechanisms involved in UVL-induced platelet activation leading to irreversible aggregation are not known. In the present investigation we have exposed platelets from normal donors and from patients with well-characterized disorders to UVL and followed the response of these cells by aggregometry.

MATERIALS AND METHODS

Reagents

Arachidonic acid as the sodium salt was obtained from NuChek Prep (Elysian, MN); injectable adrenalin and bovine thrombin were from the Parke-Davis Company (Morris Plains, NJ). Superoxide dismutase was obtained from Truett Laboratories (Dallas, TX), and prostaglandin E_1 was a gift from the UpJohn Company (Kalamazoo, MI). Unless otherwise indicated, all other chemicals were purchased from the Sigma Chemical Company (St. Louis, MO).

Platelet Preparation

Blood for these studies was obtained by informed consent from adult volunteers who had not taken any medication for at least 2 weeks prior to venesection. Blood was also provided by patients with Hermansky-Pudlak syndrome, Glanzmann's thrombasthenia, and afibrinogenemia. It was mixed immediately with citrate anticoagulant. Platelet-rich plasma (PRP) was obtained according to the method of Rao et al. (3).

Monitoring Aggregation During Irradiation

An aggregometer (Chrono-log Corporation, Havertown, PA) was modified to receive a cuvette holder built to accept a 1 x 1 cm quartz cuvette. Two quartz pencil lamps were mounted perpendicular to the light path (Spectroline, Blaen Eastern, Westbury, NY). The cell holder contained a 3-mm port at the bottom adjacent to each lamp, and the bottom of the holder was covered with aluminum foil.

Platelet Protein Profiles

Sodium dodecyl sulfate polyacrylamide gel electrophoresis (SDS-PAGE) was performed to obtain platelet protein profiles. The gel slabs were 1 mm thick, and the staining gel was 2 cm long. The samples were electrophoresed for an average of 4 hr. After staining with Coomassie brilliant blue, the gels were scanned on a densitometer (Kratos Model SD3000; Kratos Analytical, Ramsey, NJ).

RESULTS

Effect of Inhibitors on UVL-Induced Platelet Aggregation

Platelets exposed to 5 μg/ml of prostaglandin E_1 aggregated in a normal fashion when stirred under the influence of UVL (Figure 1). Cyclooxygenase inhibitors like aspirin (1 x 10^{-3} M) and indomethacin (1 x 10^{-4} M) did not block UVL-induced platelet aggregation. UVL has been known to generate free radicals. Enzymes capable of inactivating superoxide radical (superoxide dismutase) and hydrolyzing hydrogen peroxide (catalase) did not exert any inhibitory effect on platelet aggregation caused by UVL. Similarly, neither antioxidants (such as vitamin E or butylated hydroxyanisole) (BHA) nor free radical scavengers (such as tetraphenyl-cyclopentadionone and diphenyl isobenzofuran) had any inhibitory effect on UVL-induced platelet activation. Of the many known antiplatelet drugs tested, only EDTA and ascorbic acid (20 mM) prevented UVL-induced platelet aggregation.

Effect of UVL on Platelet Cytoskeletal Proteins

Densitometric tracings of the protein profiles of normal control platelets and UV-irradiated platelets are presented in Figure 2. Exposure of platelets in buffers to UVL induced the formation of high-molecular-weight proteins, which were unable to move into the gel. However, no such aggregates formed when platelets suspended in plasma were exposed to UVL. Ascorbic acid (20 mM) in the medium prevented the formation of protein aggregates as a result of UV irradiation.

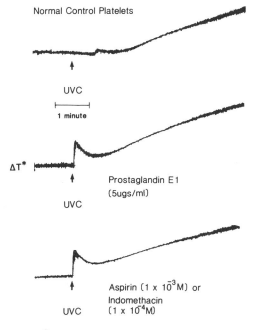

Fig. 1. Irreversible aggregation of platelets exposed to UVL. Common antiplatelet drugs did not prevent UVL-induced platelet aggregation.

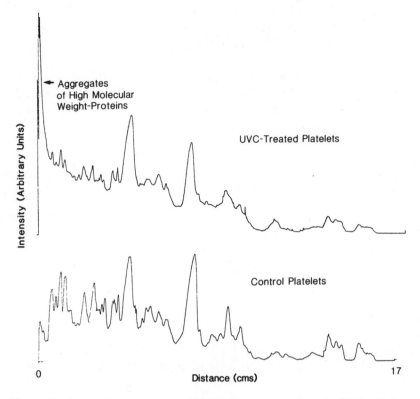

Fig. 2. Platelet proteins separated by SDS-PAGE gel electrophoresis. UV irradiation induced formation of aggregates of high-molecular-weight proteins.

Influence of UVL on Platelets of Patients With Various Disorders

Thrombosthenic platelets, which are deficient in surface membrane glycoproteins, did not aggregate when irradiated in plasma. However, platelets of a patient with afibrinogenemia aggregated normally when exposed to UVL. Platelets of a patient with storage pool defect also aggregated in response to UV irradiation.

DISCUSSION

Results of the present study suggest that UVL-induced platelet activation is independent of prostaglandin synthesis and the release reaction. Earlier studies by Dickson and associates also demonstrated the minimal role played by released granule contents. The majority of antiplatelet drugs used failed to inhibit UVL-induced platelet aggregation.

UV irradiation has been shown to induce lipid peroxidation. Therefore, the ability of antioxidants such as vitamin E, BHA, and radical scavengers to prevent UVL-induced platelet activation was followed. With the exception of high concentrations of EDTA (75 mM) and ascorbic acid (20 mM), all other chemicals and enzymes failed to inhibit UVL-induced platelet aggregation.

Studies by Lawler et al. showed that UVL induces changes in platelet proteins (2). When platelet membranes are subjected to UV radiation, a decrease in the concentration of all of the proteins occurs. In the present study, when washed platelets were exposed to UVL, the loss of several proteins and the appearance of some protein aggregates could be demonstrated.

Studies by Dickson et al. showed a facilitating role for fibrinogen in UVL-induced platelet aggregation (1). Though fibrinogen did improve the response of platelets to UVL, it was not critical for platelet activation. Similarly, released ADP was not essential because storage pool deficient platelets aggregated irreversibly when exposed to UVL. However, thrombosthenic platelets failed to aggregate when stirred under the influence of UVL, suggesting the essential nature of intrinsic surface-membrane glycoproteins for irreversible aggregation.

In conclusion, our studies confirm some of the earlier observations on the effect of UVL on platelet function. Exposure of platelets to UV radiation induces activation that is not inhibitable by antiplatelet drugs, antioxidants, or free radical scavengers. It induces alterations in proteins on short-term exposure. In view of these observations, further studies on the acute and chronic effects of UVL (UV A340 nm, B290 nm, and C250 nm) on platelet function and arachidonic acid metabolism are essential.

ACKNOWLEDGMENTS

Supported by USPHS grants HL-11880, CA-21737, GM-22167, HL-30217, a grant from the March of Dimes (MOD 1-886), and the Minnesota Medical Foundation (HDRI-38-85).

REFERENCES

1. Dickson, R. C., Doery, J. C. G., and Lewis, A. F. Induction of platelet aggregation. Sci. 172: 1140-1141, 1971.
2. Lawler, J. W., Chao, F. C., and Fang, P. The effect of ultraviolet light on platelets: Platelet activation and inhibition of thrombin-mediated functions. Thrombos. Res. 14: 489-494, 1979.
3. Rao, G. H. R., Schmid, H. H. O., Reddy, K. R., and White, J. G. Human platelet activation by an alkylacetyl analogue of phosphotidylcholine. Biochim. Biophys. Acta 715: 205-214, 1982.

EFFECTS OF EICOSANOID PRECURSORS ON TBA REACTIVE MATERIAL IN NORMAL AND MALIGNANT CELLS

M. E. Bégin, G. Ells, and D. F. Horrobin

Efamol Research Institute
Kentville, Nova Scotia, Canada, B4N 4H8

ABSTRACT

The amounts of thiobarbituric acid reactive material (TBARM) produced by human breast tumor (ZR-75-1) and normal simian kidney cells (CV-1) enriched with ethyl gammalinolenate (GLA), ethyl arachidonate (AA), and methyl eicosapentaenoate (EPA) were compared. TBARM increased in both tumor and normal cell cultures upon supplementation. TBARM measured in the culture medium and in the cells varied with the cell type and the fatty acid, as follows: (a) The tumor cells produced more TBARM than did normal cells. (b) In the tumor cells, GLA and AA produced more TBARM than did EPA. (c) In the normal cells, EPA produced more TBARM than did GLA and AA. (d) TBARM measured in the culture medium incubated without cells was proportional to the number of double bonds of the fatty acid under test. The addition of GLA, AA, and EPA killed the tumor cells but not the normal cells. The effectiveness of a given fatty acid ester in killing tumor cells correlated with the TBARM content in the cells. The addition of vitamin E acetate to the tumor cell cultures challenged with EPA reduced both cell killing and amount of TBARM.

INTRODUCTION

One basis for the differences between normal cells and cancer cells may be alteration in membrane lipid composition. Many types of cancer cells exhibit a decreased rate of lipid peroxidation compared to normal cells as a result of a lower content of polyunsaturated fatty acids (PUFA's) and/or decrease in cytochrome P450 and/or higher content of chain-breaking antioxidants (1). Since the presence of PUFA's in cell membranes makes cells susceptible to damage by lipid peroxidation, the levels of PUFA's in cancer cells may be a factor in the levels of lipid hydroperoxide levels and their degradation products. We and others (2-4) have shown that supplementation with PUFA's that are precursors of eicosanoids kill a variety of human carcinoma cells very efficiently without being lethal to normal cells. Since lipid peroxidation products may be toxic to cells, one mechanism of the effect might be an elevation of peroxidation products generated by n-3 and n-6 PUFA's as a result of providing a PUFA substrate. We have therefore compared the content of peroxidation products as measured by thiobarbituric acid reactive material (TBARM) in human breast tumor cells supplemented with ethyl gammalinolenate (18:3n-6, GLA), ethyl arachidonate (20:4n-6, AA), and methyl eicosapentaenoate (20:5n-3, EPA) and on the relationship of TBARM with cell killing in the cancer cells and in an established normal kidney cell line.

MATERIALS AND METHODS

Cell Cultures

Human breast (ZR-75-1) carcinoma and established normal simian (CV-1) kidney cell lines were seeded at 5×10^4 cells per plate in 35-mm Petri dishes as previously described (2). One day after seeding, one set of cultures was supplemented with 60 μM of GLA, AA, or EPA, a concentration that was shown to give optimal results (2). Another set of unsupplemented cultures received 0.2% final concentration of ethanol and was incubated in parallel. In some experiments, vitamin E acetate (10 μM) and indomethacin (14 μM) were added to tumor cell cultures challenged with EPA (60 μM). All experiments were done in triplicate.

Fatty Acids

EPA (90%-95% pure) was obtained from Sigma (St. Louis, MO) and the other fatty acids (>99% pure) were purchased from Nu Chek Prep, Inc (Elysian, MN). The fatty acids were prepared fresh from stock solutions (10 mg/ml of ethanol) and diluted with growth medium to give a final ethanol concentration of less than 0.2%.

Lipid Peroxidation

Medium and cells were separated and assayed for TBARM according to Gavino et al. (5). Briefly, 1 ml of 20% trichloroacetic acid (TCA) was added to 2 ml of medium or 2 ml of cells resuspended in phosphate-buffered saline, pH 7.0 (PBS). Two ml of 0.67% TBA was added. This mixture was incubated for 20 min at 90°C and centrifuged at 12,000 x g for 10 min at 4°C. The absorbance of the supernatant was measured at 532 nm using growth medium or PBS as a reference for the medium or the cell samples, respectively. Absorbance was converted to pmoles malondialdehyde-equivalent (MDA-Eq) from a standard curve generated with 1,1,3,3-tetramethoxypropane.

Extent of Cell Killing

Cell viability counts were determined by trypan blue exclusion in a Neubauer hemacytometer in parallel cultures. The percentage of dead cells was calculated as the number of stained cells in a culture/total number of cells (unstained + stained) x 100.

RESULTS

Table 1 shows the amount of TBARM found in the culture medium and in the cells obtained from tumor and normal cell cultures incubated in the presence or absence of GLA, AA, and EPA. It can be seen that both cancer and normal cell cultures challenged with PUFA's generated considerably more TBARM than did unsupplemented cell cultures. The proportion of TBARM in the culture medium differed from that obtained in the cells. TBARM found in the culture medium accounted for about 80% of the total TBARM found in tumor and normal cell cultures incubated without excess PUFA's. In contrast, in supplemented normal cell cultures, the amount of TBARM measured in the cells was, on the average, about 2-4 times greater than that found in the culture medium. In the supplemented tumor cell cultures, the proportion of TBARM found in the culture medium increased with the number of double bonds of the PUFA's so that, on the average, cells enriched in GLA, AA, and EPA contained, respectively, 2.2, 1.4, and 0.7 times more TBARM than in the culture medium. This is not surprising since TBARM formed in the culture medium incubated without cells was proportional to the number of double bonds of the fatty acid under test (data not shown).

When the average amount of the TBARM recovered from each cell type over 7 days is compared, it can be seen that both cancer and normal cells yielded the same amount of TBARM in unsupplemented cultures. Conversely, in supplemented cultures, (a) the tumor cell cultures produced more TBARM than did the normal cell cultures, (b) in the tumor cells, GLA and AA produced more TBARM than did EPA, and (c) in the normal cells, EPA produced more TBARM than did GLA and AA.

TBARM varied with time differently in the culture medium and in the cells, depending on the fatty acid and the cell type. No pattern specific for tumor or normal cells was discerned.

If selective killing of cancer cells is associated with lipid peroxidation in the cancer cells, an increase in the percentage of dead cells should correlate with an elevation of TBARM concentrations. Table 2 demonstrates that the effectiveness of GLA, AA, and EPA in inducing cancer cell death correlated with the TBARM levels found in the tumor cells but not in the normal cells. Then we checked if vitamin E and indomethacin would inhibit the PUFA-induced killing effect and if the inhibition could be correlated with a decrease in TBARM amounts. Table 3 shows that the addition of vitamin E acetate, but not indomethacin, to the tumor cell cultures challenged with EPA inhibited cell killing and reduced TBARM amounts.

DISCUSSION

We examined the contribution of lipid peroxidation in the selective killing of cancer cells by GLA, AA, and EPA, all three PUFA's being precursors of eicosanoids.

Table 1. Amount of TBARM in Growing Breast Tumor (ZR-75-1) and Normal Kidney (CV-1) Cell Cultures Incubated With Various PUFA's

pmoles MDA - Eq/10³ Cells

Treatment	Day	Medium		Cells		Total (Medium & Cell)	
		Normal	Tumor	Normal	Tumor	Normal	Tumor
Control	3	3.1 ± 0.7	5.5 ± 0.8	2.1 ± 3.5	0.3 ± 0.3	5.2 ± 4.2	5.8 ± 1.1
	4	7.4 ± 2.5	1.6 ± 0.7	0.0 ± 0.5	0.8 ± 0.5	7.4 ± 3.0	2.4 ± 1.2
	5	0.0 ± 0.8	2.0 ± 1.2	0.5 ± 0.3	1.8 ± 1.1	0.5 ± 1.1	3.8 ± 2.3
	6	1.0 ± 0.4	3.4 ± 1.5	0.1 ± 0.4	0.5 ± 0.4	1.1 ± 0.8	3.9 ± 1.9
	7	0.3 ± 0.1	0.9 ± 0.2	0.5 ± 0.3	0.2 ± 0.3	0.8 ± 0.4	1.1 ± 0.6
GLA	3	3.2 ± 1.3	12.0 ± 1.5	5.6 ± 2.8	28.0 ± 4.6	8.8 ± 4.1	40.0 ± 6.1
	4	3.9 ± 0.2	15.0 ± 1.4	5.4 ± 1.7	30.0 ± 3.2	9.3 ± 1.9	45.0 ± 4.6
	5	1.5 ± 1.3	12.2 ± 0.8	20.0 ± 6.4	22.7 ± 4.3	21.5 ± 7.7	34.9 ± 5.1
	6	2.0 ± 0.4	12.4 ± 1.4	7.0 ± 2.7	28.2 ± 7.5	9.0 ± 3.1	40.6 ± 8.9
	7	1.1 ± 0.3	10.6 ± 2.4	6.2 ± 1.1	26.5 ± 2.7	7.3 ± 1.4	37.1 ± 5.1
AA	3	3.2 ± 0.7	38.0 ± 1.4	8.3 ± 2.9	14.0 ± 3.2	11.5 ± 3.6	52.0 ± 4.6
	4	6.3 ± 1.2	31.4 ± 1.5	4.6 ± 2.4	61.0 ± 9.0	10.9 ± 3.6	92.4 ± 9.5
	5	3.1 ± 0.5	30.0 ± 1.7	10.0 ± 1.2	35.8 ± 5.3	13.1 ± 1.7	65.8 ± 7.0
	6	3.5 ± 1.0	21.7 ± 1.2	5.0 ± 2.6	32.4 ± 9.9	8.5 ± 3.6	53.1 ± 11.1
	7	2.8 ± 1.0	3.5 ± 1.6	7.1 ± 0.4	35.9 ± 10.4	9.9 ± 1.4	39.4 ± 12.0
EPA	3	4.7 ± 0.9	27.0 ± 1.6	7.4 ± 4.3	18.0 ± 3.0	12.1 ± 5.2	45.0 ± 4.6
	4	6.8 ± 1.2	28.0 ± 1.4	14.0 ± 3.5	31.3 ± 2.4	20.8 ± 4.7	40.0 ± 4.0
	5	3.1 ± 0.5	26.6 ± 3.5	24.7 ± 8.5	12.0 ± 8.1	27.8 ± 9.0	32.6 ± 11.6
	6	4.6 ± 0.6	16.7 ± 1.2	15.5 ± 3.7	15.8 ± 1.8	20.1 ± 4.2	33.5 ± 3.0
	7	3.3 ± 0.6	10.2 ± 0.6	15.0 ± 2.7	16.7 ± 1.7	18.3 ± 3.3	26.9 ± 2.3

Table 2. Cell Viability and Amount of TBARM in Growing ZR-75-1 Breast Tumor or CV-1 Normal Kidney Cells Incubated With GLA, AA, and EPA After 6 Days

Treatment	pmoles MDA-Eq/10^3 Cells[1]		Tumor/Normal	% Dead Cells	
	Normal	Tumor		Normal	Tumor
Control	0.5 ± 2.1	0.7 ± 0.9	1.4	13 ± 1	15 ± 2
GLA	9.5 ± 7.2	27.9 ± 6.1	2.9	9 ± 2	93 ± 3
AA	7.0 ± 3.3	33.8 ± 21.0	4.8	8 ± 1	99 ± 1
EPA	15.4 ± 8.2	19.2 ± 8.6	1.2	14 ± 1	64 ± 10

[1]Average of amounts detected from day 3 to day 6

Table 3. Effects of Vitamin E and Indomethacin on Cell Viability and Amount of TBARM in Growing ZR-75-1 Breast Tumor Cells Incubated with EPA

Treatment	Day	pmoles MDA-Eq/10^3 Cell	% Dead Cells
EPA	1	5.5 ± 3.9	8 ± 3
	4	9.1 ± 1.0	44 ± 15
	5	12.9 ± 3.2	83 ± 5
	6	3.6 ± 0.7	96 ± 1
EPA+ Indomethacin	1	6.6 ± 0.8	3 ± 3
	4	6.4 ± 3.3	18 ± 5
	5	15.3 ± 2.8	76 ± 2
	6	3.6 ± 0.9	92 ± 6
EPA+ Vitamin E Acetate	1	13.7 ± 6.3	6 ± 1
	4	5.1 ± 0.9	3 ± 1
	5	3.8 ± 1.1	5 ± 1
	6	0.4 ± 0.3	5 ± 4
Vitamin E Acetate	1	n.d.[1]	3 ± 1
	4	0.4 ± 0.1	1 ± 1
	5	n.d.	2 ± 1
	6	0.1 ± 0.2	4 ± 2
Indomethacin	1	3.8 ± 2.5	3 ± 1
	4	1.6 ± 0.4	3 ± 2
	5	0.5 ± 0.3	5 ± 1
	6	0.0 ± 0.3	4 ± 2

[1]Not detected

We used the TBA test as a nonspecific indicator of lipid peroxidation. Since an increase in TBARM was associated with a greater percentage of dead cells in tumor cell cultures and since inhibition of tumor cell killing was associated with a reduction in TBARM, we conclude that the effectiveness of a given PUFA in killing cancer cells correlated with the TBARM content in the cells. No correlation was found between the amount of TBARM and the percentage of dead cells in normal cell cultures.

It may be important to distinguish TBARM measured in the culture medium from TBARM measured in the cells because the extent of peroxidation and its relationship to the killing may vary, depending on the cell type, the fatty acid added, and extra- or intracellular peroxidation. We are aware of the limitations of the TBA test in whole cells. Nevertheless, consistent differences in the amount of TBARM were found relative to the effects of a given PUFA in the same cells in identical conditions. Importantly, preliminary experiments indicated that considerably more conjugated dienes were formed in the tumor cells than in the normal cells upon supplementation (unpublished data), a result supporting those obtained with the TBA test. Chemical peroxidation, as occurs in the medium without cells, is in our system directly proportional to the number of double bonds in the added PUFA (data not shown). This result agrees with those of others (6) who have consistently shown that lipid peroxide production is proportional to double bond number. This correlation is less clear in the cells, probably as a result of both enzymatic and nonenzymatic lipid peroxidation.

At this stage, it is not possible to be more specific about the precise nature of the underlying biochemical mechanism or of the material that is actually toxic to the cells. Tumor cell death may result from the production of specific peroxide species and/or derived radicals or degradation products that are found in different levels in the tumor cells, depending on the PUFA substrate. We reported earlier (7) that inhibition of eicosanoid synthesis does not block the selective killing of tumor cells by PUFA's. As shown here, neither did indomethacin, an inhibitor of endoperoxide formation, reduce the amount of TBARM. Thus it is unlikely that the endoperoxides are responsible for the tumor cell killing. A vital question is whether or not lipid peroxidation represents a coincidental outcome of radical-induced damage or if lipid peroxidation products are directly deleterious to the cells. Only a great deal of further work will enable the precise biochemical mechanism of the cell death to be determined.

The possibility that a specific mechanism of lipid peroxidation depends on the substrate may have to be considered because preferential peroxidation of GLA and AA over EPA was observed in tumor cells and, conversely, preferential peroxidation of EPA over GLA and AA in normal cells, and because a different tumor cell line revealed resistance to GLA but not to EPA (unpublished data). Preferential peroxidation of n-3 over n-6 PUFA in normal cells has also been reported previously by Wills (6).

The difference found in TBARM in the tumor cells exposed to GLA, AA, and EPA may reflect (a) differences in the rate of lipid peroxidation and in the rate of metabolism of TBARM, or (b) differences in uptake, incorporation into phospholipids or distribution of the PUFA. Experiments are under way to identify the critical steps.

The difference in the effect of PUFA-E on normal cells and tumor cells may be attributed to low levels of preventive antioxidants in these tumor cells. However, tumor cells with higher levels of antioxidants than in normal cells were also shown to be killed by essential fatty acids at comparable concentrations (8).

The fact that tumor cells are more easily killed by PUFA oxidation products than are the normal cells, even for tumor cells known to contain higher levels of antioxidants than normal cells, implies that exposure of tumor cells to adequate amounts of PUFA may be a general method to eliminate cancer cells selectively. The effective PUFA concentration has to be adjusted to overcome the inhibition exercised by chain-breaking antioxidants.

REFERENCES

1. Cheeseman, K. H., Burton, G. H., Ingold, K. H., and Slater, T. F. Lipid peroxidation and lipid antioxidants in normal and tumor cells. Toxicol. Pathol. 12: 235-239, 1984.
2. Bégin, M. E., Das, U. N., Ells, G., and Horrobin, D. F. Selective killing of human cancer cells by polyunsaturated fatty acids. Prostagl. Leukot. Med. 19: 177-186, 1985.
3. Bégin, M. E., Ells, G., Das, U. N., and Horrobin, D. F. Differential killing of human carcinoma cells by n-3 and n-6 polyunsaturated fatty acids. JNCI 77: 1053-1062, 1986.
4. Booyens, J., Engelbrecht, P., LeRoux, S., Lowrens, C. C., van der Merwe, M., and Katzeff, I. E. Some effects of the essential fatty acids linoleic acid and alpha-linolenic acid and of their metabolites gamma-linolenic acid, arachidonic acid, eicosapentaenoic acid, docosahexaenoic acid, and of prostaglandins A and E on the proliferation of human osteogenic sarcoma cells in culture. Prostagl. Leukot. Med. 15: 15-34, 1984.
5. Gavino, V. C., Miller, J. S., Ikharebha, S. O., Milo, G. E., and Cornwell, D. G. Effect of polyunsaturated fatty acids and antioxidants on lipid peroxidation in tissue cultures. J. Lipid Res. 22: 763-769, 1981.
6. Wills, E. D. The role of dietary components in oxidative stress in tissues. In: "Oxidative Stress." H. Sies, ed. Academic Press Inc., San Diego, 1985, pp. 197-218.
7. Bégin, M. E., Das, U.N., and Ells, G. Mechanism of essential fatty acid (EFA)-induced cytotoxicity in malignant cells. Abstract (#9) of 2nd International

Congress on Essential Fatty Acids, Prostaglandins and Leukotrienes. London, March 1985.

8. Tolnai, S., and Morgan, J. F. Studies on the *in vitro* antitumor activity of fatty acids. V. Unsaturated acids. Can. J. Biochem. Physiol. 40: 869-875, 1962.

DETECTION AND QUANTITATION OF
EICOSANOIDS AND LIPID PEROXIDES

DEVELOPMENT OF ENZYME-IMMUNOASSAY TECHNIQUES FOR MEASUREMENT OF EICOSANOIDS

J. Maclouf[1], J. Grassi[2], and P. Pradelles[2]

[1]U150 INSERM
LA 334 CNRS Hôpital Lariboisière
Paris, France

[2]Section de Pharmacologie et d'Immunologie
Departement de Biologie
Commissariat à l'Energie Atomique
Cedex, France

ABSTRACT

Enzyme immunoassays of eicosanoids have been developed in recent years. Although the assays are quite different in the enzyme label and the type of development, they represent a major conceptual change for the immunoanalysis of these compounds. We have coupled covalently acetylcholine esterase from electrophorus electricus to various eicosanoids, including PGE_2, PGD_2, $PGF_{2\alpha}$, 6-keto-$PGF_{1\alpha}$, TXB_2, dinor-TXB_2, 11-dehydro-TXB_2, and LTC_4. Using 96-well microtiter plates coated with the second antibody, we could obtain results equal to or better than those obtained with corresponding $[^{125}I]$-radioiodinated tracers when available. The high sensitivity of this method combined to semiautomation may prove to be a useful tool to investigate the level of these substances from various biological fluids. In addition, this enzyme has been coupled successfully to other substances, suggesting that it could be used as a general label to produce nonisotopic tracers.

Widespread interest concerning the various oxygenated metabolites of arachidonic acid has led to the development of various techniques to monitor the *in vitro* or *in vivo* production of these compounds. Two analytical techniques have been complementary in obtaining quantitative information on these compounds. The first, gas chromatography coupled to mass spectrometry (GC/MS), has allowed structural information on all these molecules (1). GC/MS has been used quantitatively with internal standards; further, extreme sensitivities by means of negative-ion chemical ionization have permitted the use of GC/MS to monitor these substances in biological fluids. Due to the specificity of the simultaneous measurement of several ions, GC/MS is used as a reference for all quantitative assays. However, the cost of equipment and the high degree of expertise required

to use this technique do not promote its widespread use (2). In addition, GC/
MS does not lend itself to the analysis of a serial number of samples.

The second technique, based on immunoanalysis of these compounds, is the more
widespread one. However, a major limitation is its validation for specificity to
ensure proper use as a reliable quantitative technique (see later). A number of
radioimmunoassays (RIA) have therefore been developed for prostaglandins,
thromboxanes, and the recently described leukotrienes. Further, the need to get
high sensitivity by increasing the specific radioactivity of the tracer has been
circumvented by the use of iodinated tracers for these molecules which, in addition
to their high specific activity, avoid quenching problems and decrease counting
time (3). However, the use of RIA entails certain disadvantages; among them are
the limited half-life of reagents susceptible to radiolysis, increasing expenses in
developed countries inherent in the rigid rules imposed for radioactive isotopes,
cost of tubes, scintillation cocktails associated with radioactive disposal, and high
personnel requirement per sample due to its limited automation.

All these reasons explain why, during these recent years, users of immunoassays
have stressed the need to obtain nonradioactive labels. Immunoassays for eicosanoids
have not escaped these evolutions, and since 1981, a consequent number of enzyme
immunoassays (EIA) have been developed. In this paper, we briefly review these
different approaches, including our own experience.

BRIEF SURVEY OF DIFFERENT EIA'S FOR EICOSANOIDS

In 1981, Yamamoto and colleagues developed the first EIA for prostaglandin
$F_{2\alpha}$ using either alkaline phosphatase (4) or beta-D-galactosidase (5) as labels to
this prostaglandin. For both systems, bound was separated from free using the
double antibody technique. Using fluorescent detection, the beta-galactosidase assay
was found to be more sensitive than the first one (Table 1). Although a good
correlation with RIA was obtained on urinary measurement, the procedure involved
laborious extraction and separation steps prior to the assay.

In 1983, the same authors (6) developed an EIA for the measurement of
thromboxane. After labeling the compounds with beta-galactosidase, the separation
of antibody-bound prostaglandin from free prostaglandin was performed using a
double antibody technique. However, this assay performed in test tubes could not
allow the direct measurement of biological samples, and although a good correlation
was obtained between this method and GC/MS results, the prerequisite for systematic
extraction and/or purification of the sample represents a serious drawback for
RIA.

More recently, Miller et al. (7) developed a heterogenous enzyme-linked
immunosorbent assay (ELISA) to measure prostaglandin, thromboxane, and
leukotriene. These authors have used an indirect assay procedure involving the
preparation of a bovine serum albumin conjugate of the leukotriene or prostaglandin

Table 1. Comparison of Sensitivity EIA's for Various Eicosanoids

Com-pound	Enzyme Label[1] (Detection)	General Procedure	Mode	IC-50 (pg)	Refer-ence
TXB$_2$	β-Galactosidase (F)	Competition	Tube	360	5
	β-Galactosidase (F)	Immobilized anti-TXB$_2$	Tube	190	9
	β-Galactosidase (F)	Immobilized Second antibody	Tube	62	8
	Alkaline phosphatase (C)	ELISA (sandwich meth.)	Plate	250	7
	Acetylcholinesterase (C)	Competition	Plate	2	12
6-keto-PGF$_{1\alpha}$	Alkaline phosphatase (C)	ELISA (sandwich meth.)	Plate	350	7
	Alkaline phosphatase (C)	ELISA	Plate	30	11
	Acetylcholinesterase (C)	Competition	Plate	2.5	12
PGE$_2$	Acetylcholinesterase (C)	Competition	Plate	2	17
PGF$_{2\alpha}$	β-Galactosidase (F)	Competition	Tube	350	4
	Alkaline phosphatase (C)	Competition	Tube	1000	3
	Acetylcholinesterase (C)	Competition	Plate	2.5	2
PGD$_2$	Peroxidase (F)	Immobilized anti-PGD$_2$	Tube	25	10
	Acetylcholinesterase (C)	Competition	Plate	2	12
LTC$_4$	Alkaline phosphatase (C)	ELISA (sandwich meth.)	Plate	125	7
	Acetylcholinesterase (C)	Competition	Plate	14	12,19
LTB$_4$	Alkaline phosphatase (C)	ELISA (sandwich meth.)	Plate	3300	7

[1]F, fluorometric detection; C, colorimetric detection
[2]This paper

subsequently adsorbed to polystyrene microtiter plates. The samples (or standards) to measure were incubated in the well with the specific rabbit antiserum. After incubation, a goat anti-rabbit labeled with fluorescence and an antifluorescein antibody linked to alkaline phosphatase were added successively. The authors could then measure leukotriene directly from various biological samples and obtain good correlation of this method with RIA, with radioisotope experiment as well as with bioassay.

Another enzyme immunoassay for thromboxane B$_2$ (TXB$_2$) has been developed recently using a beta-D-galactosidase label (8). Incubation of the reagents was performed in test tubes using either direct double antibody separation or a solid-phase procedure with glass beads coated with a second antibody. In both cases, the enzyme reaction was carried out using a fluorometric reaction with 4-methyl umbelliferyl-beta-D-galactoside as a substrate. However, this method could not be used on unextracted samples. A correlation of this method was done on urine samples with an RIA that had been correlated with GC/MS on measurements of human serum samples. A similar approach used immobilized anti-TXB$_2$-IgG bound to a polystyrene tube; after addition of the tracer and standard or unknown, the beta-galactosidase activity was revealed fluorometrically (9).

Two other EIA's have been described. The first one used a solid-phase EIA for prostaglandin D$_2$ (PGD$_2$) and a PGD$_2$-labeled horseradish peroxidase with an immobilized antibody (10). After competition between the enzymatic tracer and PGD$_2$, the activity of the enzyme bound to the antibody was assayed fluorometrically. All samples had to be purified prior to analysis. The measurement of PGD$_2$ in rat brain by EIA correlated with results obtained by RIA. The second one concerned an ELISA developed for 6-keto-prostaglandin F$_{1\alpha}$ (PGF$_{1\alpha}$) (11). After coating of microtiter plates with a 6-keto-PGF$_1$-bovine serum albumin conjugate, a competition reaction with the specific rabbit antibody and the standard (or unknown) was carried on. After washing the excess unreacted reagents, an anti-rabbit IgG labeled with alkaline phosphatase was added. The method was used successfully to estimate 6-keto-PGF$_{1\alpha}$ from unextracted endothelial cell supernatant.

The results obtained with these different EIA's have been summarized in Table 1, which shows the various IC-50's obtained with these procedures. However, due to a difference in procedure, enzyme, and antiserum, it is not possible to draw an adequate comparison of the intrinsic value of these assays. However, in most cases the sensitivity is much lower than the one currently obtained with RIA. Also, nearly all the assays require purification prior to measuring the corresponding ligand, and the use of a tube does not allow easy automation of the technique.

EIA OF EICOSANOIDS USING ACETYL CHOLINESTERASE AS LABEL

We have recently developed an EIA technique using for the first time acetylcholine esterase (AChE) from electric eel as label coupled to different eicosanoids (12). This method, which uses 96-well microtiter plates coated with a second antibody, allows sensitivity equal or superior to that obtained with RIA using (^{125}I)-radioiodinated tracers (3). In addition, the partial automation of the assay and the long shelf life of the label should promote the use of EIA as a substitute for RIA.

METHODS

The procedure described here follows that described previously (12). Briefly, AChE was purified by one-step affinity chromatography as described by Massoulie et al. (13). The preparation of enzymatic tracers was as follows: prostaglandins or thromboxanes were converted to their N-hydroxy-succinimide esters, which were subsequently incubated (1-100 nmoles) with the enzyme (50-250 μg). After 1 hour at 4°C, the reaction was stopped by the addition of a phosphate buffer containing bovine serum albumin, and purification of the conjugates from uncoupled products was performed using a Biogel A15-m column (Bio-Rad, Richmond, CA). The enzymatic activity of each fraction collected in a polystyrene tube was determined, and the fractions containing the different molecular forms of the enzyme were pooled and stored at -30°C. For the preparation of leukotriene C_4-AChE conjugate, we used the method described by Young et al. (14).

All assays were performed in a 0.1 M phosphate buffer, pH 7.4, containing 0.4 M NaCl, 1 mM EDTA, 0.1% bovine serum albumin, and 0.01% sodium azide. Each sample or buffer (50 μl) was added to each well of a 96-well microtiter plate that was previously coated with pig anti-rabbit IgG antibody (2 μg/well). The AChE conjugates (50 μl) and the specific antiserum (50 μl) were then added at appropriate dilutions. After incubation at room temperature for 4-18 hours, the plates were automatically washed with an automatic washer (Flow Laboratories, Helsinki, Finland) and 0.01 M phosphate buffer (pH 7.4) containing 0.05% Tween-20. The plates were then automatically filled with 200 μl of the enzymatic substrate [2 μg/ml acetylthiocholine and 2.15 μg/ml of 5-5'-dithiobis(2-nitrobenzoic acid) in 0.01 M phosphate buffer] (15). After 1-3 hours at room temperature, the absorbance of a yellow product in each well was measured at 414 nm using a Multiskan MC spectrophotometer (Flow Laboratories). Results are expressed in terms of B/B_0 x 100, where B and B_0 are the absorbance of eicosanoid label measured on the bound fraction in the presence or in the absence, respectively, of eicosanoid competitors. Fitting of the standard curves and calculations of the quantity of eicosanoids were done with a microcomputer (Apple II, Cupertino, CA) using a linear log-logit transformation (16). All measurements were done in duplicate.

RESULTS AND DISCUSSION

In early studies, when we performed a comparison between the performances obtained with EIA and [^{125}I]-labeled-TX-histamine conjugate, we found that both precision and sensitivity were significantly lower with the non-isotopic technique (12). These results prompted us to use EIA as a substitute for RIA since iodine-labeled tracers possess the highest specific radioactivity allowing low detection of substances. A dose-response curve of 6-keto-$PGF_{1\alpha}$ using this method is shown in Figure 1. We have now extended the labeling of eicosanoids with AChE to a variety of compounds, even though for some of them we did not possess the corresponding radioactive substance (Table 2). As can be observed for all these molecules, the mass of standard required to obtain a 50% displacement of the initial antibody-bound tracer is very low. This allowed in most cases a minimal detectable concentration of less than 15 pg/ml (i.e., 0.75 pg/well), which has been useful when measuring PGE_2 from microdissected nephron (17). As also observed in this table, the titer of the antibody (i.e., usable dilution) is extremely high. These findings are a consequence of the high specific activity of the tracers. This is further illustrated with the LTC_4 antiserum; although we used an antiserum whose binding

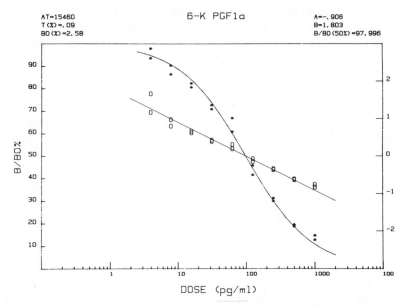

Fig. 1. Example of a standard curve obtained using the 6-keto-PGF$_{1\alpha}$ system. Open triangles correspond to values expressed as B/B (%). Squares give values after log-logit transformation. Usable calibration curve ranges from 12 to 500 pg/ml corresponding to 0.6-25 pg/well.

Table 2. Comparison of Sensitivity and Antibody Titer for Different Eicosanoids

Eicosanoid	B/B$_o$ = 50%[1] (pg/ml)	Titer (Final Dilution)
PGE$_2$	20	1/30,000
PGF$_{2\alpha}$	70	1/300,000
PGD$_2$-MO	40	1/600,000
6-keto-PGF$_{1\alpha}$	44	1/300,000
TXB$_2$	40	1/60,000
Diner TXB$_2$	48	1/60,000
11-dehydro-TXB$_2$	40	1/1,500,000
6-β-PGI$_1$	320	1/600,000
LTC$_4$	300	1/90,000

[1]Concentration of eicosanoid inducing a 50% inhibition of tracer bound in absence of unlabeled competitor (B$_o$)

properties were originally characterized with [^{3}H]LTC$_4$ (18), we obtained a substantial gain in sensitivity compared to tritium (IC-50, 390 pg/ml versus 3900 pg/ml for RIA) using 7-10 times less antiserum than for RIA (19).

In cases where extreme sensitivities are required, it is known that preincubation of the specific antibody with standard, followed by a later addition of tracer, can increase substantially the sensitivity. It was therefore of interest to see whether the preincubation technique could be applied with the second antibody-coated microtiter plates and the AChE tracer. Table 3 shows an example where this procedure has been applied to the LTC$_4$ system. As can be seen, it allowed a significant increase in sensitivity (twofold), which might be an appreciable gain whenever such sensitivity is required (i.e., a small number of cells or low volume of biological sample with a minute amount of compound). Such a procedure also has been applied successfully to other eicosanoids such as 11-dehydro-TXB$_2$, where it allowed a gain of factor 3 (to be published). However, it should be noted that such a gain in sensitivity can be obtained only when the affinity constant of the antibody for the ligand is high enough.

As mentioned above, a certain number of situations require that the biological samples be purified prior to analysis (19). Organic solvent or resin extractions before purification by thin layer chromatography have turned out to be reliable and sufficient. However, they are time-consuming and should be followed by quantitative methods such as GC/MS or RIA after corrections for losses in recoveries. In such cases, quantitation using the semi-automated analysis by EIA can be an advantage (19). We have taken advantage of this latter aspect to analyze the platelet-derived and vascular cell-derived thromboxane and prostacyclin urinary metabolites, which are believed to be good markers of cardiovascular and/or thrombotic diseases (20-22). After addition of [^{3}H]TXB$_2$ as a standard for estimating the losses, the urine was extracted by 2 x volume of diethyl ether and run on two serial 10-μm (250 x 5 mm and 150 x 5 mm) high-performance liquid chromatography columns using a nonlinear gradient (22). Each fraction was analyzed for its immunoreactivity with different antisera and corresponding tracers as well as for its content in radioactivity. The analysis allowed a profiling study of the urinary extracts for the various metabolites. Such an approach has been used to measure the urinary metabolites in normal subjects (23), and it can be applied easily to pathological situations.

Table 3. Influence of Preincubation[1] on LTC$_4$-AChE/anti-LTC$_4$ System

Normal procedure	IC-50 (pg/ml) Preincubation
220	94

[1]Incubating antibody with standard for 4-18 hours and then adding LTC$_4$-AChE during 2-3 hours

CONCLUSIONS

The technique that we have developed seems to possess a significant number of advantages over the classical RIA analysis. They are primarily semiautomation, no radioactivity handling, long shelf life of tracers, and high sensitivity. Other advantages of the technique have been discussed elsewhere (19). In addition, AChE has been used as a label for a variety of compounds, including thyronine, 3,5,3′-triiodothyronine, thyroid-stimulating hormone, thyrotropin-releasing hormone, substance P, serum factor from thymus (FTS), immunoglobulin E (specific and total), octasaccharide (heparin fragment CY 222), vasopressin metabolites (L-vasopressin and A-vasopressin), screening of monoclonal antibodies (avidin-biotin system), cyclic nucleotides, human renin, atriopeptin III, estradiol, penicilloyl, and juvenile hormone (insect) (P. Pradelles and J. Grassi, unpublished, and ref. 24). The same advantages in terms of simplicity, sensitivity, and titer of antisera were found in each situation. The intrinsic biochemical properties of this enzyme label [possessing an extremely high specific activity as well as long stability (several years frozen)] should allow more widespread use for labeling any compound to be measured by EIA analysis.

It appears today that the use of enzymatic fluorescent or luminescent probes allows the preparation of tracers having a specific activity equal to or significantly superior to that obtainable with ^{125}I. Immunoassays using nonisotopic labels can be more sensitive than corresponding RIA if high-affinity antibodies are used together with these high-specific-activity tracers. The development of such methods may increase, although there will still be a place for different labels (isotopic and nonisotopic) appropriate to the analytical problem encountered and/or the experimental environment.

ACKNOWLEDGMENTS

The authors are indebted to Dr. J. E. Pike for the kind gift of prostaglandin and thromboxane standards and to Drs. J. Rokach and R. Young for providing LTC$_4$ standard and the LTC$_4$ antibody. Part of this work was supported by a grant from INSERM/PIRMED "Lipides pharmacologiquement actifs."

REFERENCES

1. Murphy, R. C., and Harper, T. W. Mass spectrometry and icosanoid analysis. "Biochemistry of Arachidonic Acid Metabolism." W. E. M. Lands, ed. Martinus Nijhoff Publishing, Boston, 1985, pp. 417-434.
2. Granstrom, E., and Kindahl, H. "Advances in Prostaglandin and Thromboxane Research," Vol. 5. J. C. Frolich, ed. Raven Press, New York, 1978, pp. 119-210.

3. Maclouf, J., Pradel, M., Pradelles, P., and Dray, F. [125]I derivatives of prostaglandins: A novel approach in prostaglandin analysis by radioimmunoassay. Biochim. Biophys. Acta 431: 139-146, 1976.

4. Hayashi, Y., Yano, T., and Yamamoto, S. Enzyme immunoassay of prostaglandin $F_{2\alpha}$. Biochim. Biophys. Acta 663: 661-688, 1981.

5. Yano, T., Hayashi, Y., and Yamamoto, S. β-Galactosidase-linked immunoassay of prostaglandin $F_{2\alpha}$. J. Biochem. 90: 773-777, 1981.

6. Hayashi, Y., Ueda, N., Kazushige, Y., Kawamura, S., Ogushi, F., Yamamoto, Y., Yamamoto, S., Nakamura, K., Yamashita, K., Miyazaki, H., Kato, K., and Terao, S. Enzyme immunoassay of thromboxane B_2. Biochim. Biophys. Acta 75: 322-329, 1983.

7. Miller, D., Sadowski, S., De Sousa, D., Maycock, A. L., Lombardo, D. L., Young, R. N., and Hayes, E. C. Development of enzyme-linked immunosorbent assays for measurement of leukotrienes and prostaglandins. J. Immunol. Methods 81: 169-185, 1985.

8. Sawada, M., Inagawa, T., and Frolich, J. C. Enzyme-immunoassay of thromboxane B_2 at the picogram level. Prostaglandins 29: 1039-1048, 1985.

9. Shono, F., Yokota, K., and Yamamoto, S. A solid-phase enzyme immunoassay of thromboxane B_2. J. Biochem. 98: 1069-1077, 1985.

10. Hiroshima, O., Hayashi, H., Ito, S., and Hayaishi, O. Basal level of prostaglandin D_2 in rat brain by a solid-phase enzyme immunoassay. Prostaglandins 32: 63-80, 1986.

11. Doeherty, J. C., and Gerrard, J. M. An enzyme-linked immunosorbent assay for 6-keto-$PGF_{1\alpha}$. Prostaglandins 31: 375-383, 1986.

12. Pradelles P., Grassi, J., and Maclouf, J. Enzyme immunoassays of eicosanoids using acetylcholine esterase as a label: An alternative to radioimmunoassay. Anal. Chem. 57: 1170-1173, 1985.

13. Massoulie, J., and Bon, S. Affinity chromatography of acetylcholinesterase. The importance of hydrophobic interactions. Eur. J. Biochem. 68: 531-539, 1976.

14. Young, R. N., Kakushima, M., and Rokach, J. Studies on the preparation of conjugates of leukotriene C_4 with proteins for development of an immunoassay for SRS-A. Prostaglandins 23: 603-613, 1982.

15. Ellman, G. I., Courtney, K. D., Andres, V., and Featherstone, R. M. A new and rapid colorimetric determination of acetylcholinesterase activity. Biochem. Pharmacol. 7: 88-95, 1961.

16. Rodbard, D., Bridson, W., and Rayford, P. Rapid calculation of radioimmunoassays results. J. Lab. Clin. Med. 74: 770-781, 1969.

17. Farman, N., Pradelles, P., and Bonvalet, J. P. Determination of prostaglandin E_2 synthesis along rabbit nephron by enzyme immunoassay. Am. J. Physiol. 251: F238-244, 1986.

18. Hayes, E. C., Lombardo, D. L., Girard, Y., Maycock, A. L., Rokach, J., Rosenthal, A. S., Young, R. N., Egan, R. W., and Zweerink, H. J. Measuring leukotrienes of slow reacting substance of anaphylaxis: Development of a specific radioimmunoassay. J. Immunol. 131: 429-433, 1983.

19. Westcott, J. Y., Chang, S., Balazy, M., Stene, D. O., Pradelles, P., Maclouf, J., Voelkel, N. F., and Murphy, R. C. Analysis of 6-keto $PGF_{1\alpha}$, 5-HETE

and LTC_4 in rat lung: Comparison of GC/MS, RIA and EIA. Prostaglandins 32: 857-873, 1986.

20. Roberts, L. J., Sweetman, B. J., and Oates, J. A. Metabolism of thromboxane B2 in man: Identification of twenty urinary metabolites. J. Biol. Chem. 256: 8384-8393, 1981.

21. Rosenkranz, B., Fischer, C., Weimer, K. E., and Frolich, J. C. Metabolism of prostacyclin and 6-keto prostaglandin $F_{1\alpha}$ in man. J. Biol. Chem. 225: 10194-10198, 1980.

22. Patrono C., Preston, F. E., and Vermylen, J. Platelet and vascular arachidonic acid metabolites: Can they help detect a tendency towards thrombosis? Brit. J. Haematol. 57: 209-212, 1984.

23. Castagnoli, M. N., Bellucci, S., Maclouf, J., and Pradelles, P. Combined high performance liquid chromatography and enzyme immunoassay for the determination of the urinary metabolites of thromboxane and 6-keto $PGF_{1\alpha}$. Adv. Prostaglandin Thromboxane Leukotriene Res. 17: 603-607, 1987.

24. Devillier, P., Weill, B., Renoux, M., Menkes, C., and Pradelles, P. Elevated levels of tachykinin-like immunoreactivity in joint fluids from patients with rheumatic inflammatory diseases. New Engl. J. Med. 314: 1323, 1986.

DETECTION AND QUANTITATION OF EICOSANOIDS BY COMBINED GAS CHROMATOGRAPHY-MASS SPECTROMETRY

W. C. Hubbard[1], C. L. Litterst[1], M. C. Liu[2], E. R. Bleecker[2],
E. G. Mimnaugh[1], J. C. Eggleston[3], T. L. McLemore[1],
and M. R. Boyd[1]

[1]Laboratory of Experimental Therapeutics and Metabolism
National Cancer Institute
National Institutes of Health
Bethesda, Maryland 20892

[2]Pulmonary Division
Francis Scott Key Medical Center
The Johns Hopkins University School of Medicine
Baltimore, Maryland 21224

[3]Department of Pathology
The Johns Hopkins University School of Medicine
Baltimore, Maryland 21205

ABSTRACT

Mass spectrometry is an indispensable analytical technique in studies of the metabolism of arachidonic acid to prostaglandins, leukotrienes, and other related eicosanoids. In addition to providing qualitative information, mass spectrometry is recognized as the quantitative analytical technique offering the greatest degree of sensitivity, selectivity, precision, and accuracy in the measurement of very small quantities of eicosanoids in biological samples. Recent adaptations, including (a) the use of a capillary gas chromatograph interface with the mass spectrometer and (b) the use of negative-ion chemical ionization mass spectrometry of electron-capture derivatives of eicosanoids have facilitated the capability to detect 0.2 picograms (approximately $10-12$ moles) and less of these compounds in samples of biological origin. The principles and use of combined capillary gas chromatography-mass spectrometry for subpicogram detection and quantitation of eicosanoids of biological origin are presented. Presently, these techniques have permitted the detection and quantitation of prostaglandins D_2, E_2, $F_{2\alpha}$, 6-keto-prostaglandin $F_{1\alpha}$ (6KPGF$_{1\alpha}$), and thromboxane B_2 and the 15-keto-13,14-dihydro metabolites of PGE$_2$, PGF$_{2\alpha}$, and 6KPGF$_{1\alpha}$ in biological samples. Also, the use of GC-mass spectrometry for simultaneous determinations of the profile and kinetics of prostaglandin biosynthesis and metabolism in isolated tissues and in cultured cells is featured. The use of alternative interface systems of mass spectrometers as they relate to the chemistry and inherent lability of biologically derived eicosanoids is also addressed.

INTRODUCTION

Combined gas chromatography-mass spectrometry (GC-MS) is an indispensable analytical technique in studies of eicosanoid biosynthesis from arachidonic acid. In addition to providing qualitative data essential for structure elucidation of the prostaglandins, leukotrienes, and related eicosanoids, GC-MS is recognized as the quantitative analytical technique offering the greatest combined degree of selectivity, sensitivity, precision, and accuracy in the measurement of very small quantities of eicosanoids usually present in biological samples. In this report, the principles and use of recent developments in GC-MS analysis of eicosanoids are emphasized. Specifically, the use of combined capillary GC-MS of electron-capture derivatives for the detection and quantitation of eicosanoids is illustrated (1-6), and principles critical to this method of analysis are accented. The reader is directed to excellent review articles that provide detailed discussions of the principles of mass spectrometry in the analysis of eicosanoids (7) and the use of stable isotope analogs as internal standards during GC-MS analysis of these compounds (8,9). These articles provide essential background information related to the use of capillary GC-MS for the detection and quantitation of eicosanoids that cannot be included in this report.

SYNTHESIS OF ELECTRON-CAPTURE DERIVATIVES

The conversion of eicosanoids to derivatives with greater volatility is essential for gas chromatographic analysis and subsequent detection and quantitation of these compounds by mass spectrometry. The derivatization of eicosanoids to more volatile compounds is usually a multistep procedure in which the functional groups of the eicosanoids (keto, carboxyl, and hydroxyl moieties) are sequentially converted to more volatile (and usually less polar) chemical entities. Electron-capture derivatives of eicosanoids are synthesized by the use of reagents containing fluorine atoms. Table 1 gives examples of electron-capture derivatives of eicosanoids that can be synthesized with multifluorinated reagents available from commercial sources. Since all eicosanoids contain at least one carboxyl moiety, the electron-capture derivative most frequently synthesized for capillary GC-MS analysis of these compounds is the pentafluorobenzyl ester (PFBE). The formation of an electron-capture derivative of one fuctional group is usually adequate for capillary GC-MS analysis of eicosanoids. As a result, the remaining functional groups of eicosanoids (if present) are converted to either alkyl or silyl ether derivatives prior to vapor-phase analysis. The typical sequence for complete derivatization of eicosanoids is as follows: (a) formation of the methyloxime (for eicosanoids containing one or more keto functional groups), (b) synthesis of the PFBE, and (c) conversion of the hydroxyl moieties (if present) to the trimethylsilyl ether (TMS) derivative.

Table 1. Electron Capture Derivatives of Eicosanoids

Functional Group	Derivative
Keto, R-C=0	Pentafluorobenzyl oxime[1]
Carboxyl, R-COOH	Pentafluorobenzyl ester
Hydroxyl, R-OH	Pentafluorophenyl-dimethylsilyl ether

[1]Two isomers (sny- and anti-) can be formed.

Table 2. Nomenclature and Abbreviations of Derivatized Eicosanoids

Eicosanoid	Derivative	Abbreviation
$PGF_{2\alpha}$	PFBE, Tris-trimethylsilyl ether	PFBE-(TMS)$_3$
PGD_2, PGE_2	Methyloxime, PFBE, Bis-trimethylsilyl ether	MO-PFBE-(TMS)$_2$
TXB_2, $6KPGF_{1\alpha}$	Methyloxime, PFBE, Tris-trimethylsilyl ether	MO-PFBE-(TMS)$_3$
LTB_4	PFBE, Bis-trimethylsilyl ether	PFBE-(TMS)$_2$

The sequential derivatization of several eicosanoids described above results in the formation of derivatives with chemical names that are quite lengthy. In order to avoid repeated use of the chemical names, the derivative of individual eicosanoids and the abbreviations used for these compounds in this report are listed in Table 2.

The synthesis of an electron-capture derivative of eicosanoids usually results in the formation of a derivative that is less volatile than a corresponding alkyl derivative of the same compound. The decreased volatility of electron-capture derivatives of certain eicosanoids may limit the usefulness of capillary GC-MS analysis for their detection and quantitation. The use of higher temperatures for volatilization of electron-capture derivatives during capillary gas chromatography can result in the thermal degradative loss of certain eicosanoids (10). Limitations of capillary GC-MS for the analysis of eicosanoids are presented near the end of this report.

CAPILLARY GAS CHROMATOGRAPHY OF EICOSANOIDS

The use of small-diameter, fused-silica capillary columns has overcome two major limitations encountered during vapor-phase analysis of derivatized eicosanoids by conventional gas chromatography columns. First, the reduced rates of carrier gas flow with the use of capillary columns have made it possible to transfer the effluent of the column into the ion source of the mass spectrometer with minimal loss of the effluent containing the desired analytes. Second, the enhanced separation of derivatized eicosanoids by capillary GC provides resolution that is unachievable with conventional packed columns. A direct result of the use of capillary GC is improved sensitivity and resolution of eicosanoids. Prior to the use of capillary columns for vapor-phase analysis of eicosanoids, the most efficient chromatographic method for the resolution of these compounds (as either an underivatized or a partially derivatized species) in biological samples was either high-performance liquid chromatography (11-13) or thin-layer chromatography (14,15).

Prior to injection into the capillary gas chromatograph, the derivatized eicosanoids are dissolved in a hydrocarbon solvent. The criterion for the selection of a suitable solvent is discussed in detail in earlier reports (1,2,5). Two major criteria are (a) the solvent must not significantly reduce analyte resolution by the capillary column (i.e., cause broadening or tailing of the analyte peak), and (b) the solvent should have a boiling point high enough that the time required for temperature programming of the gas chromatograph for elution of the desired analytes can be minimized. A third criterion is that the solvent should have a vapor pressure that will allow its rapid removal from the mass spectrometer. The last consideration is critical for avoiding overpressure shutdown of the mass spectrometer and possible damage to the instrument.

The low levels of eicosanoids in biological samples dictate the use of an injection technique that ensures maximal transfer of derivatized analytes to the capillary column. Three injection techniques that minimize the loss of derivatized eicosanoids during injection into the gas chromatograph are splitless injection, modified (partial) splitless injection, and on-column (cool) injection. The advantages and limitations of each of these sample injection techniques are discussed extensively in an excellent monograph (16). Modified splitless injection of samples containing thermally stable (thermally stable up to 300°C) eicosanoids offers several advantages in the combined degree of sensitivity, column background interference, and previous injection carryover for capillary GC-MS detection and quantitation of these compounds.

The resolution of derivatized $PGF_{2\alpha}$, PGD_2, PGE_2, TxB_2, and $6KPGF_{1\alpha}$ by capillary GC is shown in the partial ion current chromatogram depicted in Figure 1. The identities of the more intense peaks resulting from fragmentation of the individual compounds during negative-ion MS are indicated. The derivative of the individual eicosanoids is listed in Table 2. Peak I is the signal elicited after the coelution of the less abundant oxime isomer of derivatized PGD_2 and PGE_2 from the capillary column. Peak II (dotted line) depicts the region at which the less abundant oxime isomer of derivatized $6KPGF_{1\alpha}$ elutes. This isomer of $6KPGF_{1\alpha}$ is not completely resolved from derivatized TxB_2.

NEGATIVE-ION MS FRAGMENTATION OF EICOSANOIDS

The PFBE derivatives of the prostaglandins and related eicosanoids undergo characteristic limited fragmentation during negative-ion MS analysis (1-5,7). Figure 2 shows the typical fragmentation of the methoxylamine-PFBE-(TMS)$_2$ derivative

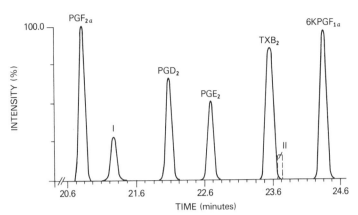

Fig. 1. Capillary gas chromatographic resolution of derivatized PGF$_{2\alpha}$, PGD$_2$, PGE$_2$, TxB$_2$, and 6KPGF$_{1\alpha}$. Approximately 1 nanogram of each compound was injected onto a 30-M SPB-5 (Supelco, Bellefonte, PA) fused silica column (ID 0.25 mm, 0.25 μm film thickness) via modified splitless injection dissolved in dodecane. GC temperature program: 200°C for 1 min following injection; 200°-240°C at 40°C/min; 240°-285°C at 1.5°C/min. Detection: Negative-ion mass spectrometry; mass range m/z 170-825. Ordinate: Reconstructed ion current depicting relative intensity of negative-ion current (%). Abscissa: time (min). Additional details in text.

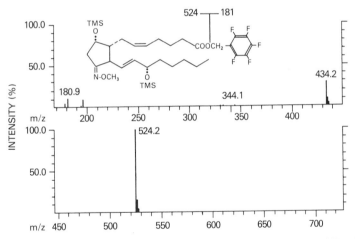

Fig. 2. Fragmentation of methoxylamine-PFBE-(TMS)$_2$ derivative of PGD$_2$ during negative-ion chemical ionization MS analysis. Instrument: Finnigan MAT 4610B. Ionizer pressure reading 0.7 torr. Ionizer temperature setting 120°C. Reagent gas, methane. Electron energy 85 eV. Emission current 0.3 mA. Ordinate = relative intensity of signal. Abscissa = mass-to-charge ratio (m/z) of fragment ion. Additional details in text.

of an eicosanoid (PGD$_2$) during negative-ion MS analysis. An intense negatively charged fragment ion at M-181 (m/z 524 for derivatized PGD$_2$) is distinct. In addition to loss of pentafluorobenzyl moiety, there is limited cleavage of the silylated hydroxyl moieties. For each hydroxyl moiety converted to the trimethylsilyl ether derivative,

Table 3. Fragment Ions of PFBE Derivatives of Eicosanoids Retaining
Eicosanoid Skeleton

Eicosanoid	Derivative[1]	Fragment Ions (m/z)
$PGF_{2\alpha}$	PFBE-(TMS)$_3$	569,479,389,299
PGE$_2$, PGD$_2$	MO-PFBE-(TMS)$_2$	524,434,344
TxB$_2$, 6KPGF$_{1\alpha}$	MO-PFBE-(TMS)$_3$	614,524,434,344
LTB$_4$	PFBE-(TMS)$_2$	479,389,299

[1]Abbreviations for derivatives are defined in Table 2.

there is usually loss of an additional 90 mass units from the loss of trimethylsilanol. Fragment ions representing the loss of one and two trimethylsilanol moieties are observed at m/z 434 and m/z 344, respectively. Other fragment ions typically present are m/z 181 (molecular weight of the pentafluorobenzyl moiety) and m/z 196 (constituency undetermined; not labeled in Figure 2).

Fragment ions of electron-capture derivatives of several eicosanoids retaining the eicosanoid skeleton are listed in Table 3. The data shown in this table indicate that isomeric eicosanoids such as PGD$_2$ and PGE$_2$ (also TxB$_2$ and 6KPGF$_{1\alpha}$) have common characteristic fragment ions. Additional coincidences in fragmentation occur when the functional groups of one or more of the eicosanoids differ only in the number of hydroxyl moieties (i.e., PGF$_{2\alpha}$ and LTB$_4$, PGD$_2$ or PGE$_2$, and TxB$_2$ or 6KPGF$_{1\alpha}$. Thus, in order to accurately detect and quantitate several eicosanoids present in biological samples having common fragment ions, complete knowledge of the fragmentation pattern of each compound is critical, and chromatographic separation prior to mass spectrometric analysis may be obligatory (i.e., LTB$_4$ analysis can be performed only if this eicosanoid is resolved from PGF$_{2\alpha}$ during capillary gas chromatography).

DETECTION AND QUANTITATION OF EICOSANOIDS

The detection and quantitation of several eicosanoids simultaneously during an analytical procedure facilitates the acquisition of the profile of eicosanoid biosynthesis in tissues and cells. The efficient resolution of derivatized eicosanoids by capillary gas chromatography (Figure 1) and the characteristic limited fragmentation of the PFBE electron-capture derivative of these compounds during negative-ion MS analysis (Figure 2, Table 3) provide the basis for profiling of eicosanoids in mammalian tissues and cells. The remaining essential item for the simultaneous detection and quantitation of several eicosanoids in biological samples by capillary GC-MS is the availability of suitable stable isotope analogs for use as internal standards and for determinations of retention times (retention indices) of eicosanoid analytes. Stable isotope analogs of PGA$_2$, PGE$_2$, PGF$_{2\alpha}$, and 6KPGF$_{1\alpha}$ are commercially available. The snythethic preparation of 5,6,8,9,11,12,14,15-octadeutero (^{2}H$_8$) arachadonic acid (17,18) facilitates the biosynthetic preparation of multi-deuterated analogs of most eicosanoids (8,19-21). In addition, methods for the preparation of ^{18}O-labeled analogs of eicosanoids that can be used as internal standards have been published (22). Another potential source of stable isotope analogs of eicosanoids and eicosanoid metabolites is custom synthesis of the desired compounds. However, the practicality of custom synthesis is limited by the expense of preparation.

In order to obtain mass spectral data that are essential for the accurate quantitation of eicosanoids, the mass spectrometer system is programmed for the acquisition of signal at a few masses instead of signal over a mass range of several hundred masses. This special use of mass spectrometry is referred to by several names: selected ion monitoring, multiple ion recording, ion current profiling, etc. In studies in which mass spectral analysis is used for the quantitation of eicosanoids, the mass spectrometer system is programmed to record signals at masses that correspond to the mass of characteristic fragment ions of the desired analyte(s) and the internal standard(s). The results of limiting acquisition of signals to a small number of masses are the greatly enhanced sensitivity and reproducibility of the signal elicited by compounds undergoing fragmentation.

The use of 0.5- to 10.0-nanogram quantities of stable isotope analogs available from commercial sources as internal standards has facilitated the detection and quantitation of subpicogram quantities by capillary GC-MS analysis (1-7). An important feature of these studies is that one deuterated analog of an eicosanoid may serve as the internal standard for more than one eicosanoid analyte. The standard curves of analytes in relation to the internal standard are linear over a wide range of concentrations to 10^{-15} molar, and the assay methods have a high degree of precision and accuracy, with intersample and intrasample variations being less than 15% at levels approaching 10^{-15} molar. An example of the detection and quantitation of femtomolar quantities of $PGF_{2\alpha}$, PGD_2, PGE_2, TxB_2, and $6KPGF_{1\alpha}$ by capillary GC-MS analysis is shown in Figure 3. Panel A depicts the signal at m/z 524, the mass of the common characteristic fragment ion of derivatized PGE_2 and PGD_2, and at m/z 528, the characteristic fragment ion of derivatized $^2H_4$-PGE_2. Panel B depicts the signal detecting derivatized $PGF_{2\alpha}$ (m/z 569) and derivatized $^2H_4$-$PGF_{2\alpha}$ (m/z 573). Derivatized TxB_2 and $6KPGF_{1\alpha}$ are detected by monitoring ions at m/z 614 while derivatized $^2H_4$-$6KPGF_{1\alpha}$ is detected by monitoring ions at m/z 618 (panel C). The peaks representing the individual prostanoid analytes and the deuterated analogs used as internal standards are indicated. Peak I (panel A) represents the signal resulting from the coelution of the less abundant isomers of derivatized PGD_2 and PGE_2. The dashed line, labeled peak II (panel C), depicts the area for the elution and detection of the less abundant isomer of derivatized $6KPGF_{1\alpha}$ and its incomplete resolution from derivatized TxB_2. The remaining peaks depicted in Figure 3 represent either unidentified compounds isolated from the biological matrix (peaks a, b, c, d, e, f, and g) or impurities present in the deuterated analogs used as internal standards (peaks x, y, and z). The horizontal line above the abscissa in the upper portions of panels A, B, and C represents the intensity of a signal that is twice the magnitude of the electronic and background "noise" of the detector and can be used to determine the signal-to-"noise" ratio of the

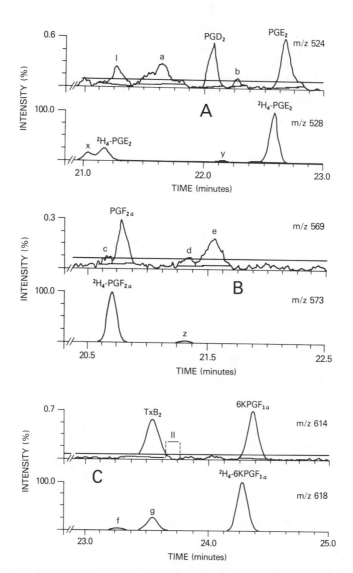

Fig. 3. Reconstructed ion current depicting detection of signal elicited by femtomolar quantities (in % of signal elicited by internal standard) of eicosanoids biosynthesized from endogenous arachidonic acid in fragments of human lung biopsy tissue (protein content 0.42 mg). Biopsy fragments were incubated in 1 ml of physiologic buffer at 37°C for 15 min. A 25-μl aliquot was placed in a silanized vial containing 0.5 ml acetone in which 0.88-1.62 ng of $^2H_4$-analogs of PGE_2, $PGF_{2\alpha}$, and 6 $KPGF_{1\alpha}$ was dissolved. After drying, sample was derivatized for vapor-phase analysis. After derivatization, sample was extracted with three 0.5-ml volumes of hexane, dried, and dissolved in 40 μl of dodecane. Refer to legends of Figures 1 and 2 for GC-MS conditions. Ions were monitored simultaneously at six different masses: m/z 524 and 528 (panel A), m/z 569 and 573 (panel B), and m/z 614 and 618 (panel C). Ordinate = reconstructed ion current depicting relative intensity of signal (%). Abscissa = time (in min). Note dissimilarities in abscissas in panels A, B, and C. Additional details in text.

Table 4. Quantitation Summary of Eicosanoid Analysis

Internal Standard[1]			Intra-Sample[2,3]		Inter-Sample[2,3]	
Name	Pg Injected	Analyte	Mean ± SD	CV(%)	Mean ± SD	CV(%)
$^2H_4$-PGE$_2$	22	PGD$_2$	94 ± 8	8.5	89 ± 9	10.1
$^2H_4$-PGE$_2$		PGE$_2$	144 ± 12	8.3	156 ± 13	8.3
$^2H_4$-PGF$_{2\alpha}$	28	PGF$_{2\alpha}$	66 ± 7	10.6	71 ± 9	12.7
$^2H_4$-6KPGF$_{1\alpha}$	41	TxB$_2$[4]	261 ± 20	7.7	250 ± 23	9.2
$^2H_4$-6KPGF$_{1\alpha}$		6KPGF$_{1\alpha}$	253 ± 22	8.7	244 ± 25	10.2

[1]Quantity of internal standard injected is based on injection of 2.5% (1/40th) of derivatized sample to which 0.88 ng $^2H_4$ -PGE$_2$, 1.12 ng $^2H_4$-PGF$_{2\alpha}$, and 1.62 ng of $^2H_4$-6KPGF$_{1\alpha}$ had been added.
[2]Mean values are expressed in femtograms (10^{-15} g).
[3]n = 4
[4]Computations include possible contribution to peak area by 6KPGF$_{1\alpha}$.

acquired signal. In these particular data, the signal-to-"noise" ratio ranges from approximately 10:1 (panel B) to approximately 20:1 (panel C).

A quantitative summary of the analysis of PGF$_{2\alpha}$, PGD$_2$, PGE$_2$, TxB$_2$, and 6KPGF$_{1\alpha}$ isolated from lung tissue (example of analysis shown in Figure 3) is presented in Table 4. These data show the quantity of each prostanoid analyte per injection, the mean and standard derivations, and the coefficients of intersample and intrasample variations. The data contained in Table 4 illustrate the high degree of sensitivity, precision, and accuracy of capillary GC-MS analysis for the detection and quantitation of prostaglandins and related eicosanoids in biological samples. The high degree of selectivity of capillary GC-MS analysis is directly attributable to the efficiency of analyte resolution achievable by capillary gas chromatography and to the limited and characteristic fragmentation of the different eicosanoids.

The detection and quantitation of eicosanoids by capillary GC-MS analysis are not limited to the specific example presented above. A number of other eicosanoids (i.e., prostaglandin metabolites, LTB$_4$, and hydroxy-eicosanoids) are amenable to vapor-phase analysis subsequent to the synthesis of electron-capture derivatives. The absence of additional specific examples of capillary GC-MS analysis for the quantitation of compounds other than those cited in references 1-7 reflects the present state of development of this type of mass spectral analysis of eicosanoids.

PURIFICATION OF BIOLOGICAL SAMPLES

The combined degree of selectivity, sensitivity, precision, and accuracy of capillary GC-MS analysis for the detection and quantitation of eicosanoids is directly related to analyte purity. Since eicosanoids are usually present in very small quantities in biological materials, the considerations of analyte purity cannot be overemphasized. The specific illustration of capillary GC-MS analysis of eicosanoids contained in this report is derived from a biological matrix which, except for the small fragment of human lung tissue (protein content 0.42 mg), was devoid of significant biological materials. The inherent sensitivity of the assay method facilitated the analysis of eicosanoids from such small quantities of biological materials. Very few specific instances can be cited (6,23) in which successful capillary GC-MS analysis of eicosanoids from biological sources has not included extraction and/or chromatographic procedures for analyte purification, but such procedures will likely be essential for most applications of GC-MS analysis of eicosanoids. Methods for the performance and use of liquid and solid-phase extraction procedures are elegantly presented in several published reports (1,2,5,14,24,25). The different liquid and solid-phase extraction and chromatographic systems are highly adaptable for specific applications and can be performed in most conventional laboratory settings.

ALTERNATIVE TECHNIQUES OF MASS SPECTRAL ANALYSIS OF EICOSANOIDS

A number of eicosanoids are not readily amenable to vapor-phase analysis. One group of compounds, certain monohydroxy-eicosatetraenoic acids and dihydroxy-eicosatetraenoic acids derived from lipoxygenase-catalyzed metabolism of arachidonic acid, may undergo extensive degradation during gas chromatographic analysis (10). The degradation of these derivatized compounds is temperature dependent and is characterized by peak broadening and/or peak tailing. Since electron-capture derivatives of eicosanoids are usually less volatile than a corresponding alkyl derivative and require higher temperatures for gas chromatographic analysis, the capillary GC-MS method for the detection and quantitation of this group of eicosanoids will require additional development. In addition to certain lipoxygenase metabolites that have a high degree of thermal lability during gas chromatographic analysis of derivatized compounds, the peptidoleukotrienes (leukotrienes C_4, D_4, and E_4) cannot be readily converted to derivatives with sufficient vapor pressure for gas chromatographic analysis (26). Two techniques have been adapted for the mass spectral analysis of thermally labile and peptide-containing eicosanoids: combined high-performance liquid chromatography(HPLC)-mass spectrometry (27,28) and fast atom bombardment (FAB) mass spectrometry (7,29,30). HPLC has been used extensively for the chromatographic resolution of eicosanoids (8,10-13,19,20,26,27). The interfacing of HPLC techniques for the separation of eicosanoids and subsequent analysis by mass spectrometry is in a very early stage of development. The principal uses of HPLC-mass spectrometry for the analysis of eicosanoids at this time are qualitative. HPLC-mass spectrometry is potentially a highly useful technique for the analysis of eicosanoids but will require additional developments for more routine use in the quantitation of biologically derived eicosanoids.

FAB is a mass spectrometric technique in which ionization and subsequent fragmentation of an analyte are independent of the vapor pressure of the sample

(31). Instead of using an electron beam for the initiation of ionization and fragmentation of compounds, surface ionization of leukotriene analytes in a glycerol matrix is achieved with the use of an intense beam of rapidly moving gaseous atoms (i.e., xenon or argon). FAB is less sensitive than GC-MS for the detection of eicosanoids and is also a technique in the very early stages of development for eicosanoid analysis.

SUMMARY AND CONCLUSIONS

Capillary GC-MS is the most sensitive method available for the detection and quantitation of eicosanoids. Femtomolar (10^{-15} molar) quantities of individual eicosanoids amenable to vapor-phase analysis can be reliably detected and quantitated. In addition to being a highly sensitive technique for the analysis of eicosanoids, capillary GC-MS is a highly selective technique with very high degrees of precision and accuracy (coefficient of variation $<15\%$). Because of the combined degree of sensitivity, selectivity, precision, and accuracy for the detection and measurement of eicosanoids, capillary GC-MS is a highly useful analytical method for studies of eicosanoid biosynthesis and metabolism.

ACKNOWLEDGMENTS

The authors express their gratitude to K. Gill and M. J. Papa for typing and organizing the text for publication.

REFERENCES

1. Waddell, K. A., Blair, I. A., and Wellby, J. Combined capillary column gas chromatography negative ion chemical ionization mass spectrometry of prostanoids. Biomed. Mass Spectrom. 10: 83-88, 1983.
2. Waddell, K. A., Robinson, C., Orchard, M. A., Barrow, S. E., Dollery, C. T., and Blair, I. A. Quantitative analysis of arachidonic acid metabolites in complex biological fluids using capillary gas chromatography/negative ion chemical ionization mass spectrometry. Int. J. Mass Spectrom. 48: 233-236, 1983.
3. Strife, R., and Murphy, R. C. Preparation of pentafluorobenzyl esters of arachidonic acid lipoxygenase metabolites. Analysis by gas chromatography and negative ion chemical ionization mass spectrometry. J. Chromatog. 305: 3-12, 1984.
4. Strife, R., and Murphy, R. C. Stable isotope labeled 5-lipoxygenase metabolites of arachidonic acid. Analysis by negative ion chemical ionization mass spectrometry. Leukotrienes, Prostaglandins Med. 13: 1-8, 1983.
5. Waddell, K. A., Barrow, S. E., Robinson, C., Orchard, M. A., Dollery, C. T., and Blair, I. A. Quantitative analysis of prostanoids in biological fluids

by chemical gas chromatography negative ion chemical ionization mass spectrometry. Biomed. Mass Spectrom. 11: 68-74, 1984.

6. Hubbard, W. C., Litterst, C. L., Liu, M. C., Bleecker, E. R., Mimnaugh, E. G., Eggleston, J. C., McLemore, T. L., and Boyd, M. R. Profiling of prostaglandin biosynthesis in biopsy fragments of human lung carcinoma and normal human lung by combined capillary gas chromatography-negative ion chemical ionization mass spectrometry. Prostaglandins, in press.

7. Murphy, R. C., and Harper, T. W. Mass spectrometry and eicosanoid analysis. In: "Biochemistry of Arachidonic Acid Metabolites." W. E. M. Lands, ed. Martinus Nijhoff, Boston, 1985, pp. 417-435.

8. Green, K., Hamberg, M., Samuelsson, B., Smigle, M., and Frolich, J. C. Measurement of prostaglandins, thromboxanes, prostacylin and their metabolites by gas chromatography-mass spectrometry. Adv. Prostaglandin Thromboxane Leukotriene Res. 5: 39-94, 1978.

9. Fischer, C., and Frolich, J. C. Analysis of prostanoids by GC/MS measurement. Adv. Lipid Res. 19: 185-202, 1982.

10. Boeynaems, J. M., and Hubbard, W. C. Preparation, purification, characterization and assay of hydroxy- and hydroperoxy-eicosatetraenoic acids. J. M. Boeynaems and A. G. Herman, eds. In: "Prostaglandin, Prostacyclin and Thromboxane Measurement." Martinus Nijhoff, Amsterdam, 1980, pp. 167-181.

11. Hubbard, W. C., Watson, J. T., and Sweetman, B. J. Prostaglandin analysis: Role in high performance liquid chromatography in sample processing. In: "Biological/Biomedical Applications of Liquid Chromatography," Vol. 10. G. L. Hawk, ed. Marcelle Dekker, New York, 1979, pp. 31-55.

12. Eling, T. E., Tainer, B., Ally, A., and Warnock, R. Separation of arachidonic acid metabolites by high-pressure liquid chromatography. In: "Methods of Enzymology," Vol. 86. W. E. M. Lands and W. L. Smith, eds. Academic Press, New York, 1982, pp. 511-517.

13. Van Rollins, M., Aveldano, M. I., Sprecher, H. W., and Horrocks, L. A. High pressure liquid chromatography of underivatized fatty acids, hydroxy acids and prostanoids having different chain lengths and double-bond positions. In: Reference 12, pp. 518-530.

14. Salmon, J. A., and Flower, R. A. Extraction and thin-layer chromatography of arachidonic acid metabolites. In: Reference 12, pp. 477-493.

15. Granstrom, E. Two-dimensional thin-layer chromatography of prostaglandins and related compounds. In: Reference 12, pp. 493-511.

16. Freeman, R. R. Instrumental consideration: In: "High Resolution Gas Chromatography." R. R. Freeman, ed. Hewlett-Packard Co., Palo Alto, CA, 1981, pp. 53-79.

17. Taber, D. F., Phillips, M. A., and Hubbard, W. C. Preparation of deuterated arachidonic acid. Prostaglandins 22: 349-352, 1981.

18. Taber, D. F., Phillips, M. A., and Rubbard, W. C. Preparation of deuterated arachidonic acid. In: Reference 12, pp. 366-369.

19. Boeynaems, J. M., Brash, A. R., Oates, J. A., and Hubbard, W. C. Preparation and assay of monohydroxy-eicosatetraenoic acids. Anal. Biochem. 104: 259-267, 1980.

20. Turk, J., Henderson, W. R., Klebanoff, S. J., and Hubbard, W. C. Iodination of arachidonic acid mediated by eosinophil peroxidase and lactoperoxidase: Identification and comparison of products. Biochim. Biophys. Acta 751: 189-200, 1983.

21. Bild, G. S., Bhat, S. G., Ramadoss, C. S., and Axelrod, B. Synthesis of 9(12)-oxy-8,11,15-trihydroxyeicosa-5, 13-dienoic acid from arachidonic acid by soybean lipoxygenase-2. Biochem. Biophys. Res. Commun. 81: 486-492, 1978.

22. Murphy, R. C., and Clay, K. L. Preparation of ^{18}O derivatives of eicosanoids for GC-MS quantitative analysis. In: Reference 12, pp. 547-551.

23. Hubbard, W. C., Mimnaugh, E. G., McLemore, T. L., and Boyd, M. R. Determination of the profile and kinetics of prostaglandin biosynthesis in microcarrier-attached human bronchioloalveolar cell lines by capillary gas chromatography-negative ion chemical ionization mass spectrometry. J. Clin. Invest., in preparation.

24. Powell, W. S. Rapid extraction of arachidonic acid metabolites from biological samples using octadecysilyl silica. In: Reference 12, pp. 467-477.

25. Ogletree, J. L., Schlesinger, K., Nettleman, M., and Hubbard, W. C. Measurement of 5-hydroxyeicosatetraenoic acid (5-HETE) in biological fluids by GC-MS. In: Reference 12, pp. 607-612.

26. Samuelsson, B., Borgeat, P., Hammarstrom, S., and Murphy, R. C. Leukotrienes: A new group of biologically active compounds. Adv. Prost. Thrombox. Res. 6: 1-18, 1980.

27. Yergey, J., and Salem, N. Thermospray HPLC-MS of prostaglandins and oxygenated metabolites of arachidonic acid (C20:4) and docosahexaenoic acid (C22:6). 32nd Annual Conference on Mass Spectrometry and Allied Topics. San Antonio, Texas, 104-105, 1984.

28. Yergey, J. A., Kim, H.-Y., and Salem, N. High performance liquid chromatography/thermospray mass spectrometry of eicosanoids and novel oxygenated metabolites of docosahexaenoic acid. Analyt. Chem. 58: 1344-1348, 1986.

29. Murphy, R. C., Mathews, W. R., Rokach, J., and Fenselau, C. Comparison of biologically-derived and synthetic leukotriene C4 by fast atom bombardment mass spectrometry. Prostaglandins 23: 201-206, 1982.

30. Taylor, G. W., Morris, H. R., Beaubien, B., and Clinton, P. M. Fast atom bombardment mass spectrometry of eicosanoids. In: "Prostaglandins, Leuko-trienes and Other Lipoxygenase Products." P. Piper, ed. John Wiley, Chichester, 1983, pp. 277-282.

31. Barber, M., Bordeli, R. S., Sedwick, R. D., and Tyler, A. N. Fast atom bombardment of solids as an ion source in mass spectrometry. Nature 293: 270-275, 1981.

LIPID PEROXIDATION AND ITS MEASUREMENT

H. Sies

Institut für Physiologische Chemie
Universität Düsseldorf
Düsseldorf, West Germany

ABSTRACT

The oxidative breakdown of polyunsaturated fatty acids in biological membranes, known as lipid peroxidation, refers to a radical-mediated reaction sequence of complex nature. The process is thought to be crucial in the expression of toxicity related to oxidative stress, as membrane integrity is challenged.

Several ways for assessing the occurrence of lipid peroxidation have been developed. Currently, major interest is in the monitoring of lipid peroxidation in intact cells and organs in order to analyze initial stages of oxidative challenge before actual cell death occurs. The introduction of several noninvasive methods has provided advances in this area.

Lipid peroxidation of ω-3 and ω-6 polyunsaturated fatty acids yields ethane and pentane as stable end products, respectively. Thus, intact animals or isolated organs and cells have been analyzed for volatile hydrocarbon evolution.

The reactions of lipid peroxy radicals according to Russell's mechanism generate singlet molecular oxygen which, in turn, may generate low-level chemiluminescence in the near-infrared. Using sensitive detection methods for these photoemissive species, studies with intact organs and cells have shown good correlation of low-level chemiluminescence with other parameters of lipid peroxidation.

Regarding invasive methods, conjugated dienes and malondialdehyde have been used in numerous studies for the detection of lipid peroxidation. Other products of interest include the assay of lipid peroxides, lipid aldehydes, lipid expoxides, and subsequent products such as fluorescent compounds that may form with cellular constituents.

INTRODUCTION

Lipid peroxidation has been much studied in recent years. The oxidative breakdown of polyunsaturated fatty acids in biological membranes, known as lipid peroxidation, is a radical-mediated reaction sequence. There are several ways in which prooxidant states in biological systems lead to an enhancement of lipid peroxidation, and conversely there is a battery of antioxidant systems. A balance in favor of prooxidant reactions is denoted as oxidative stress (see ref. 1). Lipid peroxidation is thought to be of importance in the expression of toxicity, as membrane phospholipids are essential in maintaining membrane integrity. As recently discussed by Kappus (2), lipid peroxidation has been suggested to be responsible for numerous biological effects, especially because after its initiation it concurrently proceeds by a free-radical reaction mechanism. Oxygen radicals can be responsible for lipid peroxidation. Therefore, this process has been hypothesized to explain many radical-related diseases and other states of a nonphysiological nature. An overview of methods used for detecting lipid peroxidation has been given (3,4).

This report focuses on some current areas of interest in the assessment of lipid peroxidation. Monitoring this process in intact cells and organs is of particular importance for the detection of initial stages of oxidative challenge before the actual cell death occurs. Therefore, our interest has focused on noninvasive methods.

VOLATILE HYDROCARBONS: ETHANE AND PENTANE

The exhalation of ethane by mice was first demonstrated by Riely et al. (5) in a study of carbon tetrachloride-induced lipid peroxidation in liver. Pretreatment of mice with α-tocopherol (vitamin E), a lipid antioxidant that protects against carbon tetrachloride hepatotoxicity, effectively suppressed ethane production. Similar studies were also performed regarding pentane. The method to work with intact animals *in vivo* has been described (6).

In order to eliminate contributions by extrahepatic tissues, methods using intact organs have been developed (7). As shown in Figure 1, a collection chamber around the perfused liver, for example, may be used for drawing samples for gas chromatographic analysis. Several studies using toxic agents like carbon tetrachloride, ethanol, or monoamines have been performed with this system. The rates of production of alkanes are in the order of pmol per min per gram liver wet weight. Ethane is possibly preferable to pentane, because the latter may be metabolized by cytochrome P-450-dependent hydroxylation reactions (8). Comparison of the rate of ethane production between *in vitro* (perfused liver, isolated hepatocytes, and microsomal fractions) and *in vivo* systems shows similar data for the liver *in vitro* and *in vivo,* indicating that the liver may be a major contributory organ for the ethane released from the intact animal.

As these volatile hydrocarbons are only minor end products of lipid peroxidation, changes in the pathway of lipid breakdown may also influence the amount of ethane that is released. This should be considered when analytical data are interpreted, so that caution is required in making statements on the overall process of lipid peroxidation.

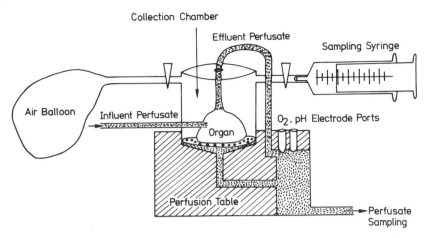

Fig. 1. System used for sampling of alkanes produced by isolated perfused rat liver (nonrecirculating perfusion). (From ref. 7)

LOW-LEVEL CHEMILUMINESCENCE

The generation of singlet molecular oxygen during lipid peroxidation has been attributed to the breakdown of lipid hydroperoxides (see ref. 9). The detection of light emission from biological samples has been shown to be a useful method for studying oxidative reactions in intact systems (10,11). The generation of electronically excited states during intracellular oxidative conditions can result from free-radical interactions that may not be associated with the peroxidation of membrane fatty acids. Alternatively, the direct generation of excited states can occur in enzyme-catalyzed reactions (12) as demonstrated in model systems using peroxidases. The excited states discussed are (a) singlet molecular oxygen as measured by photoemission during its decay to the triplet ground state in the monomol (i) or dimol (ii) reactions:

$$^{\wedge}O_2 \rightarrow \, ^3O_2 + h\nu \; (1{,}270 \text{ nm}) \tag{i}$$

$$2\,^{\wedge}O_2 \rightarrow 2 \, ^3O_2 + h\nu \; (634 \text{ and } 703 \text{ nm}) \tag{ii}$$

and (b) excited triplet carbonyls (RO*) exhibiting a weak emission in the blue-green region of the spectrum (reaction iii) or an indirect emission after energy transfer to a suitable acceptor A (reaction iv), thus eliciting sensitized photoemission (reaction v).

$$RO^* \rightarrow RO + h\nu \tag{iii}$$

$$RO^* + A \rightarrow RO + A^* \tag{iv}$$

$$A^* \rightarrow A + h\nu \qquad \text{(v)}$$

$$ROO^{\cdot} + ROO^{\cdot} \rightarrow ROH + RO + {}^{\wedge}O_2 \qquad \text{(vi)}$$

$$ROO^{\cdot} + ROO^{\cdot} \rightarrow ROH + RO^* + {}^3O_2 \qquad \text{(vii)}$$

Interactions of lipid peroxy radicals can be a source of excited states as in reactions vi and vii, as described by Russell (13) and Howard and Ingold (14). Because peroxy radicals are produced in the final stages of lipid peroxidation, they might be considered as the common mechanism for generation of chemiluminescence. They are shared by multiple types of oxidative conditions that promote lipid peroxidation, including carbon tetrachloride poisoning, iron overload, oxidative breakdown of hydroperoxides, and hyperoxia. However, it should be noted that free-radical interactions supporting redox cycling elicit chemiluminescence that is not associated with lipid peroxidation. In general, the relationship between low-level chemiluminescence and lipid peroxidation can be summarized as follows:

Chemiluminescence of systems undergoing lipid peroxidation is accompanied by an accumulation of end products such as malondialdehyde, as was observed in liver and brain homogenates (15,16), isolated hepatocytes (17), microsomal fractions, and model systems (see ref. 9 for further references). Although the amount of accumulated malondialdehyde usually correlates well with the chemiluminescence intensity observed, it should be noted that malondialdehyde and light-emitting species are formed by different pathways and at different times during the process of lipid peroxidation (18,19) (see scheme in Figure 2).

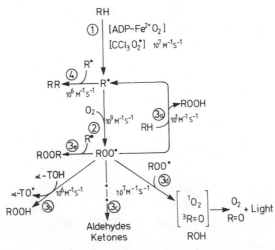

Fig. 2. Relationship between process of lipid peroxidation and generation of some products, including photoemissive species. Rate constants are given as approximate numbers, collected from literature. (From ref. 19)

INVASIVE METHODS FOR DETECTION OF LIPID PEROXIDATION

As the process of lipid peroxidation is quite ramified, there are a number of different pathways and products. Many of these have been exploited for the detection of lipid peroxidation. A most recent summary has been given by Aust (20). The most used assay seems to be that for malondialdehyde (21,22).

Lipid hydroperoxides have been used in many studies as well. Some recent methodology has been described previously (23-26). The lipid hydroperoxides in plasma are present at a concentration of 0.5 mM (25,26). In a recent study on human plasma, free fatty acid hydroperoxides and cholesterol ester hydroperoxide were present at concentrations of 0.06 μM and 0.3 μM, respectively (26). Further products of interest are the conjugated dienes (27) and, importantly, also the fluorescent products generated during lipid peroxidation (28,29).

CONCLUSIONS

It is increasingly clear that lipid peroxidation is important for the expression of toxicity and that it generates further signals to spread pathological processes to other sites within cells, organs, and the whole organism. So the future lines of research will depend on more refined techniques to evaluate lipid peroxidation. For example, the production of a number of compounds of aldehydic nature (4-hydroxynonenal being a prominent one) will have many pathophysiological consequences (see ref. 2). These toxic aldehydes have been found to be diffusible products of lipid peroxidation (30,31). Taking only one example of pathophysiological events, involvement of lipid peroxidation in the development of atherosclerotic lesions has been suggested. It was found that modified forms of human low-density lipoprotein (LDL) cause an accumulation of large amounts of cholesterol esters in macrophages, e.g., after treatment of LDL with malonaldehyde (32). 4-Hydroxynonenal also is capable of modifying LDL, and the concentrations required are 100-1,000-fold less than with malonaldehyde (33). Therefore, the aldehydes generated in lipid peroxidation must be kept low in their concentrations. One way of disposition is the formation of thioethers with glutathione, as catalyzed by glutathione transferase 4-4 (34,35) and, more recently shown, isozyme 8-8 (36). There is a subsequent need for transport of these thioethers out of the cells. In heart, the carrier system has been studied recently in our group (35-37), and there are similar carriers in other organs. Thus, there are various further defense systems related to the process lipid peroxidation.

ACKNOWLEDGMENTS

Supported by Deutsche Forschungsgemeinschaft, Bonn, and by National Foundation for Cancer Research, Washington.

REFERENCES

1. Sies, H., ed. "Oxidative Stress." Academic Press, London, 1985.
2. Kappus, H. Lipid peroxidation: Mechanisms, analysis, enzymology and biological relevance. In: Ref. 1, pp. 273-310.
3. Packer, L., ed. "Oxygen Radicals in Biological Systems. Methods in Enzymology," Vol. 105. Academic Press, Orlando, 1984.
4. Slater, T. F. Overview of methods used for detecting lipid peroxidation. In: Ref. 3, pp. 283-293.
5. Riely, C. A., Cohen, G., and Lieberman, M. Ethane evolution: A new index of lipid peroxidation. Science 183: 208-210, 1974.
6. Lawrence, G. D., and Cohen, G. Concentrating ethane from breath to monitor lipid peroxidation *in vivo*. In: Ref. 3, pp. 305-311.
7. Müller, A., and Sies, H. Assay of ethane and pentane from isolated organs and cells. In: Ref. 3, pp. 311-319.
8. Frank, H., Hintze, T., Bimboes, D., and Remmer, H. Monitoring lipid peroxydation by breath analysis: Endogenous hydrocarbons and their metabolic elimination. Toxicol. Appl. Pharmacol. 56: 337-344, 1980.
9. Cadenas, E., and Sies, H. Low-level chemiluminescence as an indicator of singlet molecular oxygen in biological systems. In: Ref. 3, pp. 221-231.
10. Boveris, A., Cadenas, E., Reiter, R., Filipkowski, M., Nakase, Y., and Chance, B. Organ chemiluminescence: Noninvasive assay for oxidative radical reactions. Proc. Natl. Acad. Sci. USA 77: 347-351, 1979.
11. Sies, H., and Cadenas, E. Oxidative stress: Damage to intact cells and organs. Philos. Trans. R. Soc. Lond. (Biol.) 311: 617-631, 1985.
12. Cilento, G. Electronic excitation in dark biological processes. In: "Chemical and Biological Generation of Excited States." W. Adam and G. Cilento, eds. Academic Press, New York, 1982, pp. 277-307.
13. Russell, G. A. Deuterium-isotope effects in the autoxidation of aralkyl hydrocarbons. Mechanism of interaction of peroxy radicals. J. Am. Chem. Soc. 79: 3871-3877, 1957.
14. Howard, J. A., and Ingold, K. U. Rate constants for self-reactions of n- and t-butyl peroxy radicals and cyclohexylperoxy radicals. The deuterium isotope effects in the termination of secondary peroxy radicals. J. Am. Chem. Soc. 90: 1058-1059, 1968.
15. De Luzio, N. R., and Stege, T. E. Enhanced hepatic chemiluminescence following carbon tetrachloride or hydrazine administration. Life Sci. 21: 1457-1464, 1977.
16. Cadenas, E., Varsavsky, A. I., Boveris, A., and Chance, B. Oxygen- or organic hydroperoxide-induced chemiluminescence of brain and liver homogenates. Biochem. J. 198: 645-654, 1981.
17. Cadenas, E., Wefers, H., and Sies, H. Low level chemiluminescence of isolated hepatocytes. Eur. J. Biochem. 119: 531-536, 1981.
18. Sugioka, K., and Nakano, M. A possible mechanism of the generation of singlet oxygen in NADPH-dependent microsomal lipoperoxidation. Biochim. Biophys. Acta 423: 203-216, 1976.

19. Noll, T., de Groot, H., and Sies, H. Distinct temporal relation between oxygen uptake, malondialdehyde formation and low-level chemiluminescence during microsomal lipid peroxidation. Arch. Biochem. Biophys., in press.

20. Aust, S. D. Lipid peroxidation. In: "Handbook of Methods for Oxygen Radical Research." R. A. Greenwald, ed. CRC Press, Boca Raton, 1985, pp. 203-207.

21. Bird, R. P., and Draper, H. H. Comparative studies on different methods of malonaldehyde determination. In: Ref. 3, pp. 299-305.

22. Esterbauer, H., Lang, J., Zadravec, S., and Slater, T. F. Detection of malondialdehyde by high-performance liquid chromatography. In: Ref. 3, pp. 319-328.

23. Pryor, W. A., and Castle, L. Chemical methods for the detection of lipid hydroperoxides. In: Ref. 3, pp. 293-299.

24. Cathcart, R., Schwiers, E., and Ames, B. N. Detection of picomole levels of lipid hydroperoxides using a dichlorofluorescin fluorescent assay. In: Ref. 3, pp. 352-358.

25. Marshall, P. J., Warso, M. A., and Lands, W. E. M. Selective micro-determination of lipid hydroperoxides. Anal. Biochem. 145: 192-199, 1985.

26. Yamamoto, Y., Brodsky, M. H., Baker, J. C., and Ames, B. N. Detection and characterization of lipid hydroperoxides at picomole levels by high performance liquid chromatography. Anal. Biochem., in press.

27. Recknagel, R. O., and Glende, E. A., Jr. Spectrophotometric detection of lipid conjugated dienes. In: Ref. 3, pp. 331-337.

28. Dillard, C. J., and Tappel, A. L. Fluorescent damage products of lipid peroxidation. In: Ref. 3, pp. 337-341.

29. Stark, W. S., Miller, G. V., and Itoku, K. A. Calibration of microphotometers as it applies to the detection of lipofuscin and the blue- and yellow-emitting fluorophores in situ. In: Ref. 3, pp. 341-347.

30. Comporti, M. Biology of disease. Lipid peroxidation and cellular damage in toxic liver injury. Lab. Invest. 53: 599-623, 1985.

31. Benedetti, A., Casini, A. F., Ferrali, M., and Comporti, M. Effects of diffusible products of peroxidation of rat liver microsomal lipids. Biochem. J. 180: 303-312, 1979.

32. Fogelman, M. A., Shechter, I., Seager, J., Hokom, M., Child, J. S., and Edwards, P. A. Malondialdehyde alteration of low density lipoproteins leads to cholesteryl ester accumulation in human monocyte-macrophages. Proc. Natl. Acad. Sci. USA 77: 2214-2218, 1980.

33. Jürgens, G., Lang, J., and Esterbauer, H. Modification of human low-density lipoprotein by the lipid peroxidation product 4-hydroxynonenal. Biochim. Biophys. Acta 875: 103-114, 1986.

34. Alin, P., Danielson, U. H., and Mannervik, B. 4-Hydroxyalk-2-enals are substrates for glutathione transferase. FEBS Lett. 179: 267-270, 1985.

35. Ishikawa, T., Esterbauer, H., and Sies, H. Role of cardiac glutathione transferase and of the glutathione S-conjugate export system in biotransformation of 4-hydroxynonenal in the heart. J. Biol. Chem. 261: 1576-1581, 1986.

36. Jenssen, H., Guthenberg, C., Alin, P., and Mannervik, B. Rat glutathione transferase 8-8, an enzyme efficiently detoxifying 4-hydroxyalk-2-enals. FEBS Lett. 203: 207-209, 1986.
37. Ishikawa, T., Zimmer, M., and Sies, H. Energy-linked cardiac transport system for glutathione disulfide. FEBS Lett. 200: 128-132, 1986.

AN *IN SITU* METHOD FOR DETECTION OF LIPID PEROXIDATION EFFECTS

T. Alvager[1] and L. Davis[2]

[1]Department of Physics
Indiana State University
Terre Haute, Indiana 47809

[2]Center for Laser Applications
University of Tennessee Space Center
Tullahoma, Tennessee 37388

ABSTRACT

An *in situ* method for observation of effects due to lipid peroxidation in biological samples has been developed by detecting the fluorescence of compounds formed in the process, such as aminoiminopropene (AIP) (excitation wavelength range 280-420 nm, emission wavelength range 380-500 nm), and by detecting the quenching of protein fluorescence. Measurements have been performed with a fiber optic light guide system coupled to either a readily available conventional fluorometer for routine sample inspections or a high-repetition-rate pulse laser system (1-psec pulses) and time resolved detection for superior sensitivity.

INTRODUCTION

Lipid peroxidation refers to the oxidative deterioration of unsaturated fatty acids in the presence of free radicals (see e.g. ref. 1). Biological material, especially membranes, contains a wide variety of substances that have polyunsaturated fatty acid side-chains that are susceptible to lipid peroxidation. These produce peroxide radicals that decompose to compounds such as malonaldehyde, a cross-linking agent whose action may result in nonfunctioning cell components and injury to the cell. Consequently there have been various attempts to understand cellular injury such as those occurring in aging and radiation damage, in terms of lipid peroxidation (see e.g. ref. 1-3).

To evaluate the various models that have been proposed regarding the effect of lipid peroxidation in cell injury, suitable detection methods are of value. In this work an *in situ* fluorescence technique is discussed that allows the detection of some of the effects of lipid peroxidation *in vivo,* thus avoiding artifacts in sample preparation. This is achieved by the use of a fiber optic light guide system.

METHOD AND INSTRUMENTATION

Fluorescence does not directly register the primary steps in lipid peroxidation but detects effects due to certain compounds formed as by-products, such as malonaldehyde. This particular molecule has the ability to cross-link other compounds, especially proteins, and form aminoiminopropene (AIP), a fluorophore with excitation and emission peak wavelengths typically in the range of 280-420 nm and 380-500 nm, respectively (see e.g. ref. 4). The mechanism involved in the formation of the fluorophore has been studied in model systems in which proteins and other cell components were incubated with malonaldehyde (4). Two main fluorescence changes were detected. Protein fluorescence decreased with half-time of approximately 2 hr. A newly formed fluorophore interpreted to be AIP reached a maximum with a much slower rate; the half-time was approximately 24 hr for pure protein and 48 hr for mitochondrial ghosts. These experiments supplemented with fluorescence lifetime measurements support the hypothesis that a two-step reaction takes place. First, malonaldehyde reacts with protein, quenching its fluorescence, and then fluorescent interprotein AIP cross-links are formed.

These general trends seen in the model system should also occur *in vivo* and could thus be used as a guide in detecting lipid peroxidation processes *in vivo.* A detection system should therefore include facilities to make it possible to (a) observe the effect *in vivo,* (b) observe the fluorescence of protein as well as that of AIP, and (c) reach high sensitivity to allow detection of low fluorescence intensity, including changes of intensity over a longer period of time.

In the work reported here, measurements were made *in vivo* by using a fiber optic light guide system to transport the excitation light and the detected fluorescence light from and to the actual site of interest. The fiber optic system was a Y-shaped Schott ultraviolet transmission guide consisting of many microfibers with approximate equal distribution of fibers in each arm of the Y-guide and shaft diameters equal to 0.3 cm. Two different light sources and detector systems were used: a conventional fluorometer and a pulsed laser system.

The conventional fluorometer consisted of a Farrand Mark 1 spectrophotofluorometer modified to couple to the fiber optic light guide system. This arrangement allows a rapid and inexpensive scan of the target but has low sensitivity (Figure 1).

The pulsed laser system consisted of a cavity dumped synchronously pumped dye laser (Quantronix 416 Nd:Yag laser in combination with a Coherent 702-1 dye laser with 7220 cavity dumper), which was frequency doubled with a 3-cm KDP crystal to produce pulses at 280-320 nm of about 1-psec duration and 7.6-MHz repetition rate. The UV-beam power into the fiber optic guide was

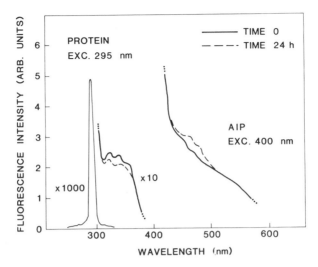

Fig. 1. *In situ* fluorescence emission spectra (conventional fluorometer) of gamma-irradiated skin tissue

approximately 10 μW. A scanning spectrometer with fast response photon-counting photomultiplier (XP2020Q) was used for fluorescence detection. The photomultiplier provided the start pulses for fluorescence decay measurements by time correlated single-photon counting, and fluorescence emission spectra were measured by recording the photon counts occurring within a specified time window as the spectrometer was scanned (Figure 2).

RESULTS AND DISCUSSION

Figure 1 shows an example of *in situ* measurements to study radiation effects in skin tissue of male Wistar rats recorded with the conventional fluorometer (T. Alvager; unpublished results). To make it possible to measure skin fluorescence from the same position over an extended period of time, a light guide holder was fixed in place in the parietal bone. This arrangement allowed observation of skin fluorescence from this area of up to several weeks. Details of this technique have been reported elsewhere (5). For the particular measurement illustrated in Figure 1, rats in the age range 3-6 months were exposed to 0.1 Gy of gamma rays from a cobalt-60 source. Measurements at excitation wavelengths 295 nm (protein) and 400 nm (AIP) were performed prior to irradiation (time 0) and 24 hours after irradiation (time 24). Two changes in the fluorescence spectra were observed: the protein fluorescence decreased slightly while the intensity at 465 nm increased in accordance with model system results (see section 2). A calibration of the protein spectrum with a serum albumin (SA) sample indicates that the skin protein fluorescence is equivalent to that of 0.5 mg/ml SA fluorescence while the AIP signal is equivalent to the effect of 2 μM of malonaldehyde on protein (1 mg/ml) observed in the model system (4).

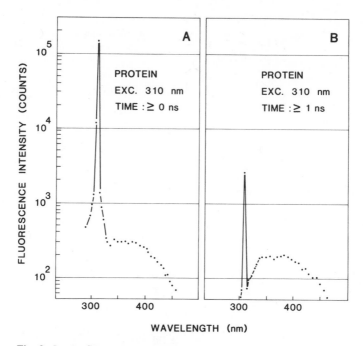

Fig. 2. *In situ* fluorescence emission spectra (laser system) of SA protein sample. A: No time analysis. B: Time analysis with 1 nsec time window to reject prompt (scattered light) signals.

The spectra in Figure 1 indicate a problem with the use of the conventional fluorometer: most applications of the light guide give large light-scattering signals that limit the overall sensitivity of the method. As illustrated in the figure, the tail of the scattering peak tends to mask the weak fluorescence signal. This signal can be enhanced over the scattering by applying time resolved fluorescence spectroscopy as demonstrated in Figure 2.

Figure 2A gives a fluorescence spectrum of a crystalline protein sample (concentration equivalent to 0.5 mg/ml SA sample) obtained with the laser system but without time analyses in operation. Figure 2B shows a similar scan of fluorescence counts versus wavelength with photon counts that arrive within the first nsec of sample excitation being rejected. The time resolution of the detector system was approximately 0.5 nsec. A small residual part of the scattering peak is still present, but by selecting a larger rejection time window, this peak could be diminished considerably. As a measure of enhancement of fluorescence over scattering, the ratio of the respective peaks with and without time analyses can be used. For the case illustrated in Figure 2, the ratio of the intensities of the scattering peaks shown in Figure 2A and 2B is approximately a factor of 100 while the corresponding fluorescence peaks are essentially equal. Thus the enhancement of the fluorescence signal over the scattering signal is in this case approximately a factor of 100.

The above results indicate that *in situ* observations of radiation effects due to lipid peroxidation reactions are feasible with fiber optic light guides. Especially promising is laser spectroscopy with its ability to enhance the fluorescence signal over the scattering signal. The relatively complex instrumentation needed for this may restrict its use to specialized studies of low radiation effects.

ACKNOWLEDGMENT

The financial support of the Center for Laser Applications, University of Tennessee Space Institute, is acknowledged.

REFERENCES

1. Slater, T. F. "Free Radical Mechanisms in Tissue Injury." Pion Limited, London, 1972.
2. Donato, H. Jr. Lipid peroxidation, cross-linking reactions, and aging. In: "Age Pigments." R. S. Sohal, ed. Elsevier/North-Holland Biomedical Press, Amsterdam, 1981, pp. 63-82.
3. Tappel, A. L. Lipid peroxidation damage to cell components. Federation Proc. 32: 1870-1874, 1973.
4. Balcavage, W., and Alvager, T. Reaction of malonaldehyde with mitochondrial membranes. Mech. of Aging and Development 19: 159-165, 1982.
5. Alvager, T., Feuquay, J., and Hamlin, G. Radiation effects studied with *in situ* fluorescence spectroscopy technique. Proc. Ind. Acad. Sci 93: 125-131, 1985.

EFFECTS OF ULTRAVIOLET LIGHT IRRADIATION, CARBON TETRACHLORIDE, AND TRICHLOROETHYLENE ON ETHANE EXHALATION IN RATS

C. D. Eskelson[1], M. Chvapil[1], J. Szebeni[1], and I. G. Sipes[2]

Departments of [1]Surgery and [2]Pharmacology/Toxicology
University of Arizona
College of Medicine
Tucson, Arizona 85724

ABSTRACT

Rats pretreated with cod liver oil (CLO) were exposed either to ultraviolet light (UVL) irradiation or to carbon tetrachloride (CCl_4) and/or trichloroethylene (TCE) treatments, and the amount of exhaled ethane was measured as an index of *in vivo* lipid peroxidation. The method of ethane collection had been modified relative to existing techniques. UVL irradiation increased ethane production with a lag period of 18-24 hr. In contrast to UVL, CCl_4 and TCE increased the exhalation of ethane without latency, irrespective of CLO pretreatment. The effects of CCl_4 and TCE showed no synergism or additivity. Our data are consistent with the established lipid peroxidation-increasing effect of chlorinated hydrocarbons, and more importantly, they demonstrate for the first time an increase in ethane production due to UVL. The considerable lag period between the irradiation and the ethane production argues against a direct lipid peroxidation-increasing effect of UVL in the skin. It is suggested that the changes are related to leukocyte infiltration of the inflamed skin surface, and that the increased lipid peroxidation results from free radical liberation during phagocytosis. A potential role of slowly released arachidonic acid from the irradiated cells is also considered.

INTRODUCTION

The measurement of exhaled ethane is a widely used method to estimate lipid peroxidation *in vivo* (1-10). In numerous studies, the hepatotoxic agent CCl_4 has been used to initiate the events resulting in lipid peroxidation (1-4). This effect of CCl_4 has been associated with the formation of trichloromethyl radical $CCl_3\cdot$, which (after combining with oxygen, forming peroxy radicals) initiates lipid peroxidation (11-15). Ethane is a decomposition product of the omega-3 unsaturated fatty acids, particularly linolenic acid (5,12,13). In the present study we addressed

393

the question of whether a different free radical-producing mechanism, i.e., UVL radiation, is reflected in ethane production. It is known that UVL irradiation triggers lipid peroxidation in various model systems (16-18) or cellular systems (19), and that this effect is in part mediated by oxygen radicals produced in the water phase (18). Regarding the complex and often harmful effects of sunlight on the skin, including cancer (20,21), an analysis of UVL-elicited lipid peroxidation in the skin seemed to be important from the medical point of view.

METHODS

Animal Manipulations

Male Sprague-Dawley rats weighing 200-300 g were divided into four groups. The first group was given orally 2 ml of CLO for 3 consecutive days in addition to the normal diet consisting of standard Purina chalks and water *ad libitum*. After the final dose of CLO, the backs of the rats were closely shaved and cleansed with ethanol under Innovar anesthesia. The animals then were placed into wire mesh cages with adjustable wire mesh floors, and irradiated by UVL (254-320 nm) at an intensity of approximately 280 mW/cm² from a distance of approximately 1 cm. The UV lamp was obtained from Fisher Scientific Co. (Pittsburgh, PA). After irradiation for 2 hr (or in one case for 1 hr), the animals were replaced in their original cages and supplied with the normal diet. The above treatment resulted in an erythema of the irradiated skin surface, which existed for several days postirradiation. The next day (18 to 24 hr after irradiation), the amount of exhaled ethane was determined as described below.

The second and third groups of rats were given an i.p. dose of either 0.1 ml per kg CCl₄ or 1 ml per kg TCE, mixed in corn oil. The fourth group was administered both CCl₄ and TCE i.p., in the above vehiculum and doses. The control animals were given only corn oil. Ethane exhalation was measured within 30 min after the administration of these agents.

Ethane Collection and Analysis

To collect expired ethane, the rats were placed in chambers made from the base of a 2-liter desiccator (Fig. 1). The desiccator was sealed with a glass plate that contained a small hole drilled in the center. The hole was plugged with a two-hole rubber stopper connected to glass tubes. These tubes served as ports for the entering and exiting air. The plate was secured airtight with silicone grease and held in place with two C clamps. The air flowed through the chamber at 200 ml/min.

A coconut charcoal trap held in a dry ice-containing Dewar flask was inserted in the pathway of exiting air to adsorb exhaled ethane. A similar trap was inserted in the pathway of the entering air to remove ambient ethane. The charcoal was

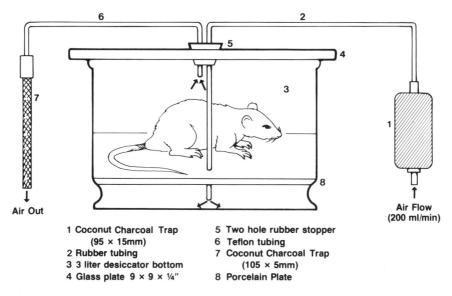

Fig. 1. Scheme of experimental setup to measure exhaled ethane by rats. Collection chamber is sealed by silicon grease at interface of desiccator bottom and 9″ x 9″ x 1/4″ glass plate. Glass plate is held in place using a wood 9″ x 9″ collar that fits around lip of upper part of desiccator. This collar is clamped to glass plate with C clamps. Other details are in Methods.

purified before use by heating to 230°C under vacuum. During this procedure, a dry-ice trap was used to prevent any vapors arising from the oil-vacuum pump from adsorbing onto the charcoal.

After flushing the system for 30 min with ethane-free air, the exhaled ethane was collected for four consecutive 30-min periods, using new traps in each cycle. The charcoal from the traps was emptied into screw-top test tubes (13.2 ml internal volume) fitted with Teflon-lined silicone septa. The tubes were then placed in a heating block, maintained at 230°-240°C, for 5-10 min to desorb the trapped ethane. The desorption was facilitated by occasional gentle shaking of the tubes.

Ethane analysis was carried out in a Hewlett-Packard gas chromatograph (Palo Alto, CA) equipped with a flame ionization detector. High-purity ethane calibration gas (99.99%) was supplied by Curtin Matheson Scientific, Inc. (Houston, TX). Five-ml air samples were taken from the headspace of the sealed vials and applied onto a 2-m long carbosive G column maintained at 180°C. The flow rate of helium carrier was 50-60 ml/min. The above method of ethane collection and analysis is a modification of the technique originally described by Lawrence and Cohen (7).

RESULTS

Methodical Observations

Figures 2 and 3 are typical GC patterns obtained under the described experimental conditions. The peak eluting at 200 sec corresponds to ethane. Two of the earlier peaks eluting between 60 and 105 sec have been identified as methane and ethylene; the rest of the peaks have not been identified. The early peaks are probably due to water reacting with the column.

When known amounts of ethane were injected into the metabolic chamber in the absence of rats, $100 \pm 10\%$ of ethane was recovered during the 30-min collection time. If a rat was also placed into the chamber, the recovery decreased to $85 \pm 10\%$; i.e., 15% of external ethane was adsorbed/metabolized by the animal.

Effect of UVL Radiation on Ethane Production

Figure 4 shows the time course of ethane exhalation by rats 18-24 hr after irradiation with UVL for 2 hr. Ethane exhalation was significantly increased over

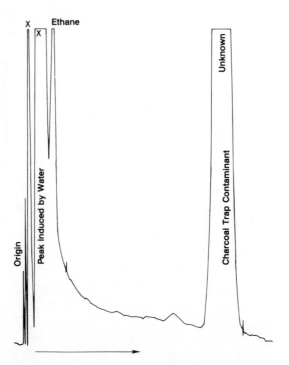

Fig. 2. Typical gas chromatogram showing elution of ethane. X represents varying mixtures of water, methane, ethylene, and column contaminants. Arrow indicates chart flow. Other details are in Methods.

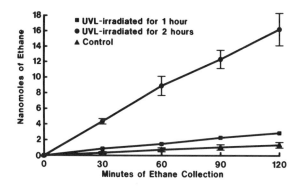

Fig. 3. Exhalation of ethane by rats irradiated with UVL. Ethane exhalation was measured 18-24 hr after irradiation. Similarly treated but nonirradiated animals served as controls. Amounts of exhaled ethane during four 30-min cycles have been cumulated. Details of animal treatments and ethane collection are described in Methods.

the control (similarly treated and CLO-primed but nonirradiated animals) by UVL irradiation. Measurement of ethane exhalation either in rats irradiated within 4-8 hr or in those irradiated 18-24 hr before but not pretreated with CLO showed no increase in ethane production (data not shown). One CLO-primed animal irradiated for 1 hr at 24 hr before the measurement did not display increased ethane production (Fig. 4). The dose-response relationship was not investigated in further detail.

Effects of Chlorinated Hydrocarbons on Ethane Production

Figure 4 shows that both CCl_4 and TCE, alone or in combination, increased the amount of exhaled ethane by the rats, compared to untreated controls. This effect was seen within 30 min of the adminisration of these solvents. The effect of TCE was significantly stronger than that of CCl_4, although the dose was also higher. The administration of TCE together with CCl_4 did not elicit a further increase in ethane exhalation relative to that caused by CCl_4. The effects of chlorinated hydrocarbons did not require CLO pretreatment of rats.

DISCUSSION

The measurement of exhaled ethane, a noninvasive method to estimate *in vivo* lipid peroxidation, has been widely used (1-10). In the present work and in a previous study from our laboratory (10) concerned with ethanol-induced lipid peroxidation, methodical modifications relating to ethane collection have been introduced. The major change relative to the earlier techniques (1-9) consists of flushing the exhaled ethane out of the collection chamber; trapping it on activated charcoal; and eliminating the CO_2, NH_3, and H_2O traps. A common feature of the previous methods is that the equilibrium concentration of ethane in a closed system was

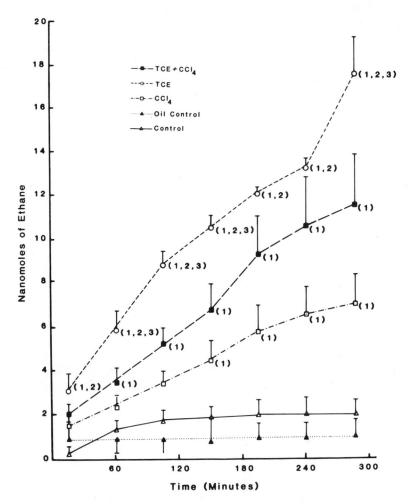

Fig. 4. Effect of haloalkanes on ethane exhalation by rats. CCl_4, carbon tetrachloride; TCE, trichloroethylene. 1, significantly different ($p < 0.05$) from control; 2, significantly different ($p < 0.05$) from CCl_4 group; 3, significantly different ($p < 0.05$) from TCE + CCl_4 group. Amounts of exhaled ethane during four 30-min cycles have been cumulated. Experimental details are in Methods.

measured using various sampling techniques. The chamber was either filled with pure oxygen [as, for example, in the study of Koster et al. (3)] or more frequently, air was circulated through the chamber in a closed system (4,8,9). In the latter cases, the consumed oxygen had to be replaced and the exhaled CO_2, H_2O, and NH_3 removed from the circulating air.

Consistent with earlier reports (4,9), we observed a slight decrease of preexisting ethane concentration in the chamber when animals were placed in it. This phenomenon originates from a partition of ethane into the tissues (predominantly the fat) and its metabolism by the hepatic cytochrome P-450 system (4,9). These processes critically depend on individual characteristics of the animal (i.e., the body volume, relative amount of adipose tissue, rate of breathing) as well as the ambient concentration of ethane. The open system applied in our experiments, by preventing the accumulation of ethane in the system, eliminates the problem of varying passive and active elimination of ethane by the animals. At the same time, the instrumentation is simpler, since there is no need for a pump to circulate the air or extra traps in addition to the charcoal.

Pure oxygen in the collection chamber was reported (3) not to have significant effect on ethane exhalation in rats. Another report (8) states that oxygen tension can affect the yield of hydrocarbon gas. We found that the exposure of rats to pure oxygen significantly increased the exhalation of ethane by the rats (Habib et al., paper in preparation). This problem is also eliminated by our method of ethane collection.

Our findings reinforced previous reports on the increased exhalation of ethane after CCl_4 administration (1,3,11). Moreover, we demonstrated that TCE, another hepatotoxic chlorinated hydrocarbon, also elicited ethane exhalation by rats. Instead of synergism or additivity between CCl_4 and TCE, we detected a reduction of the TCE-elicited ethane production when CCl_4 was also given to the animals. Since TCE and CCl_4 are probably metabolized by the same enzyme system (i.e., microsomal NADPH-cytochrome P-450 mixed function oxidase) (13-15), this phenomenon could arise from the inhibition of TCE metabolism by CCl_4. CCl_4 is known to reduce the hepatic concentration of cytochrome P-450.

The most important new finding in our study is the delayed stimulatory action of UVL irradiation on ethane production. The fact that the UVL effect was restricted to animals fed with CLO (a rich source of polyunsaturated fatty acids) proves that the ethane originates from peroxidizing lipids. Increased lipid peroxidation due to UVL irradiation has been described earlier (16-19) in several noncellular (membrane) and cellular systems. To our best knowledge, ethane evolution following UVL irradiation of the whole animal has not yet been reported.

Pathak and Stratton (20) provided direct evidence of UVL-elicited production of protein free radicals and melanin free radicals in the human skin. This proves that the preconditions for lipid peroxidation are present in the skin upon UVL irradiation. In all the cited studies (16-19), however, the increase of lipid peroxidation, as well as the free radical liberation in the human skin (20), immediately followed UVL irradiation. Since we detected increased ethane production with a latency of 18-24 hr, the direct lipid peroxidation-increasing effect of UVL is unlikely to be the main cause of ethane evolution in this system. It is known that UV irradiation of the skin precipitates an inflammation process (20,21) that involves leukocyte infiltration (20). During the process of phagocytosis, leukocytes secrete oxygen radicals, which in turn trigger lipid peroxidation with subsequent ethane evolution (6). Consequently, a possible mechanism of our observation is lipid peroxidation of phagocytic origin, associated with a protracted leukocyte infiltration of the irradiated (and inflamed) skin area. This process may be also linked to a slow release of arachidonic acid from the irradiated cells (21). Further investigations are needed to test these proposals.

ACKNOWLEDGMENT

This work was sponsored by EPA-CR812557-01 and NIEHS-NOI-ES-3-5031.

REFERENCES

1. Riely, C. A., Cohen, G., and Lieberman, M. Ethane evolution: A new index of lipid peroxidation. Science 183: 208-210, 1974.
2. Hafeman, D. G., and Hoekstra, W. G. Protection against carbon tetrachloride-induced lipid peroxidation in the rat by dietary vitamin E, selenium and methionine, as measured by ethane evolution. J. Nutr. 107: 656-665, 1977.
3. Koster, U., Albrecht, D., and Kappus, H. Evidence for carbon tetrachloride and ethanol-induced lipid peroxidation *in vivo* demonstrated by ethane production in mice and rats. Toxicol. Appl. Pharmacol. 41: 639-648, 1977.
4. Remmer, H., Hintze, T., Toranzo, E. G. D., and Frank, H. The exhaled alkanes indicative for lipid peroxidation *in vivo* quantitatively measured after CCl_4 and ethanol administration. In: 'Oxygen Radicals in Chemistry and Biology." Walter de Gruyter & Co., Berlin and New York, 1984, pp. 335-339.
5. Dumelin, E. E., and Tapel, A. L. Hydrocarbon gases produced during *in vitro* peroxidation of polyunsaturated fatty acids and decomposition of preformed hydroperoxides. Lipids 12: 894-900, 1977.
6. Ruiter, N. D., Muliawan, H., and Kappus, H. Ethane production of mouse peritoneal macrophages as indication of lipid peroxidation and the effects of heavy metals. Toxicology 17: 265-268, 1980.
7. Lawrence, G. D., and Cohen, G. Ethane exhalation as an index of *in vivo* lipid peroxidation. Anal. Biochem. 122: 283-290, 1982.
8. Lawrence, G. D., and Cohen, G. Concentrating ethane from breath to monitor lipid peroxidation *in vivo*. Methods Enzymol. 105: 305-319, 1984.
9. Wade, C. R., and Van Rij, A. M. *In vivo* lipid peroxidation in man as measured by the respiratory excretion of ethane, pentane, and other low-molecular-weight hydrocarbons. Anal. Biochem. 150: 1-7, 1985.
10. Szebeni, J., Eskelson, C. D., Mufti, S. I., Watson, R. R., and Sipes, I. G. Inhibition of ethanol induced ethane exhalation by carcinogenic pretreatment of rats 12 months earlier. Life Sci. 39: 2587-2591, 1986.
11. Rao, K. S., and Recknagel, R. O. Early onset of lipoperoxidation in rat liver after carbon tetrachloride administration. Exp. Mol. Pathol. 9: 271-278, 1968.
12. Corongiu, F. P., Lai, M., and Milia, A. Carbon tetrachloride, bromotrichloromethane and ethanol acute intoxication. Biochem. J. 212: 625-631, 1983.
13. Comporti, M. Biology of disease. Lipid peroxidation and cellular damage in toxic liver injury. Lab. Invest. 53: 599-623, 1985.

14. Link, B., Durk, H., Thiel, D., and Frank, H. Binding of trichloromethyl radicals to lipids of the hepatic endoplasmic reticulum during tetrachloromethane metabolism. Biochem. J. 223: 577-586, 1984.

15. Sipes, I. G., Krishna, G., and Gillete, J. R. Bioactivation of carbon tetrachloride, chloroform and bromotrichloromethane: Role of cytochrome P-450. Life Sci. 20: 1541-1548, 1977.

16. Orlov, S. N., Malkov, Y. A., Rebrov, V. G., and Danilov, V. S. Free radical oxidation of the lipids of biological membranes. VII. Effect of ultraviolet radiation on the free radical oxidation of the lipids in complex with protein and in biomembranes. Biofizika 21: 278-282, 1976.

17. Mandal, T. K., Ghose, S., Sur, P., and Chatterjee, S. N. Effect of ultraviolet radiation on the liposomal membrane. Int. J. Rad. Biol. 33: 75-79, 1978.

18. Das, M., Mukhtar, H., Greenspan, E. R., and Bickers, D. R. Photoenhancement of lipid peroxidation associated with the generation of reactive oxygen species in hepatic microsomes of hematoporphyrin derivative-treated rats. Cancer Res. 45: 6328-6330, 1985.

19. Roshchupkin, D. I., and Lordkipanidze, A. T. Dependence of the photodamage to erythrocytes on the intensity of ultraviolet radiation. Biofizika 29: 169-170, 1984.

20. Pathak, M. A., and Stratton, K. Free radicals in human skin before and after exposure to light. Arch. Biochem. Biophys. 123: 468-476, 1968.

21. De Leo, V. A., Hanson, D., Weinstein, I. B., and Harber, L. C. Ultraviolet radiation stimulates the release of arachidonic acid from mammalian cells in culture. Photochem. Photobiol. 41: 51-56, 1985.

A SENSITIVE AND SELECTIVE METHOD FOR QUANTITATIVE ANALYSIS OF PRODUCTS OF LIPID PEROXIDATION

M. M. Zaleska and D. F. Wilson

Department of Biochemistry and Biophysics
University of Pennsylvania
Philadelphia, Pennsylvania 19104

ABSTRACT

We report here the development of a sensitive and specific method for measuring small amounts of lipid peroxides. The method is based on the derivatization of hydroxylated fatty acid methyl esters with fluorinated anhydrides and their analysis by gas chromatograpby (GC). The peroxidation products of arachidonic (20:4) and linoleic (18:2) acids obtained by lipoxygenase treatment have been analyzed by this method. The resulting fluoroacyl derivatives are stable, can be readily separated by capillary column GC, and can be measured at picogram levels using an electron capture detector (ECD). Moreover, since fluoroacyl moieties are incorporated exclusively into hydroxylated fatty acids, this method is well suited for the measurement of low levels of peroxidation products in the presence of much larger amounts of unmodified lipid in total lipid extracts.

INTRODUCTION

Free radical-catalyzed peroxidation of membrane lipids is thought to mediate tissue damage in a number of degenerative diseases, aging, chemical toxicities, and radiation injury (1,2). Lipid peroxidation as a mechanism of toxicity of reactive oxygen species has been implicated also in the pathology of cerebral ischemia (3,4). Most of the methods currently available for measuring the products of lipid peroxidation are indirect, lack specificity, or are very laborious (for review, see reference 5). For example, measurements of total conjugated dienes or thiobarbituric acid-reactive substances are rapid but nonspecific, while analysis of the individual primary peroxidation products has required purification by high-performance liquid chromatography usually followed by GC-mass spectrometry of the silyl derivatives of the fatty acid methyl esters (6-8). Here we report a relatively rapid and specific method for measuring small amounts of lipid peroxides. The peroxides are converted to hydroxides and analyzed by capillary column gas chromatography as methyl esters of respective fatty acids. The hydroxides are specifically identified and quantitated as fluoroacyl derivatives, which can be measured with high sensitivity using an electron capture detector.

MATERIALS AND METHODS

Reagents

Analytical grade solvents were used. Fatty acids, NaBH$_4$, α',α'-dipyridyl, and soybean lipoxygenase were purchased from Sigma Chemical Co., St. Louis, MO. Trifluoroacetic acid (TFA), heptafluorobutyric acid (HFB), heptafluoro-butyrylimidazole (HFBI), and perfluorooctanoic acid (PFO) anhydrides were obtained from Pierce Chemical Co., Rockford, IL. PdCl$_2$ was from Aldrich Chemical Co., Milwaukee, WI.

Preparation of Samples

To obtain the respective fatty acid hydroperoxides, 20:4 or 18:2 acid (as ammonium salts, 0.1 or 1 mM) was incubated in 0.1 M borate buffer, pH 9.0, in the presence of 12-hydroxystearic (12OH 18:0) and heptadecanoic acids (17:0, as standards), 0.2 mM α',α'-dipyridyl, and 50 μg of soybean lipoxygenase in a total volume of 5 ml. α',α'-Dipyridyl, a potent iron chelator, was added to prevent the peroxida-tion of fatty acids due to the contamination of reagents with traces of iron salts. The following sample-handling procedure through formation of the fatty acid methyl esters is adapted from other authors, particularly Porter and co-workers (9,10). Aliquots (1 ml) of incubation mixture were withdrawn at time intervals indicated in the figure legends and acidified to pH 3.5-4.0 with 0.5 M citric acid. Fatty acids were extracted with five volumes of chloroform:methanol, 1:2. The extract was washed with saline (0.9%, 0.2 vol) and the organic phase was evaporated under nitrogen. Reduction of hydroperoxides to hydroxides and catalytic hydrogenation was achieved in one step. The dry residues were resuspended in 0.5 ml methanol in 5-ml Reacti-vials (Pierce Chemical Co.). The vials were placed in an ice bath, Teflon-coated stirring bars were added, and the ice bath placed on a magnetic stirrer. About 1 mg of PdCl$_2$ and 10 mg of NaBH$_4$ were added, vials were tightly capped, and stirring continued for 45 min followed by 10 min at room temperature. To each sample, 1 ml of saline was added and the pH adjusted to 3-4 with 1 N HCl. Hydroxylated and hydrogenated fatty acids were extracted into hexane:ether, 20:1 (2 times 3.5 ml). Following evaporation to dryness, the samples were methylated with excess of freshly prepared etheral diazomethane in the presence of methanol (10% vol) for 15 min at room temperature. The ether was removed with the stream of nitrogen, and methyl esters of fatty acids were dissolved in hexane:benzene, 1:1. To obtain fluoroacyl derivatives, the sample in 0.3-0.5 ml of hexane:benzene (1:1) was heated with 0.15 ml of 0.05 M triethylamine and 0.05 ml of TFA, HFB, HFBI, or PFO anhydride at 50° for 15 min. After cooling, the samples were shaken with 1 ml of distilled water and 1 ml of 5% NH$_4$OH. Following centrifugation at 200 x g for 5 min, the organic phase was removed for GC analysis. The above derivatizing procedures were adopted from the Pierce Handbook and Catalog.

Instrumentation

A Hewlett-Packard 5890 gas chromatograph (Hewlett-Packard, Paramus, NJ) with on column and split-splitless injectors, a flame ionization detector (FID), and an ECD were used. Gas chromatography conditions are described in the legends to Figures 1 and 2. The data were processed through a 19 bit Nelson Analytical Model 763 intelligent interface (Nelson Analytical, Cupertino, CA) and transferred to a Tandy 1200 HD microcomputer (Tandy, Fort Worth, TX). Data analysis was performed using Nelson Analytical 2600 chromatography software.

RESULTS AND DISCUSSION

In the present study we have chosen to derivatize and quantitate the lipid peroxides generated by commercial soybean lipoxygenase from 20:4 and 18:2. These particular fatty acids were chosen because they are among the most abundant unsaturated fatty acids in tissues and are very susceptible to peroxidation. Figure 1 illustrates the appearance of 15OH 20:0 (hydrogenated 15-HETE) with simultaneous disappearance of 20:0. The total amount of 20:0 injected onto the column was 6 ng or 18 pmoles (zero time incubation), and its conversion to hydroxylated product with increasing time of incubation can be clearly observed on FID chromatograms. ECD chromatograms show the time course of peroxidation even more clearly since only the fluoroacyl (hydroxylated) compounds were detected; fatty acid methyl esters appear as negative peaks. The relative response ratio of ECD/FID for HFB methyl esters of hydroxylated fatty acids approximates 10. 15-HETE and 20:4 were quantified using 12OH 18:0 and 17:0 acids as internal standards. Recovery of the standards carried through the whole procedure exceeded 95%. Under the chromatographic conditions described in the legend to Figure 1, the PFO derivative of 12OH 18:0 methyl ester was detected by FID and ECD at 5.90 and 5.57 min, respectively.

A different capillary column and on column injection were used to further increase the sensitivity of the method. Chromatograms using FID are shown in Figure 2. Volumes of 0.1-0.5 μl were injected directly onto the column. This eliminates the need to concentrate the samples and overcomes the potential losses of the sample contents that may occur with split mode injection. The appearance of the product of linoleate oxidation by lipoxygenase is shown in Figure 2. The FID response to HFB methyl ester of 13OH 18:0 (hydrogenated 13OH 18:2) was linear from 1 to at least 40 ng. Under these conditions, 12OH 18:0 methyl ester TFA and HFB derivatives were detected at 5.03 and 5.17 min, respectively. For the 15OH 20:0 methyl ester TFA and HFB derivatives, the retention times were 5.98 and 6.12 min, respectively. The above procedure was used to measure the content of HETE's in the human serum stimulated with ionophore A23187 in the presence

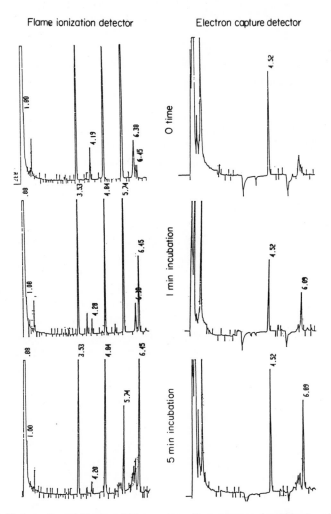

Fig. 1. Oxidation of arachidonic acid by soybean lipoxygenase. Arachidonic acid was treated with soybean lipoxygenase, and samples were withdrawn at 0, 1, and 5 min of incubation. Samples were processed as described in Materials and Methods, and aliquots of 1-3 μl were injected into a GC column. A 25-m Hewlett-Packard column (0.2-mm ID, cross-linked methyl silicone of 0.33-micron thickness) was cut in half and attached in parallel to a split-injector. One effluent end was attached to an ECD and the other to an FID. Slight difference between retention times (Rt) for same compound is due to differences in length of column and rate of carrier gas flow. Injector split ratio: 1:100; carrier gas: helium at 0.8 ml/min per column; injector temp: 280°. Temperature programming: 210°-280° at 8°/min. Derivatives were methyl esters or heptafluorobutyrate methyl esters. FID trace: attenuation = 2; standards: heptadecanoate and 12-hydroxystearate; Rt was 3.53 and 4.84 min, respectively; arachidate, Rt = 5.74 min; 15OH 20:0, Rt = 6.45 min. ECD trace: attenuation = 5; standard: 12OH 18:0, Rt = 4.52 min; 15OH 20:0, Rt = 6.09 min.

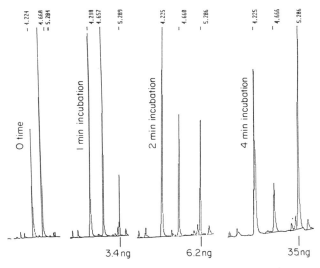

Fig. 2. Linoleic acid oxidation by soybean lipoxygenase. Samples were obtained as described in Figure 1. Column: J&W Scientific Durabond DB1 capillary; film thickness: 0.1 micron, 0.32 mm ID; length: 15 m; on column injection, 0.2 μl (50 ng of 18:0 at zero time). Temperature programming: 70°-175° at 35°/min and then to 260° at 10°/min. Derivatives were methyl esters or heptafluorobutyryl methyl esters. FID trace, standard: 17:0, Rt = 4.22 min; 18:0, Rt = 4.66 min; 13OH 18:0, Rt = 5.23 min.

of arachidonic acid. Figure 3 shows the presence of three HETE's in concentrations of 4 to 27 μM.

As shown, the HFB methyl esters of hydroxy fatty acids appear as well-resolved single peaks on capillary GC chromatograms using either FID or ECD. The thermostability of these derivatives is in contrast to the reported thermal decomposition of trifluorobutyrate methyl esters of PG's involving loss of the perfluorinated group from the allylic position 15 (11). Other authors have successfully used HFBI derivatization for the analysis of PG's in body fluids and recommend the use of this agent rather than HFB (12). In our procedure the double bonds are reduced by catalytic hydrogenation, and both derivatizing agents were found to give good results. Reduction of the double bonds adds to the stability of derivatives at high temperatures during GC analysis as well as permitting them to be stored for at least several weeks in hexane at -20°. The disadvantage of this method is that the reduction of double bonds results in loss of some structural features, and it may not be possible to determine fully the structure of the original compound. However, it should be noted that the use of deuterated sodium borohydride (NaBD₄) would make it possible to determine the positions of double bonds in original molecules using GC-mass spectrometry. The advantage of the procedure described in this study over other presently used methods is twofold: increased stability of the derivatives and specificity of the derivatizing agent toward the hydroxyl group of fatty acid. Due to the reagent specificity there is less need for purification of

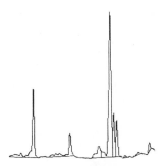

Fig. 3. Chromatogram of metabolites of arachidonic acid formed in human serum obtained from blood incubated with ionophore A23187 (15 μM) and arachidonic acid (1.5 mM) for 1 hour at 37°. GC conditions were same as in legend to Figure 2 except for slightly lower carrier gas flow. Derivatives were methyl esters or heptafluorobutyryl methyl esters. FID trace, standard: 12OH 18:0, Rt = 5.41 min; 20:0, Rt = 5.96 min; 12OH 20:0, Rt = 6.58 min; 5OH 20:0, Rt = 6.63 min; 15OH 20:0, Rt = 6.68 min.

the sample by high-performance liquid chromatography and, especially when ECD is used, small amounts of hydroxylated fatty acids can be measured in the presence of much larger amounts of unmodified lipid in total lipid extracts.

ACKNOWLEDGMENTS

This work was supported in part by Grant NS-10939. The authors thank Mr. David Nelson for his advice and technical assistance and Dr. Ronald Walenga for helpful discussion.

REFERENCES

1. Tappel, A. L. Lipid peroxidation damage to cell components. Fed. Proc. 32: 1870-1874, 1973.
2. Mead, J. F. Free radical mechanisms of lipid damage and consequences for cellular membranes. In: "Free Radicals in Biology," Vol. 1. W. A. Pryor, ed. Academic Press, New York, 1977, pp. 51-68.
3. Demopoulos, H. D., Flamm, E. S., Pietronigro, D. D., and Seligman, M. L. The free radical pathology and the microcirculation in the major central nervous system disorders. (Suppl.) Acta Physiol. Scand. 492: 91-119, 1980.

4. Kogure, K., Watson, B. D., Busto, R., and Abe, K. Potentiation of lipid peroxides by ischemia in the rat brain. Neurochem. Res. 7: 437-454, 1982.

5. Halliwell, B., and Gutteridge, J. M. C., eds. In: "Free Radicals in Biology and Medicine." Clarendon Press, Oxford, 1985, p. 161.

6. Boeynaems, J. M., Brash, A. R., Oates, J. A., and Hubbard, W. C. Preparation and assay of monohydroxy-eicosatetraenoic acids. Anal. Biochem. 104: 259-267, 1980.

7. Maclouf, J., De La Baume, H., Caen, J., Rabinovitch, H., and Rigaud, M. Complete profiling of some eicosanoids using glass capillary gas chromatography with flame ionization detection. Application to biological samples. Anal. Biochem. 109: 147-155, 1980.

8. Hughes, H., Smith, C. V., Tsokos-Kuhn, J. O., and Mitchell, J. R. Quantitation of lipid peroxidation products by gas chromatography-mass spectrometry. Anal. Biochem. 152: 107-112, 1986.

9. Funk, M. O., Isaac, R., and Porter, N. A. Preparation and purification of lipid hydroperoxides from arachidonic and γ-linolenic acids. Lipids 11: 113-117, 1976.

10. Porter, N. A., Wolf, R. A., Yarboro, E. M., and Weenen, H. The autoxidation of arachidonic acid: Formation of the proposed SRS-A intermediate. Biochem. Biophys. Res. Commun. 89: 1058-1064, 1979.

11. Levitt, M. J., Josimovich, J. B., and Broskin, K. D. Analysis of prostaglandins by electron capture chromatography. Thermal decomposition of heptafluoro-butyrate methyl esters of F_1 and F_2. Prostaglandins 1: 121-131, 1972.

12. Sugiura, M., and Hirano, K. Determination of prostaglandin F_1 and F_2 by gas-liquid chromatography. J. Chromatogr. 90: 169-177, 1974.

INDEX